STATISTICAL THEORY OF RELIABILITY AND LIFE TESTING

INTERNATIONAL SERIES IN DECISION PROCESSES

INGRAM OLKIN, Consulting Editor

A Basic Course in Statistics with Sociological Applications, 2nd ed., T. R. Anderson and M. Zelditch, Jr.
Probability Theory and Elements of Measure Theory, H. Bauer
Time Series, R. Brillinger
Decision Analysis for Business, R. Brown
Probability and Statistics for Decision Making, Ya-lun Chou
A Guide to Probability Theory and Application, C. Derman, L. J. Gleser, and I. Olkin
Statistics: Probability, Inference, and Decision, Volumes I and II and combined ed., W. L. Hays and R. L. Winkler
Introduction to Statistics, R. A. Hultquist
Introductory Statistics with Fortran, A. Kirch
Reliability Handbook, B. A. Kozlov and I. A. Ushakov (edited by J. T. Rosenblatt and L. H. Koopmans)
An Introduction to Probability, Decision, and Inference, I. H. LaValle
Elements of Probability and Statistics, S. A. Lippman
Modern Mathematical Methods for Economics and Business, R. E. Miller
Applied Multivariate Analysis, S. J. Press
Fundamental Research Statistics for the Behavioral Sciences, J. T. Roscoe, 2nd ed.
Applied Probability, W. A. Thompson, Jr.
Quantitative Methods and Operations Research for Business, R. E. Trueman
Elementary Statistical Methods, 3rd ed., H. M. Walker and J. Lev
An Introduction to Bayesian Inference and Decision, R. L. Winkler

FORTHCOMING TITLES

A Guide to Probability Theory and Application: Calculus Based, C. Derman, L. J. Gleser, and I. Olkin
Introduction to Statistics and Probability, E. Dudewicz
Nonparametric Methods for Quantitative Analysis, J. D. Gibbons
Statistics: Probability, Inference, and Decision, 2nd ed., W. L. Hays and R. L. Winkler
Fundamentals of Decision Analysis, I. H. LaValle

SERIES IN QUANTITATIVE METHODS FOR DECISION MAKING

ROBERT L. WINKLER, Consulting Editor

Decision Analysis for Business, R. Brown
Probability and Statistics for Decision Making, Ya-lun Chou
Statistics: Probability, Inference, and Decision, Volumes I and II and combined ed., W. L. Hays and R. L. Winkler
An Introduction to Probability, Decision, and Inference, I. H. LaValle
Elements of Probability and Statistics, S. A. Lippman
Modern Mathematical Methods for Economics and Business, R. E. Miller
Applied Multivariate Analysis, S. J. Press
Quantitative Methods and Operations Research for Business, R. E. Trueman
An Introduction to Bayesian Inference and Decision, R. L. Winkler

FORTHCOMING TITLES

Statistics: Probability, Inference, and Decision, 2nd ed., W. L. Hays and R. L. Winkler
Fundamentals of Decision Analysis, I. H. LaValle

Statistical Theory of Reliability and Life Testing

Probability Models

Richard E. Barlow
University of California at Berkeley

Frank Proschan
Florida State University

HOLT, RINEHART AND WINSTON, INC.

New York Chicago San Francisco Atlanta
Dallas Montreal Toronto London Sydney

To our wives Barbara and Edna, without whose whining nagging for more money we might long ago have abandoned this painful project;

To our numerous children, whose incessant bickering and generally atrocious behavior drove us to spend many long hours at the office, working on the book as the lesser of evils;

To our students, who, with malicious glee, found the many errors in earlier versions and did the dirty work of indexing, checking references, and so on, knowing that a degree and a decent recommendation would have been impossible otherwise;

To our many colleagues who were generous with suggestions—but who asked them? Any failings or errors now present in the book are undoubtedly the result of their unsolicited advice and meddlesome tampering. We accept absolutely no responsibility for errors in this book;

*And finaly, to our tipysts, whose dedic ated efforts, care^{ful} attentiun to de&ail, and skillful w*rk transmuted a rougf ilegible swrawl int^{o} a finished boo\$.*

Library of Congress Cataloguing in Publication Data

Barlow, Richard E
Statistical theory of reliability and life testing: probability models.

(International series in decision processes) (Series in quantitative methods for decision making)
Bibliography: p.
1. Reliability (Engineering)—Statistical methods.
I. Proschan, Frank, joint author. II. Title.
TS173.B37 620′.004′5 74-4480
ISBN 0-03-085853-4

Printed in the United States of America
5 6 7 8 038 9 8 7 6 5 4 3 2 1

PREFACE

This is the first of two books on the statistical theory of reliability and life testing. The present book concentrates on probabilistic aspects of reliability theory, while the forthcoming book will focus on inferential aspects of reliability and life testing, applying the probabilistic tools developed in this volume.

This book emphasizes the newer, research aspects of reliability theory. The concept of a coherent system, introduced in Chapter 1, serves as a unifying theme for much of the book. A number of new classes of life distributions arising naturally in reliability models are treated systematically: the increasing failure rate average, new better than used, decreasing mean residual life, and so on. As the names would seem to indicate, each such class of life distributions provides a realistic probabilistic description of a physical property occurring in the reliability context. Also various types of positive dependence among random variables are considered, thus permitting more realistic modeling of commonly occurring reliability situations. For example, bounds and inequalities for the prediction of system life length are obtained under the assumption that component life lengths are "associated" (see Chapter 2 and Chapter 5, Section 4) rather than under the more restrictive (and usual) assumption of component independence. It is noteworthy that these new classes of distributions and these new concepts of dependence have applications in areas other than reliability.

Even though much of the reliability theory presented is modern and at the research level, the book is intended to serve as a text, rather than as a monograph. A large number of exercises have been included to illustrate, clarify, and supplement the theory and methods presented in the body of the text. The exposition is essentially self-contained; definitions and theorems from other areas of mathematics and statistics are presented in detail as needed, rather than simply referenced. Finally, many practical examples and numerical illustrations have been included to help persuade the reader that the theoretical results presented are true not only in general but, surprisingly, even in specific cases.

Improvements of some of the results of our 1965 monograph *Mathematical Theory of Reliability* (Wiley) have been made. Some of the bounds and inequalities obtained earlier under the assumption of increasing failure rate are now shown to hold under weaker assumptions, such as increasing failure rate average or new better than used. Similarly, earlier results for systems of independent components are now shown to hold when component life lengths are merely associated.

The readers of our 1965 monograph should note that a number of new topics are treated in the present volume. These include recent work in coherent structure theory, the theory of extreme value distributions, fault tree analysis, availability theory, shock models, the Marshall–Olkin multivariate exponential distribution, and positive dependence of various kinds among random variables. On the other hand, some of the topics treated in the 1965 monograph are not included in the present volume.

Who should use this book? Having resisted our initial greedy impulse to respond "every person and every library," we more modestly and realistically suggest that the book would be of value to reliability analysts, operations researchers, and statisticians interested in seeing how statistics can be applied to an engineering science. The text is suitable for a one- or two-quarter engineering, operations research, or statistics course for students who have had a basic course in mathematical statistics. Although a number of relatively advanced mathematical and statistical tools are used, the reader with only a modest background in probability and statistics need not fear; the more advanced mathematical and statistical notions and tools are defined and explained when introduced to the extent needed for their application to reliability problems. The reader interested in possible engineering applications may first wish to read the examples considered in the appendix.

What topics are covered in this book? Chapters 1 and 2 discuss deterministic and probabilistic properties of coherent structures, since practically every system encountered in applications is a coherent structure. The approach taken is often more general than the usual one, since the components of the system are not necessarily assumed stochastically independent. Chapter 3 introduces the notion of failure rate and shows how it may be used to describe various types of aging. The important class of increasing (decreasing) failure rate distributions is described. Next are presented various well-known parametric families of life distributions. Many basic properties of the exponential distribution are developed, since this distribution plays such a fundamental role in reliability and life testing, and since many of these properties will be used in the statistical inference of the forthcoming companion book.

Chapter 4 presents a systematic study of various classes of life distributions based on notions of aging. Life distributions of coherent systems of increasing failure rate components are shown essentially to constitute the class of distributions with increasing failure rate average. Certain shock models in which damage cumulates also lead to this same class of life distribution, as shown next. After this, the effects of performing various reliability operations on distributions within a class (such as convolving, mixing, forming coherent systems, and so on) are studied. Finally, system reliability bounds are obtained based on partial information concerning component life distributions.

Chapter 5 studies aspects of nonindependence among components. First, the important Marshall–Olkin multivariate exponential distribution is treated. Next, various concepts of positive dependence among random variables are treated, and the relationship among them developed. Positive dependence among subsystem life lengths is often observed in actual practice as a result of exposure to a common environment, dependence on a common power source, load sharing, and so forth.

In Chapter 6 some basic tools are presented which are useful in analyzing maintenance policies, such as renewal theory. We also give a comprehensive treatment of certain classes of life distributions, such as new better than used, which are basic in the study of maintenance policies. In Chapter 7 we consider the effects of maintenance policies on such basic system parameters as availability. We also show how to compute system reliability when both spares are provided and repair of failed parts occurs. Finally, we present a general optimization model and a general algorithm for its solution, which can be used to solve a variety of practical problems in determining optimum spares kits or optimum component redundancy.

Chapter 8 obtains limit distributions for various types of systems as the number of components increases indefinitely. A practical implication of these results is that for a large system, one can often approximate system life by an appropriate extreme value distribution.

The appendix departs from the mathematical style of the first eight chapters to present selected engineering applications of the preceding theory. Coherent systems are represented as event trees. This representation is useful in translating a complex system into the mathematical language used in this book. It is also useful for computer implementation.

An introductory reliability course can be based on the chapter sequence 1, 2, 3, 4, 6, and 7. For readers interested in the mathematical tools useful in designing systems and improving the reliability of systems, we recommend the chapter sequence 1, 2, 3, and the appendix. For readers interested mainly in maintenance problems, we recommend the chapter sequence 1, 2, 3, 6, and 7. For readers interested in multivariate distributions and concepts of dependence, we recommend the chapter sequence 1, 2, 3, and 5. Sophisticated readers interested in new fields of research will find the chapter sequence 1, 2, 4, 5, and 8 useful.

Equation numbers are governed by the section in which the equations appear. Thus Equation (2.3) is the third equation of Section 2. Definitions, theorems, corollaries, lemmas, results, remarks, and examples, when numbered, are numbered in sequence within a section. Thus in Section 1 of Chapter 8, 1.1, 1.2, 1.3, and 1.4 are examples, followed by 1.5, which is a definition, followed by 1.6, which is a lemma, and so on. Figures are numbered by chapter as well as by section; thus Figure 3.5.2 is the second figure in Section 5 of Chapter 3. Starred exercises are more difficult than unstarred; starred sections may be omitted upon first reading of the text.

We wish to acknowledge the support provided by the Air Force Office of Scientific Research under AFOSR Grant 71-2058, by the Office of Naval Research under Contract N00014-69-A-0200-1036, and by the Boeing Scientific Research Laboratories. In particular, we wish to thank Bruce McDonald and Seymour Selig of the ONR, David Osteyee of the AFOSR, and Burton H. Colvin of the BSRL for their encouragement and support. We are also grateful for the help and advice given by James D. Esary, Myles Hollander, George Kimeldorf, Howard Lambert, Albert W. Marshall, Ingram Olkin, Sheldon Ross, I. Richard Savage, Nozer Singpurwalla, Erwin Straub, and Ronald Wolff. We thank Linda Betters, Barbara Brewer, Noreen Revelli, and Karen Stine for their skillful typing.

Berkeley, California — Richard E. Barlow
Tallahassee, Florida — Frank Proschan

CONTENTS

1

STRUCTURAL PROPERTIES OF COHERENT SYSTEMS

In this chapter we consider the structural relationship between a system and its components, that is, those relationships that are *deterministic*. In the next chapter we consider basic aspects of the reliability of systems, which are, of course, *probabilistic* in nature. In both chapters, except for result 3.4 of Chapter 1, we consider the system at a fixed moment of time (say, the present moment); the present state of the system is assumed to depend only on the present states of the components. In later chapters we study dynamic models in which time plays a crucial role. In the appendix we consider problems involved in computer implementation of coherent structure theory for complex systems. Also, additional applications of coherent structures are presented in the appendix. Throughout we use the terms "system" and "structure" interchangeably.

1. SYSTEMS OF COMPONENTS

Throughout Chapters 1 and 2 we will distinguish between only two states: a functioning state and a failed state. This dichotomy applies to the system as well as to each component. To indicate the state of the ith component, we assign a binary indicator variable x_i to component i:

$$x_i = \begin{cases} 1 & \text{if component } i \text{ is functioning,} \\ 0 & \text{if component } i \text{ is failed,} \end{cases}$$

for $i = 1, \ldots, n$, where n is the number of components in the system.

Similarly, the binary variable ϕ indicates the state of the system:

$$\phi = \begin{cases} 1 & \text{if the system is functioning,} \\ 0 & \text{if the system is failed.} \end{cases}$$

(Unless otherwise specified, the term *binary variable* will refer to a variable taking on the values 0 or 1. Similarly, a *binary function* will be a function taking on the values 0 or 1.)

We assume that the state of the system is determined completely by the states of the components, so that we may write

$$\phi = \phi(\mathbf{x}),$$

where

$$\mathbf{x} = (x_1, \ldots, x_n).$$

The function $\phi(\mathbf{x})$ is called the *structure function* of the system. Since a knowledge of the structure function is equivalent to a knowledge of the structure, we will often use the phrase "structure ϕ" in place of "structure having structure function ϕ." The number of components n in the system is called the *order* of the system.

1.1. Example. A *series* structure

Figure 1.1.1.
Series structure.

functions if and only if each component functions. The structure function is given by $\phi(\mathbf{x}) = \prod_{i=1}^{n} x_i = \min(x_1, \ldots, x_n)$.

1.2. Example. A *parallel* structure functions if and only if at least one

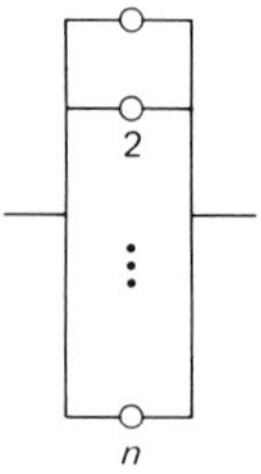

Figure 1.1.2.
Parallel structure.

component functions. The structure function is given by

$$\phi(\mathbf{x}) = \coprod_{i=1}^{n} x_i = \max(x_1, \ldots, x_n)$$

where

Notation.

$$\coprod_{i=1}^{n} x_i \equiv 1 - \prod_{i=1}^{n} (1 - x_i),$$

$$x_1 \amalg x_2 = 1 - (1 - x_1)(1 - x_2).$$

Note that $\coprod$ and $\amalg$ bear the same relation to each other as $\sum$ and $+$; $\amalg$ and $+$ operate on pairs, while $\coprod$ and $\sum$ operate over sets of indices.

1.3. Example. A *k-out-of-n* structure functions if and only if at least k of the n components function. The structure function is given by

$$\phi(\mathbf{x}) = \begin{cases} 1 & \text{if } \sum_{i=1}^{n} x_i \geq k, \\ 0 & \text{if } \sum_{i=1}^{n} x_i < k, \end{cases}$$

or equivalently,

$$\phi(\mathbf{x}) = \prod_{1}^{n} x_i \qquad \text{for} \quad k = n,$$

while

$$\begin{aligned} \phi(\mathbf{x}) &= (x_1 \cdots x_k) \amalg (x_1 \cdots x_{k-1} x_{k+1}) \amalg \cdots \amalg (x_{n-k+1} \cdots x_n) \\ &\equiv \max\{(x_1 \cdots x_k), (x_1 \cdots x_{k-1} x_{k+1}), \ldots, (x_{n-k+1} \cdots x_n)\} \end{aligned}$$

for $1 \leq k \leq n$, where every choice of k out of the n x's appears once exactly.

1.4. Remark. Note that a series structure (Example 1.1) is an n-out-of-n structure, and a parallel structure (Example 1.2) is a 1-out-of-n structure.

To illustrate the k-out-of-n structure further, consider the special case of the 2-out-of-3 structure, with structure function given by

$$\begin{aligned} \phi(\mathbf{x}) &= x_1 x_2 \amalg x_1 x_3 \amalg x_2 x_3 \\ &\equiv x_1 x_2 x_3 + x_1 x_2 (1 - x_3) + x_1 (1 - x_2) x_3 + (1 - x_1) x_2 x_3. \end{aligned}$$

Note that we have replicated each component for purposes of analysis; physically each component appears once only.

An airplane which is capable of functioning if and only if at least two of its three engines are functioning is an example of a 2-out-of-3 system.

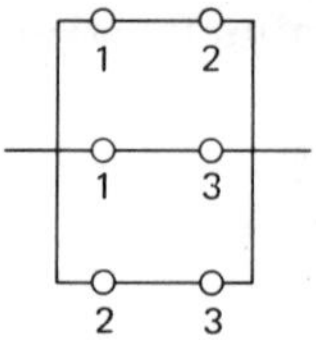

Figure 1.1.3.
2-out-of-3 structure.

1.5. Example. A more elaborate example is a stereo hi-fi system with the following components:

(a) FM tuner.
(b) Record changer.
(c) Amplifier.
(d) Speaker A.
(e) Speaker B.

We consider the system functioning if we can obtain music (monaural or stereo) through FM or records. The system diagram is illustrated in Figure 1.1.4.

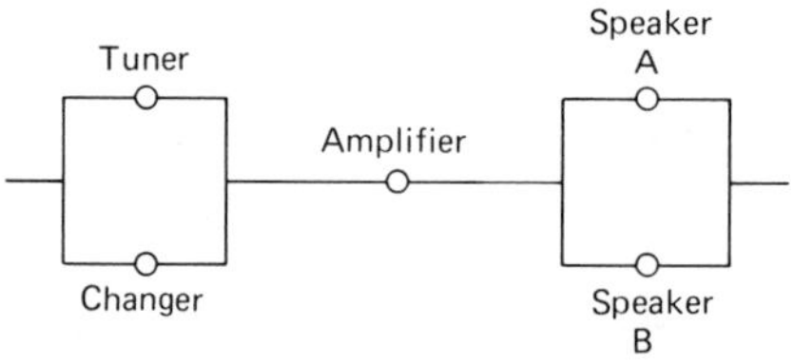

Figure 1.1.4.
Hi-fi system.

The structure function is given by

$$\phi(\mathbf{x}) = (x_1 \amalg x_2)x_3(x_4 \amalg x_5).$$

Each component of a physical system serves some sort of useful function. This motivates the following:

1.6. Definition. The ith component is *irrelevant* to the structure ϕ if ϕ is constant in x_i; that is, $\phi(1_i, \mathbf{x}) = \phi(0_i, \mathbf{x})$ for all $(\cdot_i, \mathbf{x})$.[1] Otherwise the ith component is *relevant* to the structure.

[1] *Notation.*

$$(1_i, \mathbf{x}) \equiv (x_1, \ldots, x_{i-1}, 1, x_{i+1}, \ldots, x_n),$$
$$(0_i, \mathbf{x}) \equiv (x_1, \ldots, x_{i-1}, 0, x_{i+1}, \ldots, x_n),$$
$$(\cdot_i, \mathbf{x}) \equiv (x_1, \ldots, x_{i-1}, \cdot, x_{i+1}, \ldots, x_n).$$

For example, component 2 is irrelevant to the structure pictured in Figure 1.1.5.

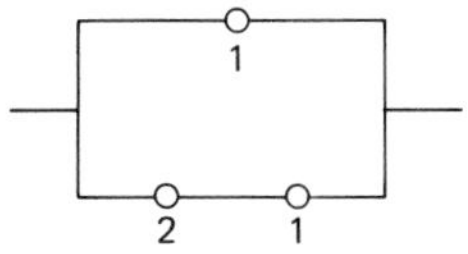

Figure 1.1.5.
Example of irrelevant component.

We will find the following *pivotal decomposition* of the structure function a fundamental tool in carrying through inductive proofs. See Hohn (1966).

1.7. Lemma. The following identity holds for any structure function ϕ of order n:

$$\phi(\mathbf{x}) = x_i\phi(1_i, \mathbf{x}) + (1 - x_i)\phi(0_i, \mathbf{x}) \qquad \text{for all} \quad \mathbf{x}\ (i = 1, \ldots, n). \qquad (1.1)$$

The proof is left to Exercise 1.

Note that Lemma 1.7 permits us to express a structure function of order n in terms of structure functions of order $n - 1$. By repeated applications we obtain the representation:

$$\phi(\mathbf{x}) = \sum_{\mathbf{y}} \prod_{j=1}^{n} x_j^{y_j}(1 - x_j)^{1-y_j}\phi(\mathbf{y}), \qquad (1.2)$$

where the sum is extended over all binary vectors $\mathbf{y}$ of order n, and $0^0 \equiv 1$.

1.8. Example. Let $\phi(x_1, x_2) = x_1x_2$ be the structure function of a series system of two components. Then the representation (1.2) is explicitly

$$\phi(x_1, x_2) = x_1(1 - x_1)^0x_2(1 - x_2)^0 \cdot 1 + x_1(1 - x_1)^0x_2{}^0(1 - x_2) \cdot 0$$
$$+ x_1{}^0(1 - x_1)x_2(1 - x_2)^0 \cdot 0 + x_1{}^0(1 - x_1)x_2{}^0(1 - x_2) \cdot 0 = x_1x_2.$$

1.9. Definition. Given a structure ϕ, we define its *dual* ϕ^D by

$$\phi^D(\mathbf{x}) = 1 - \phi(\mathbf{1} - \mathbf{x}),$$

where $\mathbf{1} - \mathbf{x} = (1 - x_1, \ldots, 1 - x_n)$.

The concept of dual structure is useful in analyzing systems of components subject to two kinds of failure, such as relay systems and safety monitoring systems. (See Section 3.)

Example. The dual of a series (parallel) system of n components is a parallel (series) system of n components. More generally, the dual of a k-out-of-n structure is an $(n - k + 1)$-out-of-n structure.

EXERCISES

1. Prove Lemma 1.7.
2. Prove that a structure function ϕ is increasing[2] if and only if $\phi(1_i, \mathbf{x})$ and $\phi(0_i, \mathbf{x})$ are increasing and $\phi(1_i, \mathbf{x}) \geq \phi(0_i, \mathbf{x})$ for all $(\cdot_i, \mathbf{x})$.
3. Prove (1.2).

2. COHERENT STRUCTURES

A physical system would be quite unusual (or perhaps poorly designed) if improving the performance of a component (that is, replacing a failed component by a functioning component) caused the system to deteriorate (that is, to change from the functioning state to the failed state). Thus we restrict consideration to structure functions that are monotonically increasing in each argument. Also, to avoid trivialities we will, wherever possible, eliminate consideration of any system whose state does not depend on the state of its components.

2.1. Definition. A system of components is *coherent* if (a) its structure function ϕ is increasing and (b) each component is relevant.

In Exercise 1 the reader is asked to show that ϕ coherent implies $\phi(\mathbf{0}) = 0$ and $\phi(\mathbf{1}) = 1$.

In most discussions, the only pertinent aspect of the coherent system is its structure; in such cases we say "a coherent system ϕ." In a few cases we need to keep track of the particular set C of components comprising the coherent system; in such cases we say the coherent system (C, ϕ). The set C is a set of integers designating the components.

2.2. Example. All distinct coherent structures (up to permutations of components) of order 1, 2, and 3 are displayed in Figure 1.2.1.

Some basic results follow immediately for coherent structures.

2.3. Theorem. Let ϕ be a coherent structure of n components. Then

$$\prod_{i=1}^{n} x_i \leq \phi(\mathbf{x}) \leq \coprod_{i=1}^{n} x_i. \tag{2.1}$$

[2] *Terminology.* Throughout the book we use the term "increasing" in place of "nondecreasing" and "decreasing" in place of "nonincreasing." Also, we say $f(x_1, \ldots, x_n)$ is increasing if f is increasing in each argument.

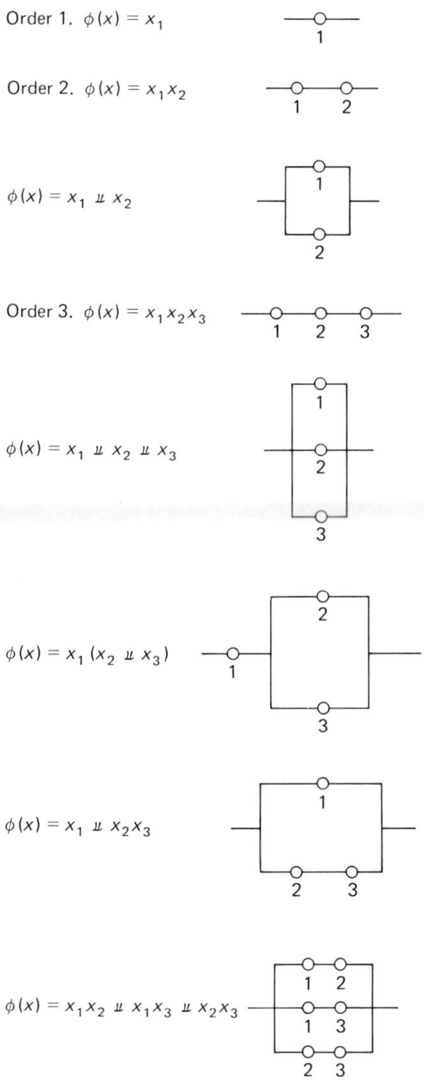

Figure 1.2.1.
Coherent structures of order 1, 2, 3.

PROOF Suppose $\prod_{i=1}^{n} x_i = 1$. Then $x_1 = \cdots = x_n = 1$. Thus $\phi(\mathbf{x}) = 1$ by Exercise 1. The left-hand inequality follows.

Suppose $\coprod_{i=1}^{n} x_i \equiv 1 - \prod_{i=1}^{n} (1 - x_i) = 0$. Then $x_1 = \cdots = x_n = 0$. Thus $\phi(\mathbf{x}) = 0$ by Exercise 1. The right-hand inequality follows. ‖

Theorem 2.3 tells us that the performance of a coherent system is bounded below by the performance of a series system and above by the performance of a parallel system.

We use the notation

$$\mathbf{x} \amalg \mathbf{y} \equiv (x_1 \amalg y_1, \ldots, x_n \amalg y_n),$$

and

$$\mathbf{x} \cdot \mathbf{y} \equiv (x_1 y_1, \ldots, x_n y_n).$$

2.4. Theorem. Let ϕ be a coherent structure. Then

(a) $\phi(\mathbf{x} \amalg \mathbf{y}) \geq \phi(\mathbf{x}) \amalg \phi(\mathbf{y})$,
(b) $\phi(\mathbf{x} \cdot \mathbf{y}) \leq \phi(\mathbf{x})\phi(\mathbf{y})$.

Equality holds in (a) for all $\mathbf{x}$ and $\mathbf{y}$ if and only if the structure is parallel. Equality holds in (b) for all $\mathbf{x}$ and $\mathbf{y}$ if and only if the structure is series.

PROOF. $x_i \amalg y_i \geq x_i \ \forall i$, so that $\phi(\mathbf{x} \amalg \mathbf{y}) \geq \phi(\mathbf{x})$ since ϕ is increasing. Similarly, $x_i \amalg y_i \geq y_i \ \forall i$, so that $\phi(\mathbf{x} \amalg \mathbf{y}) \geq \phi(\mathbf{y})$. It follows that

$$\phi(\mathbf{x} \amalg \mathbf{y}) \geq \max[\phi(\mathbf{x}), \phi(\mathbf{y})] \equiv \phi(\mathbf{x}) \amalg \phi(\mathbf{y}).$$

A similar argument proves (b).

If the system is series, $\phi(\mathbf{x} \cdot \mathbf{y}) = \prod_{i=1}^n x_i y_i = \prod_{i=1}^n x_i \prod_{i=1}^n y_i = \phi(\mathbf{x})\phi(\mathbf{y})$, yielding equality in (b). If the system is parallel, a similar argument yields equality in (a). The proof of necessity for equality is left to Exercise 8 in Section 3. ‖

Theorem 2.4(a) states that redundancy at the component level is more effective than redundancy at the system level. This principle is well known among design engineers.

EXERCISES

1. Prove that a structure function ϕ increasing in each argument has at least one relevant component if and only if

$$\phi(\mathbf{0}) = 0 \quad \text{and} \quad \phi(\mathbf{1}) = 1.$$

2. Show that the dual of a coherent structure is coherent.

3. Let ϕ be a coherent structure and $\mathbf{x}^{(1)}, \ldots, \mathbf{x}^{(k)}$ be component state vectors. Then

   (a) $\phi(\coprod_{i=1}^k \mathbf{x}^{(i)}) \geq \coprod_{i=1}^k \phi(\mathbf{x}^{(i)})$,
   (b) $\phi(\prod_{i=1}^k \mathbf{x}^{(i)}) \leq \prod_{i=1}^k \phi(\mathbf{x}^{(i)})$.

   (Generalization of Theorem 2.4.)

3. REPRESENTATION OF COHERENT SYSTEMS IN TERMS OF PATHS AND CUTS

We will find it very useful to represent a coherent structure in several alternate ways. We will need the following terminology.

3.1. Definitions. Let $\mathbf{x}$ indicate the states of the set of components $C = \{1, \ldots, n\}$. Then we define $C_0(\mathbf{x}) = \{i \mid x_i = 0\}$ and $C_1(\mathbf{x}) = \{i \mid x_i = 1\}$.

Assume the structure (C, ϕ) is coherent.

A *path vector is a vector* $\mathbf{x}$ such that $\phi(\mathbf{x}) = 1$. The corresponding *path set* is $C_1(\mathbf{x})$.

A *minimal path vector* is a path vector $\mathbf{x}$ such that $\mathbf{y} < \mathbf{x} \Rightarrow \phi(\mathbf{y}) = 0$.[3] The corresponding *minimal path set* is $C_1(\mathbf{x})$. Physically, a minimal path set is a minimal set of elements whose functioning insures the functioning of the system.

A *cut vector* is a vector $\mathbf{x}$ such that $\phi(\mathbf{x}) = 0$. The corresponding *cut set* is $C_0(\mathbf{x})$.

A *minimal cut vector* is a cut vector $\mathbf{x}$ such that $\mathbf{y} > \mathbf{x} \Rightarrow \phi(\mathbf{y}) = 1$. The corresponding *minimal cut set* is $C_0(\mathbf{x})$. Physically, a minimal cut set is a minimal set of elements whose failure causes the system to fail.

Note that if $\mathbf{x}$ is a path vector and $\mathbf{y} \geq \mathbf{x}$, then $\phi(\mathbf{y}) = 1$, where $\mathbf{y} \geq \mathbf{x}$ means $y_i \geq x_i$ $(i = 1, \ldots, n)$. Similarly, if $\mathbf{x}$ is a cut vector and $\mathbf{y} \leq \mathbf{x}$, then $\phi(\mathbf{y}) = 0$.

We illustrate these concepts below.

3.2. Example. The *bridge* structure is shown in the following diagram:

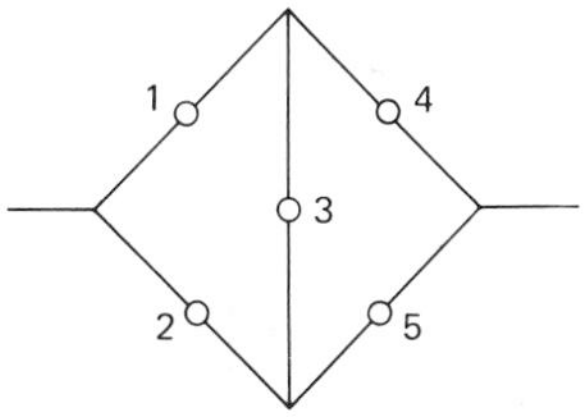

Figure 1.3.1.
Bridge structure.

The minimal path sets are:

$$P_1 = \{1, 4\}, \quad P_2 = \{2, 5\}, \quad P_3 = \{1, 3, 5\}, \quad P_4 = \{2, 3, 4\}.$$

[3] *Notation.* $\mathbf{y} < \mathbf{x}$ means $y_i \leq x_i$ $(i = 1, \ldots, n)$, with $y_i < x_i$ for some i.

The minimal cut sets are:

$$K_1 = \{1, 2\}, \quad K_2 = \{4, 5\}, \quad K_3 = \{1, 3, 5\}, \quad K_4 = \{2, 3, 4\}.$$

With the jth minimal path set P_j of a coherent structure ϕ, we may associate a binary function with arguments x_i, $i \in P_j$:

$$\rho_j(\mathbf{x}) = \prod_{i \in P_j} x_i, \tag{3.1}$$

which takes the value 1 if all components in the jth minimal path set function, and 0 otherwise ($j = 1, \ldots, p$, where p is the number of minimal path sets of ϕ). We refer to the structure ρ_j as the jth *minimal path series structure*, since clearly ρ_j is the structure function of a series arrangement of the components of the jth path set.

Since the underlying structure functions if and only if at least one of the minimal path structures is functioning, we may write the identity

$$\phi(\mathbf{x}) \equiv \coprod_{j=1}^{p} \rho_j(\mathbf{x}) \equiv 1 - \prod_{j=1}^{p} [1 - \rho_j(\mathbf{x})], \tag{3.2}$$

representing the underlying structure as a parallel arrangement of the minimal path series structures.

Similarly, with the jth minimal cut set K_j of a coherent structure ϕ, we may associate a binary function with arguments x_i, $i \in K_j$:

$$\kappa_j(\mathbf{x}) = \coprod_{i \in K_j} x_i, \tag{3.3}$$

which takes the value 0 if all the components in the jth minimal cut set fail, and 1 otherwise ($j = 1, \ldots, k$, where k is the number of minimal cut sets of ϕ). We refer to the structure κ_j as the jth *minimal parallel cut structure*, since clearly κ_j is the structure function of a parallel arrangement of the components of the jth cut set.

Since the underlying structure fails if and only if at least one of the minimal cut structures fails, we may write the identity

$$\phi(\mathbf{x}) \equiv \prod_{j=1}^{k} \kappa_j(\mathbf{x}), \tag{3.4}$$

representing the underlying structure as a series arrangement of the minimal cut parallel structures.

In the appendix we present an algorithm, suitable for computer implementation, for finding all minimal cuts and/or minimal paths for coherent structures.

3.3 Representation of Bridge Structure. In Example 3.2 we displayed the structure of the bridge and wrote down its minimal path sets and minimal cut sets. Using (3.2) and the minimal path sets, we may represent the bridge as a parallel-series structure:

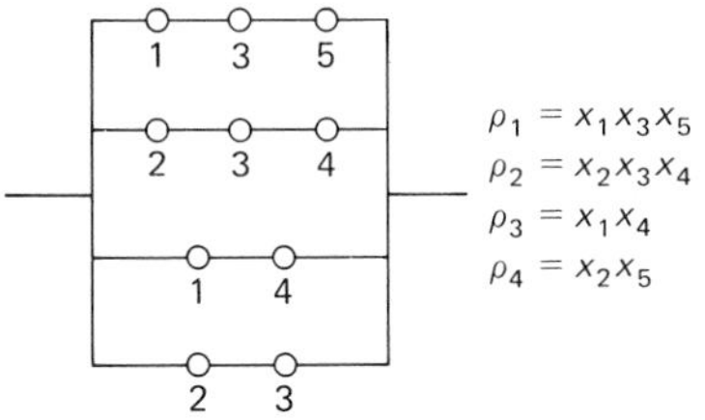

Figure 1.3.2.
Minimal path representation for bridge.

Similarly, using (3.4) and the minimal cut sets, we may represent the bridge as a series-parallel structure:

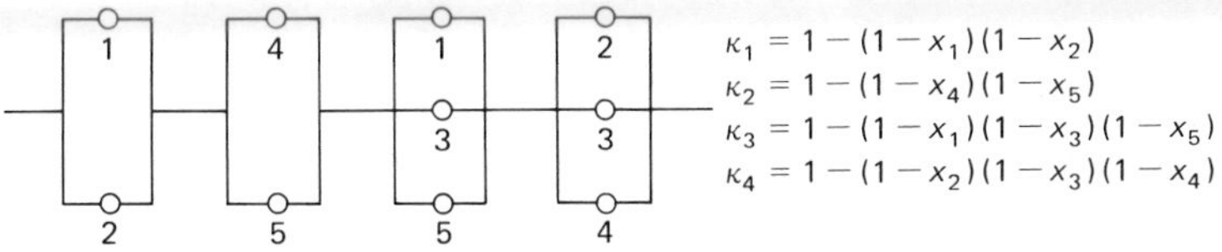

Figure 1.3.3.
Minimal cut representation for bridge.

Given the minimal cut (or minimal path) sets, any coherent system can be represented as in the above "circuit" diagrams. An alternative representation as an *event tree* with AND and OR gates is presented in the appendix. This latter representation is used in *fault tree analysis*.

3.4. Time to Failure. Thus far we have been considering systems of components at a fixed point in time. Now we consider systems of components operating in time; each component (and the system itself) operates until it fails at some time. No repair is performed. We will show that for this simple model, the domain and range of the identity (3.2) may be extended from binary values to the positive half of the axis. In Chapter 3 we discuss random times of failure.

Recall that the following identities hold for binary variables $x_1, \ldots, x_k$ (but not necessarily for general variables):

$$\prod_{i=1}^{k} x_i \equiv \min(x_1, \ldots, x_k),$$

$$\coprod_{i=1}^{k} x_i \equiv \max(x_1, \ldots, x_k).$$

Thus we may rewrite the identity (3.2) as

$$\phi(\mathbf{x}) \equiv \max_{1 \le j \le p} \rho_j(\mathbf{x}) \equiv \max_{1 \le j \le p} \min_{i \in P_j} x_i, \tag{3.5}$$

and (3.4) as

$$\phi(\mathbf{x}) \equiv \min_{1 \le j \le k} \kappa_j(\mathbf{x}) \equiv \min_{1 \le j \le k} \max_{i \in K_j} x_i. \tag{3.6}$$

Identities (3.5) and (3.6) have a meaningful interpretation not only in the case of binary variables indicating component states but also in the case of component times of failure. Let t_i be the time of failure of the ith component for $i = 1, \ldots, n$. Let $\tau_\phi(\mathbf{t})$ be the time of failure of a structure ϕ as a function of component times of failure.

3.5. Theorem. If ϕ is a coherent structure with minimal path sets $P_1, \ldots, P_p$ and minimal cut sets $K_1, \ldots, K_k$, then

$$\max_{1 \le j \le p} \min_{i \in P_j} t_i \equiv \tau_\phi(\mathbf{t}) \equiv \min_{1 \le j \le k} \max_{i \in K_j} t_i. \tag{3.7}$$

PROOF. The first identity follows from: (a) A coherent system fails when the last minimal path series structure in it fails. (b) A series structure fails at the time of the earliest component failure.

The second identity follows from: (c) A coherent system fails when the first minimal cut parallel structure in it fails. (d) A parallel structure fails at the time of the last component failure. ||

Theorem 3.5 tells us that the domain and range of the function ϕ as defined in (3.5) or (3.6) for binary arguments may be extended to the real line.

3.6. Dual Coherent Structures. Given a structure ϕ, in Definition 1.8 we defined its dual by

$$\phi^D(\mathbf{x}) = 1 - \phi(\mathbf{1} - \mathbf{x}).$$

Clearly, if $\mathbf{x}$ is a path vector for ϕ, then $\mathbf{1} - \mathbf{x}$ is a cut vector for ϕ^D, and vice versa. Also minimal path sets for ϕ are minimal cut sets for ϕ^D and vice versa (Exercise 9). These observations are useful in analyzing switching systems and certain types of safety monitoring systems.

Example: Relay Structures. Relays are subject to two kinds of failure: failure to close and failure to open. Similarly, circuits constructed from these relays are subject to the same kinds of failure. Consider the schematic diagram for a relay circuit shown in Figure 1.3.4.

Let $x_i = 1$ if the ith relay responds correctly to a command to close (that is, closes), and 0 otherwise. Let $\phi = 1$ if the circuit responds correctly to a command to close (that is, closes), and 0 otherwise. Then $\phi(\mathbf{x}) = x_1(x_2 \amalg x_3)$

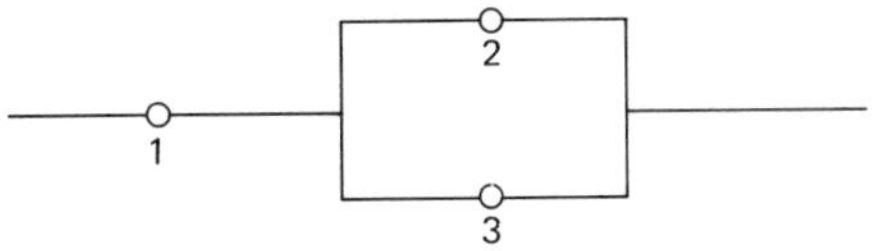

Figure 1.3.4.
Relay circuit.

represents the structure function representing the correct system response to a command to close.

Next let $y_i = 1$ if the ith relay responds correctly to a command to open (that is, opens), and 0 otherwise. Let $\psi = 1$ if the circuit responds correctly to a command to open (that is, opens), and 0 otherwise. Then $\psi(\mathbf{y}) = y_1 \amalg (y_2 y_3) = \phi^D(\mathbf{y})$, the dual of ϕ, represents the structure function representing the correct system response to a command to open.

It is interesting and useful to note that both failure to close and failure to open can be analyzed using the same structure function ϕ.

3.7. Relative Importance of Components. For a given coherent system, some components are more important than others in determining whether the system functions or not. For example, if a component is in series with the rest of the system, then it would seem to be at least as important as any other component in the system. It is clearly of value to the designer and reliability analyst to have a quantitative measure of the importance of the individual components in the system.

How important is component i in determining whether the system functions or not? First, suppose we are given the state of each of the remaining components, $(\cdot_i, \mathbf{x})$. Then if $\phi(1_i, \mathbf{x}) = 1$ while $\phi(0_i, \mathbf{x}) = 0$, that is, if

$$\phi(1_i, \mathbf{x}) - \phi(0_i, \mathbf{x}) = 1, \tag{3.8}$$

we would consider component i more important than if $\phi(1_i, \mathbf{x}) = 1 = \phi(0_i, \mathbf{x})$ or $\phi(1_i, \mathbf{x}) = 0 = \phi(0_i, \mathbf{x})$. In the first case, (3.8), the state of component i determines whether the system functions or not, whereas in the alternative cases [(3.8) not true], the state of component i is of no consequence. When (3.8) holds, we call $(1_i, \mathbf{x})$ a *critical path vector for* i and $C_1(1_i, \mathbf{x})$ the corresponding *critical path set for* i. We let

$$n_\phi(i) = \sum_{\{\mathbf{x} \mid x_i = 1\}} [\phi(1_i, \mathbf{x}) - \phi(0_i, \mathbf{x})] \tag{3.9}$$

denote the total number of critical path vectors for i (or equivalently, the total number of critical path sets for i).

This suggests the following plausible measure of the *structural importance of component* i:

$$I_\phi(i) = \frac{1}{2^{n-1}} n_\phi(i) = \frac{1}{2^{n-1}} \sum_{\{\mathbf{x} \mid x_i = 1\}} [\phi(1_i, \mathbf{x}) - \phi(0_i, \mathbf{x})], \tag{3.10}$$

the proportion of the 2^{n-1} outcomes having $x_i = 1$ which are critical path vectors for i.

Thus for any given structure ϕ, we may order the components as to structural importance by ordering the values $I_\phi(1), \ldots, I_\phi(n)$.

Examples. (a) Let ϕ be a 2-out-of-3 structure. Then $I_\phi(1) = 1/2^2 \cdot 2 = 1/2$, since among the four outcomes 100, 101, 110, 111, there are two critical path vectors for component 1 (101 and 110).

By symmetry, $I_\phi(2) = I_\phi(3) = 1/2$ also.

(b) Let $\phi(\mathbf{x}) = x_1(x_2 \amalg x_3)$. Then $I_\phi(1) = 1/2^2 \cdot 3 = 3/4$, since among the four outcomes 100, 101, 110, 111, there are three critical path vectors for component 1 (101, 110, and 111).

However, $I_\phi(2) = 1/2^2 \cdot 1 = 1/4$, since among the four outcomes 010, 011, 110, 111, there is only one critical path vector for component 2—namely, 110. By symmetry, $I_\phi(3) = 1/4$ also.

Note that component 1 is distinctly more important than component 2 or 3. This is to be expected, since component 1 is in series with the rest of the system.

A similar analysis of structural importance may be performed in terms of cut vectors and cut sets. Note that (3.8) implies simultaneously that (a) $(1_i, \mathbf{x})$ is a path vector and (b) $(0_i, \mathbf{x})$ is a cut vector. We may thus call $(0_i, \mathbf{x})$ a *critical cut vector for* i, and $C_0(0_i, \mathbf{x})$ the corresponding *critical cut set for* i. It is thus apparent that *the number of critical cut sets for* i *is the same as the number of critical path sets for* i. Thus the structural importance of i, computed from (3.10), may be considered either as (a) the proportion of the 2^{n-1} outcomes having $x_i = 1$ which are critical path vectors for i, or (b) the proportion of the 2^{n-1} outcomes having $x_i = 0$ which are critical cut vectors for i.

EXERCISES

Note: Difficult exercises are marked with an asterisk.

1. Find all the path vectors and cut vectors of the bridge structure of Example 3.2.

2. Obtain the minimal path and cut representations of the following coherent structures:

   (a) The 2-out-of-3 system;
   (b) The 3-out-of-4 system;
   (c) The series system of three components;
   (d) The parallel system of three components.

3. Prove that a minimal path (cut) set of a coherent system is a minimal cut (path) set of the dual system.

4. (a) Let $P_1, \ldots, P_r$ be the minimal path sets of a coherent structure (C, ϕ), with each component relevant. Then $C = \bigcup_{j=1}^{r} P_j$.
   (b) Let $K_1, \ldots, K_s$ be the minimal cut sets of a coherent structure (C, ϕ), with each component relevant. Then $C = \bigcup_{j=1}^{s} K_j$.

*5. (a) Let $P_1, \ldots, P_r$ be nonempty sets of components no one of which is a subset of any other. Let $C = \bigcup_{j=1}^{r} P_j$. Finally, let $\phi(\mathbf{x}) = 1$ if and only if $C_1(\mathbf{x}) \supseteq P_j$[4] for some $j = 1, \ldots, r$; otherwise $\phi(\mathbf{x}) = 0$. Then (C, ϕ) is a coherent structure with minimal path sets $P_1, \ldots, P_r$.
   (b) Let $K_1, \ldots, K_s$ be nonempty sets of components no one of which is a subset of any other. Let $C = \bigcup_{j=1}^{s} K_j$. Finally, let $\phi(\mathbf{x}) = 0$ if and only if $C_0(\mathbf{x}) \supseteq K_j$ for some $j = 1, \ldots, s$. Then ϕ is a coherent structure consisting of the set C of components with minimal cut sets $K_1, \ldots, K_s$.

*6. (a) Show that P a minimal path set and $Q \subset P$ imply Q^c (the set complementary to Q) is a cut set.
   (b) Show that K a minimal cut set and $L \subset K$ imply L^c is a path set.

7. Find all distinct coherent structure functions of order 4.
[Hint: Use Exercise 5(a) or (b).]

8. Prove that equality holds in Theorem 2.4(a) only if the structure is parallel. Prove that equality holds in Theorem 2.4(b) only if the structure is series.

9. Let $K_1, K_2, \ldots, K_k (P_1, P_2, \ldots, P_p)$ be the min cut (path) sets for coherent structure ϕ and f any real-valued function of integers $1, 2, \ldots, n$. Show that

$$\min_{1 \le j \le k} \max_{i \in K_j} f(i) = \max_{1 \le j \le p} \min_{i \in P_j} f(i).$$

10. Suppose a coherent structure has min path sets

$$\begin{array}{cccc} \{1, 2\} & \{2, 3, 4\} & \{2, 4, 5\} & \{3, 4, 5\} \\ \{1, 3\} & \{2, 3, 5\} & \{2, 4, 6\} & \{3, 4, 6\} \\ & \{2, 3, 6\} & \{2, 5, 6\} & \{3, 5, 6\} \\ & & & \{4, 5, 6\}. \end{array}$$

Determine the number of critical paths for each component. Show that $I_\phi(1) < I_\phi(2)$ even though 1 appears in more two-component min path sets than does 2.

11. Show that if component i is itself a one-component min path set, then it is

[4] *Notation.* $A \subseteq B$ means the set A is contained in the set B. $A \subset B$ means A is *properly* contained in B.

structurally more important than any component which is not itself a one-component min path set.

12. Compute the structural importance of components in the bridge structure of Example 3.2.

4. MODULES OF COHERENT SYSTEMS

In practical reliability analysis, the procedure often followed is to compute first the reliability of each of the disjoint subsystems comprising a system, and then compute the overall system reliability from these subsystem reliabilities. To discuss such procedures and related computing procedures which yield bounds, we need to define precisely the concept of a module, which in engineering terms refers to a "package" of components which can be replaced as a whole.

4.1. Definition. The coherent system (A, χ) is a *module* of the coherent system (C, ϕ) if $\phi(\mathbf{x}) = \psi[\chi(\mathbf{x}^A), \mathbf{x}^{A^c}]$,[5] where ψ is a coherent structure function and $A \subseteq C$. The set $A \subseteq C$ is called a *modular set* of (C, ϕ). If $A \subset C$, then (A, χ) is a *proper module* of (C, ϕ).

Intuitively, a module (A, χ) of (C, ϕ) is a coherent subsystem that acts as if it were just a component. Knowing whether χ is 1 or 0 is as informative as knowing the value of x_i for each i in A, in determining the value of ϕ. In the usual performance diagram of a system, we can identify a module by the fact that it is a cluster of components with one wire leading into it and one wire leading out of it.

It is immediate that each component and its indicator function constitute a module.

4.2. Example. Consider the coherent system (C, ϕ), where $\phi(\mathbf{x}) = x_1(x_2 \amalg x_3)(x_4 \amalg x_5)$ and $C = \{1, \ldots, 5\}$. Then a module of (C, ϕ) is (A, χ),

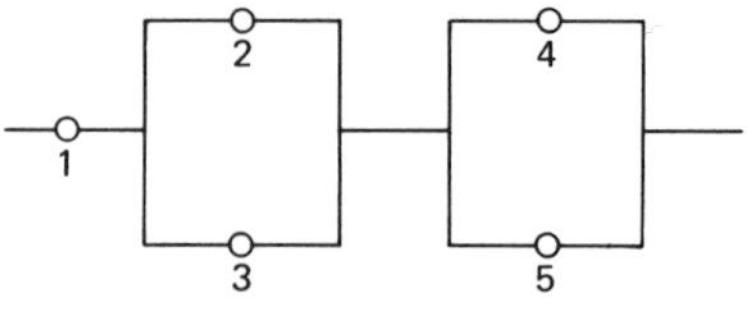

Figure 1.4.1.
Hi-fi example.

[5] *Notation.* Let A be a subset of C. Then (a) $\mathbf{x}^A$ denotes the vector with elements x_i, $i \in A$, and (b) A^c denotes the subset of C complementary to A.

where $A = \{2, 3\}$ is a modular set, and $\chi(\mathbf{x}^A) = x_2 \amalg x_3$ is the modular coherent structure function. We may write

$$\phi(\mathbf{x}) = \psi(\chi(\mathbf{x}^A), \mathbf{x}^{A^c}) = x_1 \cdot \chi \cdot (x_4 \amalg x_5).$$

Other modules are:

$$(\{4, 5\}, x_4 \amalg x_5)$$

$$(\{2, 3, 4, 5\}, (x_2 \amalg x_3) \cdot (x_4 \amalg x_5))$$

$$(\{1, 2, 3\}, x_1 \cdot (x_2 \amalg x_3))$$

$$(\{1, 4, 5\}, x_1 \cdot (x_4 \amalg x_5)).$$

The remaining modules are simply the individual units, and the system itself.

In practice, the process of designing a system automatically yields a modular decomposition of the system, that is, decomposition of the system into disjoint modules.

4.3. Definition. A *modular decomposition* of a coherent system (C, ϕ) is a set of disjoint modules $\{(A_1, \chi_1), \ldots, (A_r, \chi_r)\}$ together with an *organizing structure* ψ

$$\text{(a) } C = \bigcup_{i=1}^{r} A_i, \text{ where } A_i \bigcap A_j = E \text{ (the empty set) for } i \neq j;$$

$$\text{(b) } \phi(\mathbf{x}) \equiv \psi[\chi_1(\mathbf{x}^{A_1}), \chi_2(\mathbf{x}^{A_2}), \ldots, \chi_r(\mathbf{x}^{A_r})]. \tag{4.1}$$

Often, in practice, given a modular decomposition of a coherent system, we may find it helpful to obtain a *refinement* of the decomposition. That is, we may decompose each of the modules of the original decomposition into smaller modules. Clearly, a refinement of a decomposition of a coherent system (C, ϕ) is itself a decomposition of (C, ϕ). In practice, refinements arise as follows. A system is decomposed into its major subsystems. Each major subsystem is decomposed into components. Each component is decomposed into parts. The reliability of each part is determined from a handbook or observed data. From this information the reliability of each component is computed. Knowing the reliability of each component, we may compute the reliability of each subsystem. Finally, from the subsystem reliabilities we compute system reliability. In Section 4 of Chapter 2 we show how to obtain lower bounds on system reliability of successively greater accuracy using successive refinements of the original structure.

EXERCISES

1. If A is a modular set, is A^c necessarily a modular set?
2. Let (A, χ) be a module of (C, ϕ). Suppose $\chi(\mathbf{x}_1{}^A) = 1$, $\chi(\mathbf{x}_0{}^A) = 0$. Then $\phi(\mathbf{x}_1{}^A, \mathbf{x}) \equiv \phi(\mathbf{1}^A, \mathbf{x})$, $\phi(\mathbf{x}_0{}^A, \mathbf{x}) \equiv \phi(\mathbf{0}^A, \mathbf{x})$.

3. (a) Find all the modules of Figure 1.4.2.

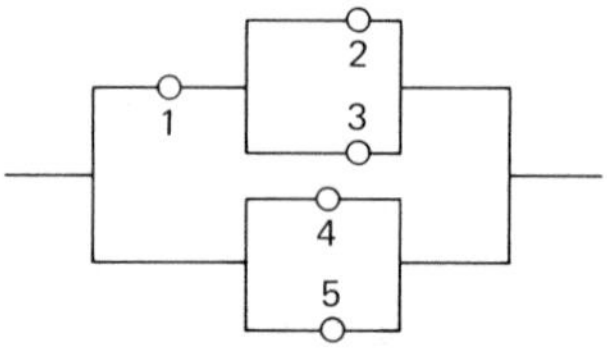

Figure 1.4.2.
Example.

(b) Find all the modules of Example 3.2.

4. Form a modular decomposition of the systems of Exercise 3 and successive refinements of it.

5. Show that the only modules in a k-out-of-n structure, where $1 < k < n$, are the individual components and the structure. Is this result true when $k = 1$ or $k = n$?

6. Assume that the structure (C, ϕ) is coherent.
   (a) Let κ_j be the jth minimal cut structure with corresponding cut set K_j $(j = 1, \ldots, k)$. Suppose $K_1 \bigcap \bigcup_{j=2}^{k} K_j = E$. Then (K_1, κ_1) is a module of (C, ϕ). (E is the empty set.)
   (b) Let ρ_j be the jth minimal path structure with corresponding path set P_j $(j = 1, \ldots, p)$. Suppose $P_1 \bigcap \bigcup_{j=2}^{p} P_j = E$. Then (P_1, ρ_1) is a module of (C, ϕ).

7. Show that the dual of a module of a coherent structure is a module of the dual structure.

*8. Let (A, χ), (B, λ) be modules of a coherent system (C, ϕ), with $B \subset A$. Then (B, λ) is a module of (A, χ).

5. NOTES AND REFERENCES

Section 1. What we have called a *relevant* component in Definition 1.6 has been called in previous literature [Esary-Proschan (1962), Esary-Proschan (1963a), Birnbaum-Esary (1965), B-P (1965), and so on] an *essential* component. We prefer to avoid this latter term, since components "essential" according to the usage in previous literature need not be essential to the functioning of a system. However, such components are *relevant* in determining whether the system functions or not.

Section 2. The formulation of the notion of a coherent system and the development of its basic properties are due to Birnbaum-Esary-Saunders

(1961). In B-P (1965), the term "monotonic structure" is used in place of the term "coherent structure" of Birnbaum-Esary-Saunders (1961).

The result formally stated in Theorem 2.4(a) has long been known in an intuitive way among design engineers concerned with reliability. The simple proof presented appears in Esary-Marshall-Proschan (1970).

Section 3. The minimal path and cut representations of Section 3 appear in Birnbaum-Esary-Saunders (1961). Theorem 3.5 is presented in Esary and Marshall (1970). Structural importance is defined in Birnbaum (1969).

Section 4. The mathematical formulation of the concept of a module and the development of its properties are due to Birnbaum and Esary (1965). The main theorem in their paper is the Three Modules Theorem:

Let (C, ϕ) be a coherent system. Suppose A_1, A_2, and A_3 are nonempty disjoint sets of components such that $A_1 \cup A_2$ and $A_2 \cup A_3$ are modular sets. Then

(a) A_1, A_2, and A_3 are modular sets.
(b) $A_1 \cup A_2 \cup A_3$ is a modular set.
(c) Modules χ_{A_1}, χ_{A_2}, χ_{A_3} appear either in series or in parallel.
(d) $A_1 \cup A_3$ is modular.

2

RELIABILITY OF COHERENT SYSTEMS

In Chapter 1 we discussed the deterministic aspects of structures, particularly coherent structures. In the present chapter we relate the reliability of the system to the reliabilities of the components. Methods for evaluating exact system reliability are discussed. Since it is often difficult or impossible to do this for very complex systems, we also obtain useful bounds on system reliability in terms of the minimal path or minimal cut reliabilities.

After Section 1 we do not restrict ourselves to the case of statistically independent components. Rather, we consider the more general case in which components are statistically "associated" in a sense to be made precise. This statistical model applies in particular to the familiar situation in which components are subjected to common stresses.

1. RELIABILITY OF SYSTEMS OF INDEPENDENT COMPONENTS

In this section *we assume that components are statistically independent.* Suppose that the state X_i of the ith component is random with

$$P[X_i = 1] = p_i = EX_i{}^{1} \qquad \text{for} \quad i = 1, \ldots, n. \tag{1.1}$$

Also, as in Chapter 1, $\mathbf{X} = (X_1, X_2, \cdots, X_n)$. We refer to p_i, the probability that i functions, as the *reliability* of i. Similarly, the reliability of the system is given by

$$P[\phi(\mathbf{X}) = 1] = h = E\phi(\mathbf{X}). \tag{1.2}$$

[1] *Notation.* EX denotes the expected value of the random variable X.

Under the assumption of independent components, we may represent system reliability as a function of component reliabilities:

$$h = h(\mathbf{p}).$$

When $p_1 = \cdots = p_n = p$, we will use the symbol $h(p)$. We will refer to $h(\mathbf{p})$ [or, in the case of like components, $h(p)$] as the *reliability function* of the structure ϕ. Note that if components are not independent, system reliability may not be a function of $\mathbf{p}$ alone; in this case, $h(\mathbf{p})$ will not be used.

As illustrations, the series structure $\phi(\mathbf{x}) = \prod_{i=1}^{n} x_i$ of Example 1.1 of Chapter 1 has reliability function $h(\mathbf{p}) = \prod_{i=1}^{n} p_i$, the parallel structure $\phi(\mathbf{x}) = \coprod_{i=1}^{n} x_i$ of Example 1.2 of Chapter 1 has reliability function $h(\mathbf{p}) = \coprod_{i=1}^{n} p_i \equiv 1 - \prod_{i=1}^{n} (1 - p_i)$. The hi-fi system of Example 1.5 of Chapter 1 has reliability function $h(\mathbf{p}) = (p_1 + p_2 - p_1 p_2)p_3(p_4 + p_5 - p_4 p_5)$. Assuming $p_1 = p_2 = \cdots = p_n = p$, the k-out-of-n system $\phi(\mathbf{x}) = 1$ if and only if $\sum_{i=1}^{n} x_i \geq k$ of Example 1.3 of Chapter 1 has reliability function

$$h(p) = \sum_{i=k}^{n} \binom{n}{i} p^i (1 - p)^{n-i}.$$

Graphical Example. In Figure 2.1.1 we plot the reliability function $h(p)$ for the following systems of three identical components.

(a) Series system: $h(p) = p^3$.
(b) 2-out-of-3 system: $h(p) = 3p^2(1 - p) + p^3$.
(c) Parallel system: $h(p) = 1 - (1 - p)^3$.

Some Basic Properties of System Reliability

From Lemma 1.7 of Chapter 1 we immediately obtain the corresponding pivotal decomposition of the reliability function.

1.1. Lemma. The following identity holds for the reliability function:

$$h(\mathbf{p}) = p_i h(1_i, \mathbf{p}) + (1 - p_i) h(0_i, \mathbf{p}) \qquad \text{for} \quad i = 1, \ldots, n. \tag{1.3}$$

PROOF. $h(\mathbf{p}) = E\phi(\mathbf{X}) = EX_i E\phi(1_i, \mathbf{X}) + (1 - EX_i)E\phi(0_i, \mathbf{X})$, from (1.1) of Chapter 1 and the independence of components. Equation (1.3) follows immediately. ‖

From (1.3), we note that $h(\mathbf{p})$ is multilinear, that is, linear in each p_i. Moreover, when $p_1 = p_2 = \cdots = p_n = p$, $h(p)$ is a polynomial in p.

Recall that coherent structures have increasing structure functions. The corresponding monotonicity property for reliability functions is given in Theorem 1.2 below.

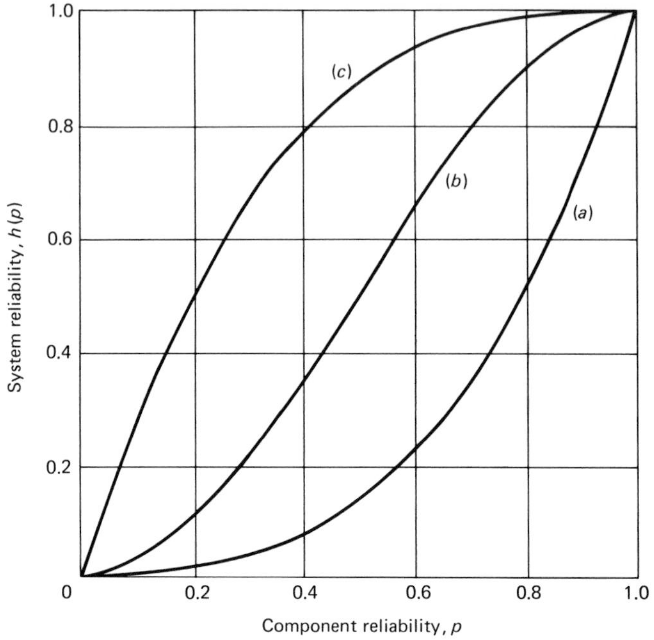

Figure 2.1.1.
System reliability versus component reliability for systems of three components.

1.2. Theorem. Let $h(\mathbf{p})$ be the reliability function of a coherent structure. Then $h(\mathbf{p})$ is strictly increasing in each p_i for $\mathbf{0} \ll \mathbf{p} \ll \mathbf{1}$.[2]

PROOF. From (1.3),

$$\frac{\partial h}{\partial p_i} = h(1_i, \mathbf{p}) - h(0_i, \mathbf{p}), \tag{1.4}$$

so that

$$\frac{\partial h}{\partial p_i} = E[\phi(1_i, \mathbf{X}) - \phi(0_i, \mathbf{X})]. \tag{1.5}$$

Since ϕ is increasing, then $\phi(1_i, \mathbf{x}) - \phi(0_i, \mathbf{x}) \geq 0$. In addition, $\phi(1_i, \mathbf{x}^0) - \phi(0_i, \mathbf{x}^0) = 1$ for some $\mathbf{x}^0$ since each component is relevant. Since $\mathbf{0} \ll \mathbf{p} \ll \mathbf{1}$, $\mathbf{x}^0$ has positive probability of occurring. Thus $E[\phi(1_i, \mathbf{X}) - \phi(0_i, \mathbf{X})] > 0$, and the desired result follows. ||

In Theorem 2.4 of Chapter 1 it was shown for structures that redundancy at the component level is more effective than redundancy at the system level. A corresponding result holds for reliability functions.

[2] *Notation.* $\mathbf{a} \ll \mathbf{b} \Leftrightarrow a_i < b_i$ for all i.

1.3. Theorem. Let h be the reliability function of a coherent system. Then

(a) $h(\mathbf{p} \amalg \mathbf{p}') \geq h(\mathbf{p}) \amalg h(\mathbf{p}')$,
(b) $h(\mathbf{p} \cdot \mathbf{p}') \leq h(\mathbf{p})h(\mathbf{p}')$,

for all $\mathbf{0} \leq \mathbf{p} \leq \mathbf{1}$, $\mathbf{0} \leq \mathbf{p}' \leq \mathbf{1}$. Equality holds in (a) for all $\mathbf{p}$, $\mathbf{p}'$ if and only if the system is parallel. Equality holds in (b) for all $\mathbf{p}$, $\mathbf{p}'$ if and only if the system is series.

PROOF. Let $X_1, \ldots, X_n$, $X_1', \ldots, X_n'$ be mutually independent binary random variables with $P[X_i = 1] = p_i$, $P[X_i' = 1] = p_i'$. Then from Theorem 2.4(a) of Chapter 1,

$$h(\mathbf{p} \amalg \mathbf{p}') - h(\mathbf{p}) \amalg h(\mathbf{p}')$$
$$= \sum_{\mathbf{x}} \sum_{\mathbf{x}'} [\phi(\mathbf{x} \amalg \mathbf{x}') - \phi(\mathbf{x}) \amalg \phi(\mathbf{x}')]P[\mathbf{X} = \mathbf{x}]P[\mathbf{X}' = \mathbf{x}'] \geq 0,$$

so that (a) above holds. A similar argument yields (b) above.

In both (a) and (b), the sufficient condition for equality follows from the corresponding sufficient condition for equality in Theorem 2.4 of Chapter 1. If $\mathbf{0} \ll \mathbf{p} \ll \mathbf{1}$ and $\mathbf{0} \ll \mathbf{p}' \ll \mathbf{1}$, then $P[\mathbf{X} = \mathbf{x}] > 0$ for all $\mathbf{x}$ and $P[\mathbf{X}' = \mathbf{x}'] > 0$ for all $\mathbf{x}'$. Thus in both (a) and (b), the necessary condition for equality follows from the corresponding necessary condition for equality in Theorem 2.4 of Chapter 1. ||

Example. Given the system of like components shown in Figure 2.1.2(a), we compute the corresponding reliability function, $h(p) = p[2p - p^2]$. Replicating the whole system yields a reliability function

$$h(p) \amalg h(p) = 1 - [1 - p(2p - p^2)]^2.$$

Replicating components yields a reliability function

$$h(p \amalg p) = [2p - p^2][1 - (1 - p)^4].$$

The two system reliability functions are plotted in Figure 2.1.2(b); of course $h(p \amalg p)$ dominates $h(p) \amalg h(p)$, as proved in general in Theorem 1.3.

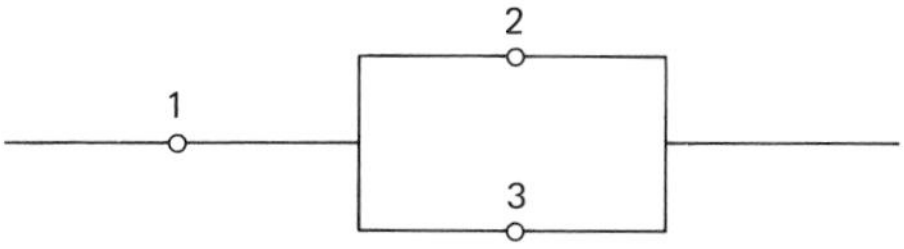

Figure 2.1.2(a).
Three-component system.

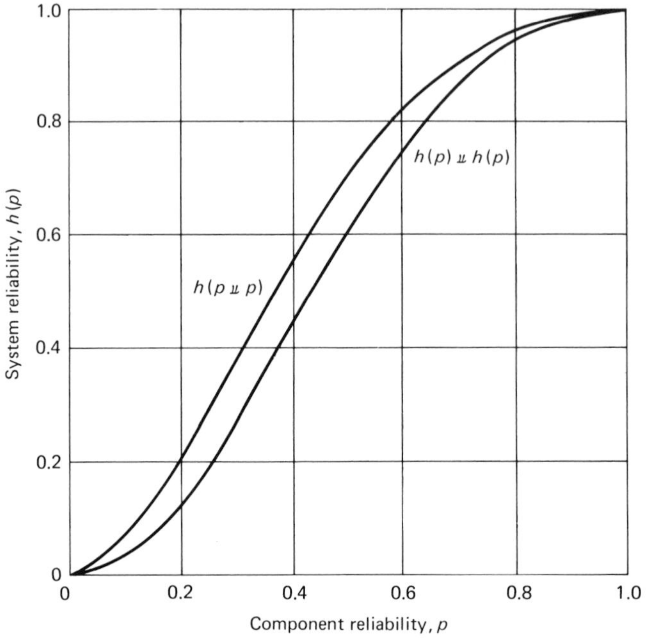

Figure 2.1.2(b).
Redundancy at component level versus redundancy at system level.

Computing Exact System Reliability

The minimal cut and minimal path representations developed in Chapter 1 provide a means for systematically computing the reliability of any coherent system. By (3.2) and (3.4) of Chapter 1, we have

$$\phi(\mathbf{x}) = \coprod_{j=1}^{p} \prod_{i \in P_j} x_i \tag{1.6}$$

and

$$\phi(\mathbf{x}) = \prod_{j=1}^{k} \coprod_{i \in K_j} x_i. \tag{1.7}$$

Expanding the right-hand sides into multinomial expressions in the x_i's and using the idempotency of x_i (that is, $x_i^2 = x_i$), we can compute system reliability simply by taking the expectation; that is,

$$h(\mathbf{p}) = E \coprod_{j=1}^{p} \prod_{i \in P_j} X_i$$

and

$$h(\mathbf{p}) = E \prod_{j=1}^{k} \coprod_{i \in K_j} X_i.$$

Example. If we take the structure of Example 4.2 of Chapter 1, then

$$\begin{aligned}\phi(\mathbf{x}) &= x_1(x_2 \amalg x_3)(x_4 \amalg x_5)\\ &= x_1(x_2 + x_3 - x_2x_3)(x_4 + x_5 - x_4x_5)\\ &= x_1x_2x_4 + x_1x_2x_5 - x_1x_2x_4x_5\\ &\quad + x_1x_3x_4 + x_1x_3x_5 - x_1x_3x_4x_5\\ &\quad - x_1x_2x_3x_4 - x_1x_2x_3x_5 + x_1x_2x_3x_4x_5.\end{aligned}$$

Hence

$$\begin{aligned}h(\mathbf{p}) = E\phi(\mathbf{X}) &= p_1p_2p_4 + p_1p_2p_5 - p_1p_2p_4p_5 + p_1p_3p_4 + p_1p_3p_5\\ &\quad - p_1p_3p_4p_5 - p_1p_2p_3p_4 - p_1p_2p_3p_5 + p_1p_2p_3p_4p_5.\end{aligned}$$

Alternatively, we can compute system reliability by pivoting about the ith component as in (1.3). This is the method of taking successive conditional expectations. We can also compute $h(\mathbf{p})$ by summing over all 2^n vectors $\mathbf{x}$ with 0 or 1 coordinates as

$$h(\mathbf{p}) = \sum_{\mathbf{x}} \phi(x_1, x_2, \ldots, x_n) \prod_i p_i^{x_i} q_i^{1-x_i}, \tag{1.8}$$

where $q_i = 1 - p_i$, and $0^0 \equiv 1$. See (1.2) of Chapter 1.

It is clear from the previous example that the exact computation of system reliability is usually a formidable task for complex systems. For this reason we spend considerable time on bounds on system reliability in Section 3. The following method together with the min path algorithm in the appendix can be used to approximate the reliability of complex systems.

The Inclusion-Exclusion Method

The following method provides successive upper and lower bounds on system reliability which converge to the exact system reliability. Let E_r be the event that all components in min path set P_r work. Then

$$P(E_r) = \prod_{i \in P_r} p_i.$$

System success corresponds to the event $\bigcup_{r=1}^{p} E_r$ if the system has p min path sets. Then

$$h = P\left[\bigcup_{r=1}^{p} E_r\right].$$

Let

$$S_k = \sum_{1 \le i_1 < i_2 < \cdots < i_k \le p} P[E_{i_1} \cap E_{i_2} \cap \ldots \cap E_{ik}].$$

By the inclusion–exclusion principle (Feller, 1968, pp. 98–101),

$$h = \sum_{k=1}^{p} (-1)^{k-1} S_k, \tag{1.9}$$

and

$$h \le S_1 = \sum_{r=1}^{p} \prod_{i \in P_r} p_i,$$

$$h \ge S_1 - S_2,$$

$$h \le S_1 - S_2 + S_3,$$

$$h \ge S_1 - S_2 + S_3 - S_4,$$

and so on. Although it is not true in general that the upper bounds decrease and the lower bounds increase, in practice it may be necessary to calculate only a few S_k's to obtain a close approximation.

Of course similar formulas for computing system *unreliability* in terms of min cut sets and component unreliabilities can be given.

Reliability Importance of Components

In 3.7 of Chapter 1 we developed a measure of the *structural importance* of each component of a coherent system. This measure was based on a knowledge of the *structure* of the system only. Next we develop a measure of the *reliability importance* of each component, which takes into account *component reliabilities* as well as system structure. Such a measure can be very useful in system analysis in determining those components on which additional research and development effort can be most profitably expended.

Intuitively, it would seem reasonable to measure the importance of a component in contributing to system reliability by the rate at which system reliability improves as the reliability of the component improves. Specifically, we present the following definition.

1.4. Definition. The *reliability importance* $I_h(j)$ of component j is given by

$$I_h(j) = \frac{\partial h(\mathbf{p})}{\partial p_j}. \tag{1.10a}$$

By (1.4), an equivalent definition is

$$I_h(j) = h(1_j, \mathbf{p}) - h(0_j, \mathbf{p}), \tag{1.10b}$$

or more explicitly,

$$I_h(j) = E[\phi(1_j, \mathbf{X}) - \phi(0_j, \mathbf{X})]. \tag{1.10c}$$

Remarks. (a) From (1.10c), we see that if $p_i = \frac{1}{2}$ for $i \neq j$, the *reliability importance*, $I_h(j) \equiv I_\phi(j)$, the *structural importance*, given in (3.10) of Chapter 1. Thus, if $p_i = \frac{1}{2}$ for $i = 1, \ldots, n$, then reliability importance and structural importance coincide for each component of the system.

(b) Using the arguments of Theorem 1.2, it is clear that for a coherent system with each component reliability in (0, 1), component reliability importance satisfies

$$0 < I_h(j) < 1 \qquad \text{for} \quad j = 1, \ldots, n \qquad \text{and} \quad n > 1. \tag{1.11}$$

(c) In the case of *statistically dependent* components, which will be considered in the next section, we would use (1.10c) to define the reliability importance of component j.

The reliability importance of components may be used to evaluate the effect of an improvement in component reliability on system reliability, as follows. By the chain rule for differentiation,

$$\frac{dh}{dt} = \sum_{j=1}^{n} \frac{\partial h}{\partial p_j} \frac{dp_j}{dt},$$

where t is a common parameter, say, the time elapsed since system development began. Using (1.10a), we have

$$\frac{dh}{dt} = \sum_{j=1}^{n} I_h(j) \frac{dp_j}{dt}. \tag{1.12}$$

Thus the rate at which system reliability grows is a weighted combination of the rates at which component reliabilities grow, where the weights are the reliability importance numbers.

From (1.12) we may also obtain

$$\Delta h \cong \sum_{j=1}^{n} I_h(j)\, \Delta p_j, \tag{1.13}$$

where Δh is the perturbation in system reliability corresponding to perturbations Δp_j in component reliabilities. As in (1.12), the reliability importance numbers enter as weights. Thus small improvements Δp_j in component reliabilities lead to a corresponding improvement Δh in system reliability in accordance with (1.13).

Examples. Assume components have been labeled so that component reliabilities are ordered as follows

$$p_1 \leq p_2 \leq \cdots \leq p_n.$$

(a) *Series System.* If $h(\mathbf{p}) = \prod_{i=1}^{n} p_i$, then

$$I_h(j) = \prod_{i \neq j} p_i$$

and $I_h(1) \geq I_h(2) \geq \cdots \geq I_h(n)$, so that the component with lowest reliability is the most important to the system. This reflects the well-known principle that "a chain is as strong as its weakest link."

(b) *Parallel System.* If $h(\mathbf{p}) = \coprod_{i=1}^{n} p_i$, then

$$I_h(j) = \prod_{i \neq j} (1 - p_i)$$

and $I_h(1) \leq I_h(2) \leq \cdots \leq I_h(n)$, so that the component with highest reliability is the most important to the system. This, too, is intuitively reasonable, since if just *one* component functions, the system functions.

(c) *2-out-of-3 System.* For a 2-out-of-3 system, $h(\mathbf{p}) = p_1 p_2 + p_1 p_3 + p_2 p_3 - 2p_1 p_2 p_3$. Then

$$I_h(1) = p_2 + p_3 - 2p_2 p_3$$
$$I_h(2) = p_1 + p_3 - 2p_1 p_3$$

and

$$I_h(3) = p_1 + p_2 - 2p_1 p_2.$$

If $p_i \geq \frac{1}{2}$ for $i = 1, 2, 3$, then

$$I_h(3) \geq I_h(2) \geq I_h(1),$$

and the component with highest reliability is the most important to the system. If $p_i \leq \frac{1}{2}$ for $i = 1, 2, 3$, then

$$I_h(1) \geq I_h(2) \geq I_h(3),$$

and the component with lowest reliability is now the most important to the system.

EXERCISES

1. Assume $\phi_1(\mathbf{x}) \leq \phi_2(\mathbf{x})$ for all $\mathbf{x}$. Then $h_1(\mathbf{p}) \leq h_2(\mathbf{p})$ for all $\mathbf{0} \leq \mathbf{p} \leq \mathbf{1}$. Under what conditions does strict inequality hold?
2. Assume $h_1(\mathbf{p}) \leq h_2(\mathbf{p})$ for all $\mathbf{0} \ll \mathbf{p} \ll \mathbf{1}$. Then $\phi_1(\mathbf{x}) \leq \phi_2(\mathbf{x})$ for all $\mathbf{x}$.
3. Compute the exact system reliability of the bridge in Example 3.2 of Chapter 1.
4. The reliability importance of components can also be calculated using the concept of modules presented in Section 4 of Chapter 1. Show that if component j is in a module with structure function χ, which is in turn part of a coherent system ϕ, then $I_h(j)$ may be computed as the product of the importance of j in structure χ, multiplied by the importance of the module considered as a component. We remind the reader that components are assumed to be statistically independent.

5. Compute the reliability importance of components in the bridge of Example 3.2 of Chapter 1.

2. ASSOCIATION OF RANDOM VARIABLES

In a great many reliability situations, the random variables of interest are *not* independent, but rather are "associated." As examples, consider:

(a) Minimal path structures of a coherent system having components in common;
(b) Components subjected to the same set of stresses;
(c) Structures in which components share the load, so that failure of one component results in increased load on each of the remaining components.

Note that in each case the random variables of interest tend to act similarly. Thus in case (a), failure of a component will tend to degrade the performance of all the minimal path structures containing it. In case (b), a high stress will affect all the components adversely. In case (c), the functioning (failure) of a component contributes to the functioning (failure) of the remaining components.

We wish now to formulate a definition of association appropriate to reliability situations. If we had just two random variables S and T, we could consider them associated if $\mathrm{cov}[S, T] \geq 0$.[3] A stronger requirement would be $\mathrm{cov}[f(S), g(T)] \geq 0$ for all increasing f and g. Finally, if $\mathrm{cov}[f(S, T), g(S, T)] \geq 0$ for all f and g increasing in each argument, we would have a still stronger version of association.

The strongest of these three criteria has a natural multivariate generalization which serves as a useful definition of association, especially for reliability applications, as we will see. See Esary-Proschan-Walkup (1967) and Esary-Proschan (1970).

2.1. Definition. Random variables $T_1, \ldots, T_n$ (not necessarily binary) are *associated* if

$$\mathrm{cov}[\Gamma(\mathbf{T}), \Delta(\mathbf{T})] \geq 0$$

for all pairs of increasing binary functions Γ, Δ.

By restricting our test functions Γ, Δ to be binary and increasing, no loss of generality is suffered, since we will see in Exercise 5 that this implies

[3] *Notation.* $\mathrm{cov}[S, T] = E[(S - ES)(T - ET)]$ denotes the covariance of S and T.

nonnegative covariance for all increasing test functions for which the covariance exists.

Association of random variables satisfies the following desirable multivariate properties.

(P_1) Any subset of associated random variables are associated. We leave the proof to Exercise 1.

(P_2) The set consisting of a single random variable is associated. We leave the proof to Exercise 2.

(P_3) Increasing functions of associated random variables are associated.

PROOF. Let $T_1, \ldots, T_n$ be associated, f_i increasing, and $S_i \equiv f_i(\mathbf{T})$ for $i = 1, \ldots, m$. If Γ and Δ are increasing binary functions, then $\Gamma(f_1(\mathbf{T}), \ldots, f_m(\mathbf{T}))$ and $\Delta(f_1(\mathbf{T}), \ldots, f_m(\mathbf{T}))$ are increasing binary functions of $\mathbf{T}$. By definition of association,

$$\text{cov}_{\mathbf{S}}[\Gamma(\mathbf{S}), \Delta(\mathbf{S})] = \text{cov}_{\mathbf{T}}[\Gamma(\mathbf{f}(\mathbf{T})), \Delta(\mathbf{f}(\mathbf{T}))] \geq 0. \quad \|$$

(P_4) If two sets of associated random variables are independent of one another, then their union is a set of associated random variables.

PROOF. Let $\mathbf{X}$ be associated, $\mathbf{Y}$ be associated, and $\mathbf{X}$ and $\mathbf{Y}$ be independent of each other. Let Γ and Δ be binary increasing functions. Writing Γ for $\Gamma(\mathbf{X}, \mathbf{Y})$ and Δ for $\Delta(\mathbf{X}, \mathbf{Y})$, we have

$$\begin{aligned}
\text{cov}(\Gamma, \Delta) &= E_{\mathbf{X},\mathbf{Y}}\Gamma\Delta - E_{\mathbf{X},\mathbf{Y}}\Gamma E_{\mathbf{X},\mathbf{Y}}\Delta \\
&= [E_{\mathbf{X}}E_{\mathbf{Y}}\Gamma\Delta - E_{\mathbf{X}}(E_{\mathbf{Y}}\Gamma \cdot E_{\mathbf{Y}}\Delta)] \\
&\quad + [E_{\mathbf{X}}(E_{\mathbf{Y}}\Gamma E_{\mathbf{Y}}\Delta) - E_{\mathbf{X}}E_{\mathbf{Y}}\Gamma \cdot E_{\mathbf{X}}E_{\mathbf{Y}}\Delta] \\
&= E_{\mathbf{X}} \,\text{cov}_{\mathbf{Y}}(\Gamma, \Delta) + \text{cov}_{\mathbf{X}}(E_{\mathbf{Y}}\Gamma, E_{\mathbf{Y}}\Delta),
\end{aligned}$$

where for any vector of random variables $\mathbf{Z}$, $E_{\mathbf{Z}}$ denotes expectation over the distribution of $\mathbf{Z}$. In the second equality we have used the fact that $E_{\mathbf{X},\mathbf{Y}} = E_{\mathbf{X}}E_{\mathbf{Y}}$ since $\mathbf{X}$ and $\mathbf{Y}$ are independent of each other.

Since $\text{cov}_{\mathbf{Y}}[\Gamma(\mathbf{x}, \mathbf{Y}), \Delta(\mathbf{x}, \mathbf{Y})] \geq 0$ for each fixed $\mathbf{x}$, then $E_{\mathbf{X}} \,\text{cov}_{\mathbf{Y}}(\Gamma, \Delta) \geq 0$. Since $E_{\mathbf{Y}}\Gamma(\mathbf{x}, \mathbf{Y})$ and $E_{\mathbf{Y}}\Delta(\mathbf{x}, \mathbf{Y})$ are increasing functions of $\mathbf{x}$,

$$\text{cov}_{\mathbf{X}}(E_{\mathbf{Y}}\Gamma, E_{\mathbf{Y}}\Delta) \geq 0,$$

using Exercise 5. ‖

Properties P_2 and P_4 immediately imply the following theorem.

2.2. Theorem. Independent random variables are associated.

In the reliability setting, property P_3 tells us that coherent systems built out of associated (say, independent) components are themselves associated. Thus, returning to the example illustrated by case (a), the minimal path structure functions of a coherent system are increasing in the component performance indicator variables $X_1, \ldots, X_n$. Thus if $X_1, \ldots, X_n$ are associated, the minimal path indicator variables are associated. We will use this fact to obtain convenient bounds on system reliability in Theorem 3.3.

In the special case in which the associated random variables are *binary*, we obtain the following theorem.

2.3. Theorem. If $X_1, \ldots, X_n$ are associated binary random variables, then $1 - X_1, \ldots, 1 - X_n$ are also associated binary random variables.

PROOF. Let Γ, Δ be binary increasing functions. Then the dual functions

$$\Gamma^D(\mathbf{x}) = 1 - \Gamma(\mathbf{1} - \mathbf{x}), \ \Delta^D(\mathbf{x}) = 1 - \Delta(\mathbf{1} - \mathbf{x})$$

are also binary increasing.

We may write

$$\begin{aligned} \text{cov}[\Gamma(\mathbf{1} - \mathbf{X}), \Delta(\mathbf{1} - \mathbf{X})] &= \text{cov}[1 - \Gamma^D(\mathbf{X}), 1 - \Delta^D(\mathbf{X})] \\ &= \text{cov}[\Gamma^D(\mathbf{X}), \Delta^D(\mathbf{X})] \geq 0 \end{aligned}$$

by P_3. ‖

EXERCISES

1. Prove property P_1 of association.
2. Prove property P_2 of association.
3. Show that if X and Y are binary random variables, then $\text{cov}(X, Y) \geq 0$ implies X and Y are associated. Does this hold for general random variables?

*4. Let S and T be random variables with finite expectations ES, ET, EST. Define $X_S(s) = 1$ if $S > s$, 0 otherwise. Then

$$\text{cov}[S, T] = \int_{-\infty}^{\infty} \int_{-\infty}^{\infty} \text{cov}[X_S(s), X_T(t)]\, ds\, dt. \tag{2.1}$$

5. Using Exercise 4, show that if $\mathbf{T}$ are associated and f, g are increasing functions, then $\text{cov}\,[f(\mathbf{T}), g(\mathbf{T})] \geq 0$, assuming the covariance exists.
6. Show that if S and T are associated and $\text{cov}[S, T] = 0$, then S and T are independent. Does $\text{cov}[S, T] = 0$ imply the independence of S and T in general?

7. For a coherent system of independent components, show that the lifetimes corresponding to the minimal path (minimal cut) structure functions are associated.
8. Show that if X and Y are jointly distributed with bivariate normal distribution with correlation coefficient $\rho \geq 0$, then X and Y are associated.
9. Show that a set of order statistics $X_{1:n} \leq X_{2:n} \leq \cdots \leq X_{n:n}$ is associated.
10. Give examples of sets of random variables that are *not* associated.
11. Let $T_1, \ldots, T_n$ be associated. Then $aT_1, \ldots, aT_n$ are associated, for any real valued a.

3. BOUNDS ON SYSTEM RELIABILITY

If we calculate the reliability of a series system assuming the components independent when in fact they are associated but not independent, we will underestimate system reliability. The reverse is true for parallel systems. This follows from Theorem 3.1:

3.1. Theorem. If $X_1, \ldots, X_n$ are associated binary random variables, then

$$P\left[\prod_{i=1}^{n} X_i = 1\right] \geq \prod_{i=1}^{n} P[X_i = 1], \tag{3.1}$$

$$P\left[\coprod_{i=1}^{n} X_i = 1\right] \leq \coprod_{i=1}^{n} P[X_i = 1]. \tag{3.2}$$

PROOF OF (3.1). X_1 and $\prod_2^n X_i$, being increasing functions of $\mathbf{X}$, are associated. Thus

$$\operatorname{cov}\left[X_1, \prod_2^n X_i\right] = EX_1 \prod_2^n X_i - EX_1 E \prod_2^n X_i \geq 0.$$

Repeated applications of this argument yield

$$E \prod_1^n X_i \geq \prod_1^n EX_i,$$

which is equivalent to (3.1).

We leave the proof of (3.2) to Exercise 1. ||

A generalization of Theorem 3.1 is discussed in Esary-Proschan-Walkup (1967).

Remark. An extreme example of association occurs if $X_1 = X_2 = \cdots = X_n$ with probability 1. Then

$$P\left[\prod_1^n X_i = 1\right] = P[X_1 = 1],$$

whereas

$$\prod_{i=1}^n P[X_i = 1] = P^n[X_1 = 1],$$

so that inequality (3.1) clearly holds. Similarly,

$$P\left[\coprod_{i=1}^n X_i = 1\right] = 1 - P[X_1 = 0],$$

whereas

$$\coprod_{i=1}^n P[X_i = 1] = 1 - P^n[X_1 = 0],$$

so that (3.2) clearly holds.

The following results will be useful later.

3.2. Theorem. If $T_1, T_2, \ldots, T_n$ are associated random variables (not necessarily binary), then

$$P[T_1 > t_1, \ldots, T_n > t_n] \geq \prod_{i=1}^n P[T_i > t_i] \tag{3.3}$$

and

$$P[T_1 \leq t_1, \ldots, T_n \leq t_n] \geq \prod_{i=1}^n P[T_i \leq t_i]. \tag{3.4}$$

PROOF. Let $X_i(t_i) = 1$ if $T_i > t_i$ and $X_i(t_i) = 0$ if $T_i \leq t_i$. Then $X_i(t_i)$ is nondecreasing in T_i, and so by property P_3 of association, $X_1(t_1), \ldots, X_n(t_n)$ are associated. Using (3.1) and (3.2) we obtain (3.3) and (3.4). ||

3.3. Corollary. If $T_1, T_2, \ldots, T_n$ are associated, then

$$P[\min_{1 \leq i \leq n} T_i > t] \geq \prod_{i=1}^n P[T_i > t] \tag{3.5}$$

and

$$P[\max_{1 \leq i \leq n} T_i > t] \leq \coprod_{i=1}^n P[T_i > t]. \tag{3.6}$$

We leave the proof for Exercise 6.

We obtain crude bounds on the reliability of a coherent system by comparing it with a series system (for the lower bound) and a parallel system (for the upper bound).

3.3. Theorem. Let ϕ be a coherent structure of associated components with reliabilities $p_1, \ldots, p_n$. Then

$$\prod_{i=1}^{n} p_i \leq P[\phi(\mathbf{X}) = 1] \leq \coprod_{i=1}^{n} p_i. \tag{3.7}$$

PROOF. Taking expectations in (2.1) of Chapter 1, we obtain

$$E \prod_{i=1}^{n} X_i \leq E\phi(\mathbf{X}) \leq E \coprod_{i=1}^{n} X_i.$$

Since $E \prod X_i \geq \prod EX_i$ by Theorem **3.1**, (3.7) follows. ||

Next we obtain improved bounds on system reliability using additional information, namely, the minimal path and minimal cut representations of the structure given in (3.2) and (3.4) of Chapter 1, respectively. These bounds are generally more readily computed than the exact system reliability. See Example 3.7 below.

3.4. Theorem. Let ϕ be a coherent structure of associated components. Let $\rho_1(\mathbf{x}), \ldots, \rho_p(\mathbf{x})$ be the minimal path series structures, and $\kappa_1(\mathbf{x}), \ldots, \kappa_k(\mathbf{x})$ be the minimal cut parallel structures of ϕ. Then

$$\prod_{j=1}^{k} P[\kappa_j(\mathbf{X}) = 1] \leq P[\phi(\mathbf{X}) = 1] \leq \coprod_{j=1}^{p} P[\rho_j(\mathbf{X}) = 1]. \tag{3.8}$$

PROOF. $\kappa_1(\mathbf{x}), \ldots, \kappa_k(\mathbf{x})$ are increasing functions of $\mathbf{x}$. Since $\mathbf{X}$ are associated by hypothesis, then $\kappa_1(\mathbf{X}), \ldots, \kappa_k(\mathbf{X})$ are associated by P_3. Thus

$$\begin{aligned} P[\phi(\mathbf{X}) = 1] &= P\left[\prod_{j=1}^{k} \kappa_j(\mathbf{X}) = 1\right] && \text{[by (3.4) of Chapter 1]} \\ &\geq \prod_{j=1}^{k} P[\kappa_j(\mathbf{X}) = 1] && \text{[by (3.1)],} \end{aligned}$$

establishing the lower bound. Similarly,

$$\begin{aligned} P[\phi(\mathbf{X}) = 1] &= P\left[\coprod_{j=1}^{p} \rho_j(\mathbf{X}) = 1\right] && \text{[by (3.2) of Chapter 1]} \\ &\leq \coprod_{j=1}^{p} P[\rho_j(\mathbf{X}) = 1] && \text{[by (3.2)],} \end{aligned}$$

establishing the upper bound. ||

We can obtain more explicit bounds on system reliability in terms of component reliabilities when components are independent.

3.5. Corollary. Let ϕ be a coherent system of independent components. Then

$$\prod_{j=1}^{k} \coprod_{i \in K_j} p_i \leq P[\phi(\mathbf{X}) = 1] \leq \coprod_{j=1}^{p} \prod_{i \in P_j} p_i. \tag{3.9}$$

PROOF. For the jth minimal path series structure, $P[\rho_j(\mathbf{X}) = 1] = \prod_{i \in P_j} p_i$. Thus from (3.8), $P[\phi(\mathbf{X}) = 1] \leq \coprod_{j=1}^{p} \prod_{i \in P_j} p_i$.

For the jth minimal cut parallel structure $P[\kappa_j(\mathbf{X}) = 1] = \coprod_{i \in K_j} p_i$. Thus from (3.8), $P[\phi(\mathbf{X}) = 1] \geq \prod_{j=1}^{k} \coprod_{i \in K_j} p_i$. ‖

Note that the lower (upper) bound can be obtained by replacing each set of replicated components in the minimal cut (path) representation of ϕ by a corresponding set of independently operating components of the same reliability, and then assuming all components in the structure statistically independent. It is easy to demonstrate the following properties.

3.6. Properties of the Bounds for the Case of Independent Components. Let $l(\mathbf{p}) \equiv \prod_{j=1}^{k} \coprod_{i \in K_j} p_i$ and $u(\mathbf{p}) \equiv \coprod_{j=1}^{p} \prod_{i \in P_j} p_i$. Then

(a) $l(\mathbf{p})$ and $u(\mathbf{p})$ are increasing functions.

(b) $l(\mathbf{p}) < h(\mathbf{p}) < u(\mathbf{p})$ for $\mathbf{0} \ll \mathbf{p} \ll \mathbf{1}$ if at least two minimal cut sets overlap and at least two minimal path sets overlap. (See Exercise 2.)

3.7. Example. When components are independent with high (low) reliabilities, the lower (upper) bound in Corollary 3.5 yields a good approximation to system reliability.

As an example, consider the two-terminal network shown in Figure 2.3.1. This example appears in B–P (1965), Chapter 7.

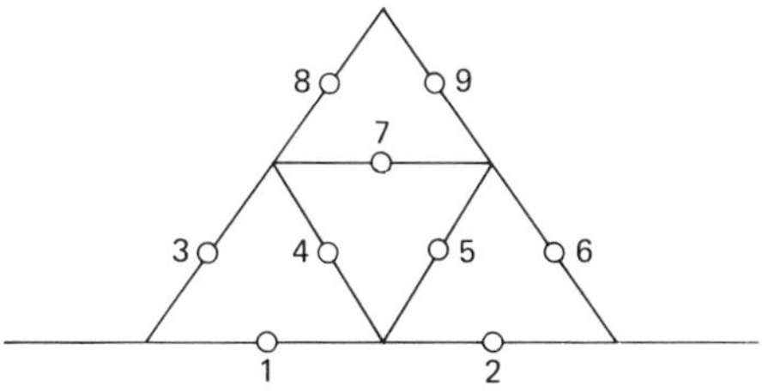

Figure 2.3.1.
Two-terminal network.

The minimal path sets are readily seen to be

12	156	1476	14896	342
3456	376	3752	3896	38952.

Similarly, the minimal cut sets are

$$\begin{array}{cccc} 13 & 1478 & 1479 & 1456 \\ 2543 & 2578 & 2579 & 26. \end{array}$$

Assuming independent components with common reliability p, the lower bound given in (3.9) is

$$l(p) = (1 - q^2)^2(1 - q^4)^6,$$

where $q = 1 - p$. The upper bound in (3.9) is

$$u(p) = 1 - (1 - p^2)(1 - p^3)^3(1 - p^4)^4(1 - p^5)^2.$$

Figure 2.3.2 presents a graphical comparison of the true system reliability $h(p)$ with the upper and lower bounds $u(p)$ and $l(p)$.

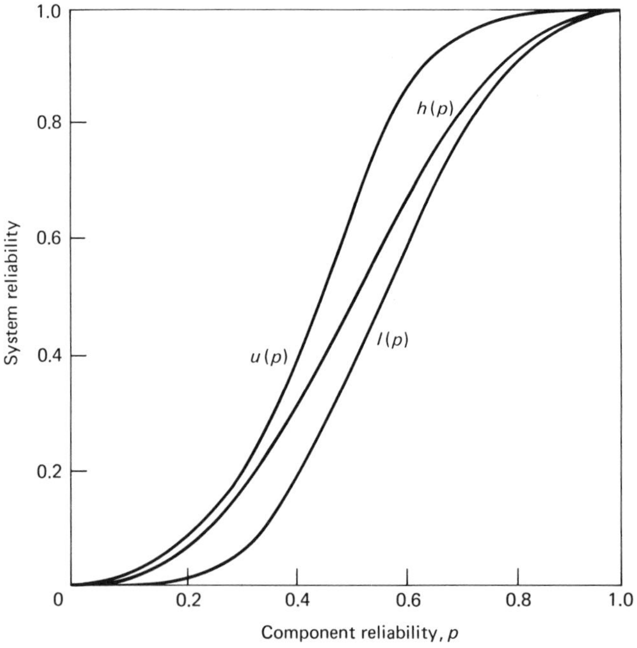

Figure 2.3.2.
Bounds on structure reliability for a two-terminal network.

3.8. Comparison of Minimal Cut Lower Bound and the Series Lower Bound. It might be conjectured that

$$l_\phi(\mathbf{p}) = \prod_{j=1}^{k} \coprod_{i \in K_j} p_i \geq \prod_{i=1}^{n} p_i \tag{3.10}$$

is always true since

$$\phi(\mathbf{x}) = \prod_{i=1}^{k} \kappa_i(\mathbf{x}) \geq \prod_{i=1}^{n} x_i$$

by (2.1) of Chapter 1. The following example shows that (3.10) is *not* always true. Let ϕ be the structure function for a 3-out-of-4 system. The minimal cut schematic diagram is

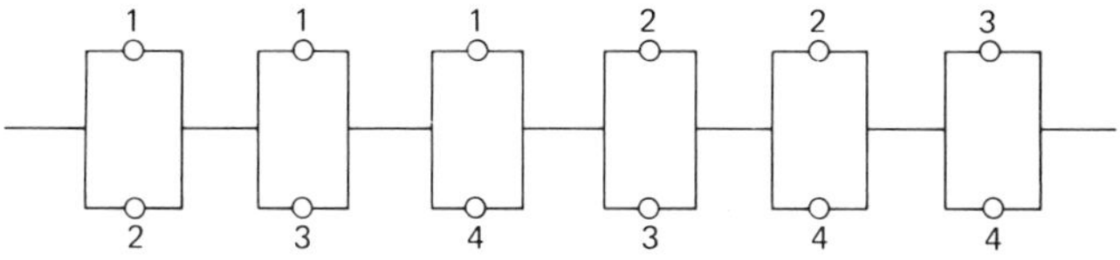

and

$$l_\phi(p) = (2p - p^2)^6,$$

where $p_i = p$ for $i = 1, 2, 3, 4$. Then $l_\phi(\mathbf{p}) \geq \prod_{i=1}^{4} p_i = p^4$ if and only if $(2 - p)^6 p^2 \geq 1$, which is false for small p.

The following *min-max* bounds will *always* improve on the trivial bounds of (3.7).

Min-Max Bounds on System Reliability

Using (3.5) and (3.6) of Chapter 1 we obtain *upper* bounds which depend on min *cuts* and *lower* bounds which depend on min *paths*, in contrast to the bounds of Theorem 3.4.

3.9. Theorem (Min-Max Bounds). Let ϕ be a coherent structure. Let $P_1, P_2, \ldots, P_p$ be the component min path sets corresponding to ϕ, and let $K_1, K_2, \ldots, K_k$ be the component min cut sets corresponding to ϕ. Then the following bounds always hold:

$$\max_{1 \leq r \leq p} P[\min_{i \in P_r} X_i = 1] \leq P[\phi(\mathbf{X}) = 1] \leq \min_{1 \leq s \leq k} P[\max_{i \in K_s} X_i = 1]. \quad (3.11)$$

If, in addition, components are associated, then the more explicit bounds hold:

$$\max_{1 \leq r \leq p} \prod_{i \in P_r} p_i \leq P[\phi(\mathbf{X}) = 1] \leq \min_{1 \leq s \leq k} \coprod_{i \in K_s} p_i. \quad (3.12)$$

PROOF. By (3.5) and (3.6) of Chapter 1,

$$\phi(\mathbf{X}) = \max_{1 \leq r \leq p} \min_{i \in P_r} X_i = \min_{1 \leq s \leq k} \max_{i \in K_s} X_i.$$

Thus

$$\min_{i \in P_r} X_i \leq \phi(\mathbf{X}) \leq \max_{i \in K_s} X_i.$$

As a consequence,

$$P[\min_{i \in P_r} X_i = 1] \leq P[\phi(\mathbf{X}) = 1] \leq P[\max_{i \in K_s} X_i = 1]$$

for $1 \leq r \leq p$ and $1 \leq s \leq k$. Equation (3.11) now follows by maximizing over r and minimizing over s.

Equation (3.12) follows from (3.11), (3.1), and (3.2). ||

Remark. Note that when components are associated,

$$\prod_{i=1}^{n} p_i \leq \max_{1 \leq r \leq p} \prod_{i \in P_r} p_i \leq \min_{1 \leq s \leq k} \coprod_{i \in K_s} p_i \leq \coprod_{i=1}^{n} p_i, \tag{3.13}$$

by (3.1) and (3.2). Thus the bounds in (3.12) are always better than the bounds in (3.7).

It follows from (3.13) that the min-max lower bound will be better than the $l_\phi(\mathbf{p})$ lower bound in the example of 3.8 when p is very small. In general, for *independent* components we have the improved bounds

$$\max\left[l_\phi(\mathbf{p}),\ \max_{1 \leq r \leq p} \prod_{i \in P_r} p_i\right] \leq h_\phi(\mathbf{p}) \leq \min\left[u_\phi(\mathbf{p}),\ \min_{1 \leq s \leq k} \coprod_{i \in K_s} p_i\right].$$

For "large" p_i we conjecture that $l_\phi(\mathbf{p})$ will provide the better lower bound while $\min_{1 \leq s \leq k} \coprod_{i \in K_s} p_i$ will provide the better upper bound.

EXERCISES

1. Prove (3.2) of Theorem 3.1. (Hint: Use Theorem 2.3.)
2. Prove properties 3.6(a) and (b).
3. (a) Find the minimal path sets and minimal cut sets of the following two-terminal network:

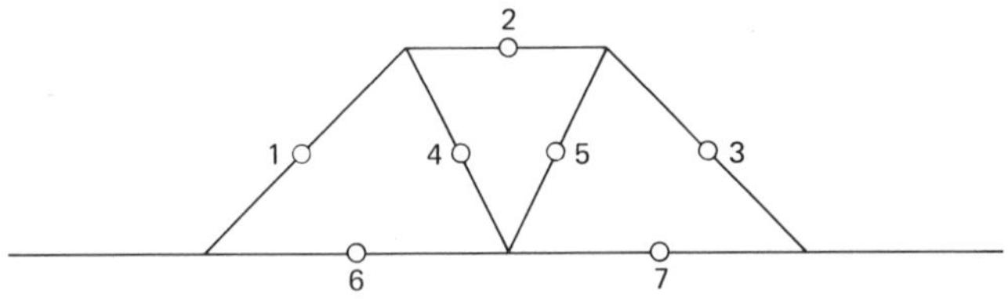

(b) Assuming independent components with common reliability p, compute the upper bound $u(p)$ and the lower bound $l(p)$.

4. A television set has an audio subsystem and a video subsystem which have certain components in common. Assuming all components independent, show that the probability that the set produces both a picture and sound is at least as great as the product of the probability that the set produces a picture by the probability that the set produces sound.

5. Suppose $\phi_1(\mathbf{x}) \leq \phi_2(\mathbf{x})$ for all $\mathbf{x}$. Then is it true that $l_{\phi_1}(\mathbf{p}) \leq l_{\phi_2}(\mathbf{p})$ and $u_{\phi_1}(\mathbf{p}) \leq u_{\phi_2}(\mathbf{p})$ for $\mathbf{0} \leq \mathbf{p} \leq \mathbf{1}$?

6. Let $T_1, T_2, \ldots, T_n$ be associated. Prove

$$P[\min_{1 \leq i \leq n} T_i > t] \geq \prod_{i=1}^{n} P[T_i > t]$$

and

$$P[\max_{1 \leq i \leq n} T_i > t] \leq \coprod_{i=1}^{n} P[T_i > t].$$

7. For the following systems of associated components, compute the bounds on system reliability given in (3.12):

(a) 2-out-of-3 system;
(b) Hi-fi system shown in Figure 1.1.4;
(c) Bridge system described in Example 3.2 of Chapter 1.

8. For a k-out-of-n system of associated components, express the bounds of (3.12) as explicitly as possible:

(a) When components have respective reliabilities $p_1, \ldots, p_n$;
(b) When each component reliability is the same value p.

9. A 2-out-of-3 system has independent components, each with reliability p. Compare the bounds of (3.9) with those of (3.12), determining the values of p for which (3.9), or alternatively (3.12), is preferable.

4. IMPROVED BOUNDS ON SYSTEM RELIABILITY USING MODULAR DECOMPOSITIONS

It seems intuitively plausible that the minimal cut lower bound presented in Theorem 3.4 can be improved by making use of modular decomposition.

Assume the components of (C, ϕ) are independent, and that (C, ϕ) has the modular decomposition $\{(A_1, \chi_1), \ldots, (A_r, \chi_r)\}$, as described in Definition 4.3 of Chapter 1. It follows that

$$h_\phi(\mathbf{p}) = h_\psi[h_{\chi_1}(\mathbf{p}), \ldots, h_{\chi_r}(\mathbf{p})], \tag{4.1}$$

using the following notation.

Notation. $h_\phi(\mathbf{p})$ denotes the reliability of structure ϕ having components of respective reliabilities $p_1, \ldots, p_n$.

Equation (4.1) states symbolically that system reliability can be computed by first computing the reliability of the individual modules $\chi_1, \ldots, \chi_r$, and then computing the reliability of the organizing structure ψ composed of these modules. Theorem 4.1 below shows us that the lower bound of Theorem 3.4 may be improved upon by using either (a) the minimal cut lower bounds for the modules $\chi_1, \ldots, \chi_r$, or (b) the minimal cut lower bound for the organizing structure ψ, or (c) both.

4.1. Theorem. Let (C, ϕ) be a coherent structure of independent components with modular decomposition $\{(A_i, \chi_i)\}_{i=1,\ldots,r}$ and organizing structure ψ, so that both (4.1) of Chapter 1 and (4.1) hold. Then

$$h_\phi(\mathbf{p}) \geq \left\{ \begin{matrix} l_\psi[h_{\chi_1}(\mathbf{p}), \ldots, h_{\chi_r}(\mathbf{p})] \\ h_\psi[l_{\chi_1}(\mathbf{p}), \ldots, l_{\chi_r}(\mathbf{p})] \end{matrix} \right\} \geq l_\psi[l_{\chi_1}(\mathbf{p}), \ldots, l_{\chi_r}(\mathbf{p})] \geq l_\phi(\mathbf{p}). \quad (4.2)$$

Before presenting a formal proof of the theorem, we give an example to help clarify the concepts involved.

Example. Consider the coherent structure presented in the schematic diagram shown in Figure 2.4.1.

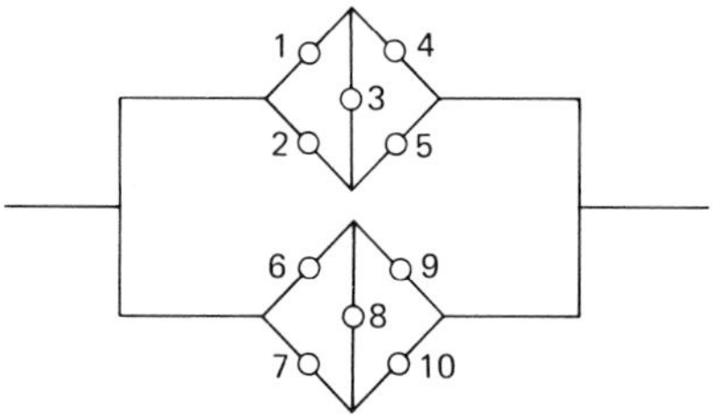

Figure 2.4.1.
Coherent structure of 10 components.

A modular decomposition of the system consists of modules (A_1, χ_1) and (A_2, χ_2), where

$$A_1 = \{1, 2, 3, 4, 5\},$$

$$\chi_1 = (x_1 \amalg x_2)(x_1 \amalg x_3 \amalg x_5)(x_2 \amalg x_3 \amalg x_4)(x_4 \amalg x_5),$$

$$A_2 = \{6, 7, 8, 9, 10\},$$

$$\chi_2 = (x_6 \amalg x_7)(x_6 \amalg x_8 \amalg x_{10})(x_7 \amalg x_8 \amalg x_9)(x_9 \amalg x_{10}),$$

with organizing structure $\psi(\chi_1, \chi_2) = \chi_1 \amalg \chi_2$, a parallel arrangement of the two modules.

Using the minimal cut representation of χ_1 and χ_2 given above, we may represent the system as shown in Figure 2.4.2.

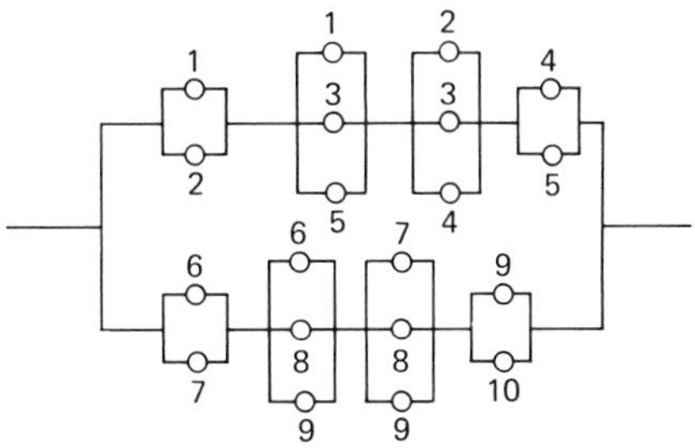

Figure 2.4.2.
Minimal cut representation.

Now suppose all components are mutually independent with common probability p of functioning, and common probability $q \equiv 1 - p$ of failing. It follows that

$$h_\psi[l_{\chi_1}(p), \dots, l_{\chi_r}(p)] = 2(1 - q^2)^2(1 - q^3)^2 - (1 - q^2)^4(1 - q^3)^4.$$

Since ψ is a parallel structure, the minimal cut representation of ψ coincides with ψ, so that

$$\begin{aligned} l_\psi[l_{\chi_1}(p), \dots, l_{\chi_r}(p)] &\equiv h_\psi[l_{\chi_1}(p), \dots, l_{\chi_r}(p)] \\ &= 2(1 - q^2)(1 - q^3)^2 - (1 - q^2)^4(1 - q^3)^4. \end{aligned}$$

Table 4.1 compares these lower bounds on system reliability for a range of values of component reliability p.

Table 4.1. Comparison of Lower Bounds

p	$h_\phi(p)$	$h_\psi[l_{\chi_1}(p), \dots, l_{\chi_r}(p)] = l_\psi[l_{\chi_1}(p), \dots, l_{\chi_r}(p)]$	$l_\phi(p)$
.99	.9999999	.9999999	.9999999
.95	.9999727	.9999725	.9999724
.90	.999536	.999522	.999516
.75	.9807	.9779	.9758
.50	.750	.675	.562
.25	.258	.123	.011
.10	.042	.005	$.533 \times 10^{-6}$
.01	$.40 \times 10^{-3}$	$.698 \times 10^{-6}$	$.941 \times 10^{-21}$

Next we prove Theorem 4.1. We will find it helpful to first prove the following lemma.

4.2. Lemma. Assume that the coherent structure ϕ consists of r modules in parallel, that is, $\phi(\mathbf{x}) = \coprod_{i=1}^{r} \chi_i(\mathbf{x})$, and that all components are statistically independent. Then

$$\coprod_{i=1}^{r} l_{\chi_i}(\mathbf{p}) \geq l_\phi(\mathbf{p}).$$

PROOF. In terms of its modules, the system appears as follows:

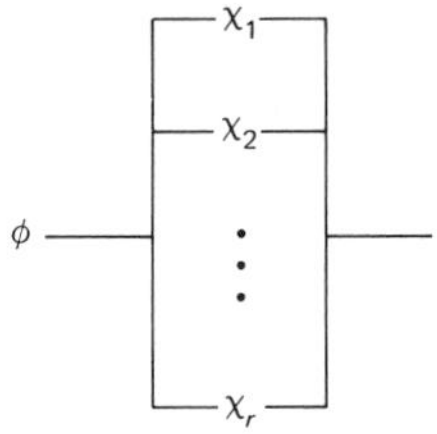

We may represent χ_i in terms of its minimal cut structure functions $\lambda_{i1}, \ldots, \lambda_{im_i}$:

$$\chi_i(\mathbf{x}) = \prod_{j=1}^{m_i} \lambda_{ij}(\mathbf{x}), \qquad i = 1, \ldots, r.$$

It follows that $l_{\chi_i}(\mathbf{p}) = \prod_{j=1}^{m_i} P[\lambda_{ij}(\mathbf{X}) = 1]$, and hence

$$\coprod_{i=1}^{r} l_{\chi_i}(\mathbf{p}) = \coprod_{i=1}^{r} \prod_{j=1}^{m_i} P[\lambda_{ij}(\mathbf{X}) = 1].$$

If we replace replicated components in the minimal cut representation for $\chi_i(\mathbf{x})$ ($i = 1, 2, \ldots, r$) by identical but mutually statistically independent components we will have a new coherent system with structure function, say, ϕ^*. Schematically, the new structure will look as follows:

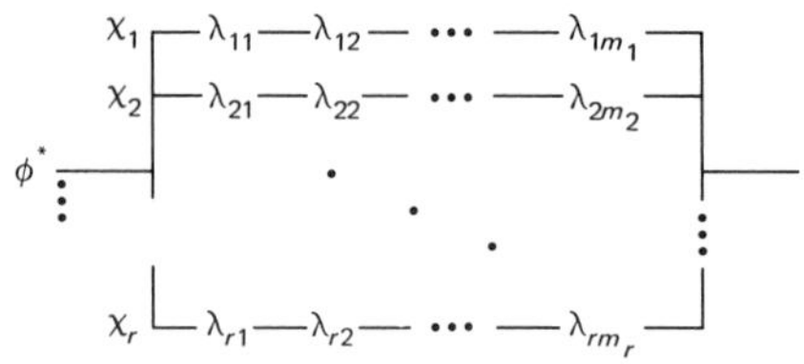

It follows that

$$h_{\phi^*}(\mathbf{p}) = \coprod_{i=1}^{r} l_{\chi_i}(\mathbf{p}).$$

Next note that any minimal cut structure β_j of ϕ can be formed by connecting in parallel $\lambda_{1j_1}, \lambda_{2j_2}, \ldots, \lambda_{rj_r}$, for some choice $j_1, j_2, \ldots, j_r$, and conversely, any such parallel connection corresponds to a β_j; moreover, the correspondence is one-to-one. It follows that

$$l_{\phi^*}(\mathbf{p}) = l_\phi(\mathbf{p}).$$

But by Corollary 3.4, $h_{\phi^*}(\mathbf{p}) \geq l_{\phi^*}(\mathbf{p})$, so that $\coprod_{i=1}^{r} l_{\chi_i}(\mathbf{p}) = h_{\phi^*}(\mathbf{p}) \geq l_{\phi^*}(\mathbf{p}) = l_\phi(\mathbf{p})$. ||

Now we may prove Theorem 4.1.

PROOF OF THEOREM 4.1. Clearly

$$h_\phi(\mathbf{p}) = h_\psi[h_{\chi_1}(\mathbf{p}), \ldots, h_{\chi_r}(\mathbf{p})] \geq l_\psi[h_{\chi_1}(\mathbf{p}), \ldots, h_{\chi_r}(\mathbf{p})],$$

since l_ψ is the minimal cut lower bound corresponding to the organizing structure ψ. Also

$$h_\psi[h_{\chi_1}(\mathbf{p}), \ldots, h_{\chi_r}(\mathbf{p})] \geq h_\psi[l_{\chi_1}(\mathbf{p}), \ldots, l_{\chi_r}(\mathbf{p})],$$

since l_{χ_i} is the minimal cut lower bound corresponding to modular structure χ_i $(i = 1, \ldots, r)$, and since h_ψ is an increasing function. In a similar fashion, we may verify

$$\begin{matrix} l_\psi[h_{\chi_1}(\mathbf{p}), \ldots, h_{\chi_r}(\mathbf{p})] \\ h_\psi[l_{\chi_1}(\mathbf{p}), \ldots, l_{\chi_r}(\mathbf{p})] \end{matrix} \geq l_\psi[l_{\chi_1}(\mathbf{p}), \ldots, l_{\chi_r}(\mathbf{p})].$$

Finally, we prove $l_\psi[l_{\chi_1}(\mathbf{p}), \ldots, l_{\chi_r}(\mathbf{p})] \geq l_\phi(\mathbf{p})$. Let $\eta_1, \ldots, \eta_t$ denote the minimal cut structure functions of ψ, let $\phi_j(\mathbf{x}) = \eta_j[\chi_1(\mathbf{x}), \ldots, \chi_r(\mathbf{x})]$, and let $\mu_{j1}, \ldots, \mu_{jm_j}$ denote the minimal cut structure functions for ϕ_j $(j = 1, \ldots, t)$.

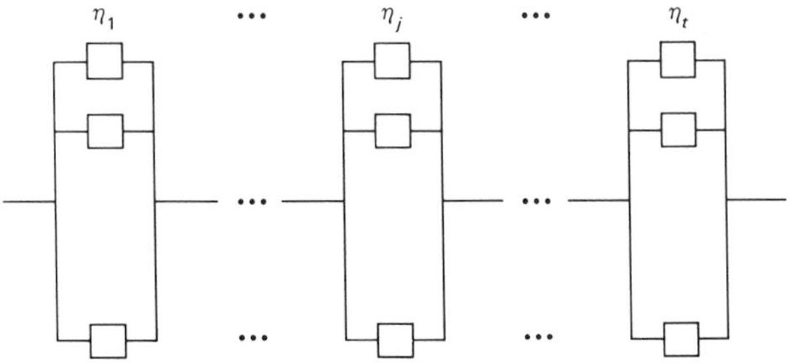

Figure 2.4.3.
Minimal cut structure functions.

We claim that

$$\{\mu_{jk}\}_{j=1,\ldots,t}^{k=1,\ldots,m_j}$$

constitute the set of minimal cut structure functions of ϕ. To verify this claim, note that

(a) The μ_{jk} are distinct since the modules in the modular decomposition of (C, ϕ) do not overlap.
(b) $\mu_{jk} = 0 \Rightarrow \eta_j = 0 \Rightarrow \psi = 0 \Rightarrow \phi = 0 \Rightarrow \mu_{jk}$ is a cut structure of ψ; moreover, it is easy to see that μ_{jk} is minimal.

It follows that

$$l_\phi(\mathbf{p}) = \prod_{j=1}^{t} \prod_{k=1}^{m_j} h_{\mu_{jk}}(\mathbf{p}). \tag{4.3}$$

Since the modular components of η_j are connected in parallel we may apply Lemma 4.2 to obtain

$$h_{\eta_j}[l_{\chi_1}(\mathbf{p}), \ldots, l_{\chi_r}(\mathbf{p})] \geq l_{\phi_j}(\mathbf{p}). \tag{4.4}$$

Finally, using (4.3) and (4.4), we have

$$\begin{aligned} l_\psi[l_{\chi_1}(\mathbf{p}), \ldots, l_{\chi_r}(\mathbf{p})] &= \prod_{j=1}^{t} h_{\eta_j}[l_{\chi_1}(\mathbf{p}), \ldots, l_{\chi_r}(\mathbf{p})] \geq \prod_{j=1}^{t} l_{\phi_j}(\mathbf{p}) \\ &= \prod_{j=1}^{t} \prod_{k=1}^{m_j} h_{\mu_{jk}}(\mathbf{p}) = l_\phi(\mathbf{p}), \end{aligned}$$

yielding the desired result. ||

In practice, Theorem 4.1 may be used to advantage as follows. The design and reliability analysis of a large complex system is undertaken by decomposing the system into separate functional subsystems, decomposing each subsystem into components, and finally decomposing each component into its individual parts. Using Theorem 3.3, the minimal cut lower bound may be obtained for each component reliability from information on part reliability. Similarly, from these component reliability lower bounds, using Theorem 3.3 again, a lower bound on subsystem reliability may be obtained. Finally, from the subsystem reliability lower bounds, using Theorem 3.3, a lower bound on system reliability may be computed. Theorem 4.1 tells us that the system reliability lower bound computed in this fashion is an improved lower bound over that obtained from the minimal cut representation of the entire system. See Exercise 2.

EXERCISES

1. Construct an example for which

$$h_\phi(\mathbf{p}) = h_\psi[l_{\chi_1}(\mathbf{p}), \ldots, l_{\chi_r}(\mathbf{p})] > l_\psi[l_{\chi_1}(\mathbf{p}), \ldots, l_{\chi_r}(\mathbf{p})] > l_\phi(\mathbf{p}).$$

*2. Let $\{(A_i, \chi_i)\}_{i=1,\ldots,r}$ be a modular decomposition of the coherent system (C, ϕ) with organizing structure ψ, so that $\phi(\mathbf{x}) = \psi[\chi_1(\mathbf{x}), \ldots, \chi_r(\mathbf{x})]$. Let $\{(B_{ij}, \Omega_{ij})\}_{j=1,\ldots,s_i}$ be a modular decomposition of (A_i, χ_i) with organizing structure σ_i, so that $\chi_i(\mathbf{x}) = \sigma_i[\Omega_{11}(\mathbf{x}), \ldots, \Omega_{is_i}(\mathbf{x})]$ for $i = 1, \ldots, r$. Thus

$$(B_{ij}, \Omega_{ij}) \begin{Bmatrix} j = 1, \ldots, s_i \\ i = 1, \ldots, r \end{Bmatrix}$$

is a refinement of $\{(A_i, \chi_i)\}_{i=1,\ldots,r}$, as described in Section 4 of Chapter 1. Finally, let $\theta(\Omega) = \psi[\sigma_1(\Omega), \ldots, \sigma_r(\Omega)]$, the composition of the organizing structure functions. Show that

(a) $l_\psi[h_{\chi_1}(\mathbf{p}), \ldots, h_{\chi_r}(\mathbf{p})] \geq l_\theta[h_{\Omega_{11}}(\mathbf{p}), \ldots, h_{\Omega_{rs_r}}(\mathbf{p})]$,

(b) $h_\psi[l_{\chi_1}(\mathbf{p}), \ldots, l_{\chi_r}(\mathbf{p})] \leq h_\theta[l_{\Omega_{11}}(\mathbf{p}), \ldots, l_{\Omega_{rs_r}}(\mathbf{p})]$.

5. SHAPE OF THE SYSTEM RELIABILITY FUNCTION

In this section we study system reliability as a function of component reliabilities. Our main result is that for independent identically distributed components, the reliability $h(p)$ of a coherent system without path sets or cut sets of size 1 is an S-shaped function of component reliability p; that is, there exists a value p_0, $0 < p_0 < 1$, such that $h(p_0) = p_0$, $h(p) \leq p$ for $0 \leq p \leq p_0$, and $h(p) \geq p$ for $p_0 \leq p \leq 1$. The practical implication of this result is that for a coherent system with redundancy, when all the components have attained sufficiently high reliability, then system reliability is greater than component reliability (when component reliabilities are not all alike, use the lowest component reliability for a lower bound).

To establish the S-shapedness of the reliability function, we first prove the following lemma.

5.1. Lemma. Let ϕ be a coherent structure of n associated components. Then

$$\operatorname{cov}\left[\phi(\mathbf{X}), \sum_{i=1}^{n} X_i\right] \geq \operatorname{var} \phi(\mathbf{X}). \tag{5.1}$$

PROOF. It is sufficient to prove that

$$\operatorname{cov}\left[\phi(\mathbf{X}), \sum_{i=1}^{n} X_i - \phi(\mathbf{X})\right] \geq 0.$$

But since $\phi(\mathbf{X})$ and $\sum_{i=1}^{n} X_i - \phi(\mathbf{X})$ are increasing functions of $\mathbf{X}$, the last inequality follows from property P_3 of associated random variables and Exercise 5 of Section 2. ‖

Actually strict inequality holds in (5.1) when the components are independent, as stated in Lemma 5.2.

5.2. Lemma. Let ϕ be a coherent structure of $n \geq 2$ independent components. Then

$$\operatorname{cov}\left[\phi(\mathbf{X}), \sum_{i=1}^{n} X_i\right] > \operatorname{var} \phi(\mathbf{X}). \tag{5.2}$$

We leave the proof to Exercise 3.

Next we prove Lemma 5.3.

5.3. Lemma. Let ϕ be a coherent structure of independent components with respective reliabilities $p_1, \ldots, p_n$, and system reliability $h(\mathbf{p})$. Then

$$\operatorname{cov}\left[\phi(\mathbf{X}), \sum_{i=1}^{n} X_i\right] = \sum_{i=1}^{n} p_i q_i \frac{\partial h}{\partial p_i}, \tag{5.3}$$

where $q_i = 1 - p_i$.

PROOF. $\operatorname{cov}[\phi(\mathbf{X}), X_i] = E\phi(\mathbf{X})X_i - E\phi(\mathbf{X})EX_i = p_i h(1_i, \mathbf{p}) - p_i h(\mathbf{p}) = p_i h(1_i, \mathbf{p}) - p_i[p_i h(1_i, \mathbf{p}) + q_i h(0_i, \mathbf{p})]$ [by (1.3)] $= p_i q_i[h(1_i, \mathbf{p}) - h(0_i, \mathbf{p})] = p_i q_i(\partial h/\partial p_i)$, by (1.4).

Since $\operatorname{cov}[\phi(\mathbf{X}), \sum_{i=1}^{n} X_i] = \sum_{i=1}^{n} \operatorname{cov}[\phi(\mathbf{X}), X_i]$, (5.3) follows. ||

From Lemmas 5.2 and 5.3, we obtain our main result, as shown in Theorem 5.4.

5.4. Theorem. Let $h(\mathbf{p})$ be the reliability function of a coherent structure of $n \geq 2$ independent components. Then

(a) $$\sum_{i=1}^{n} p_i q_i \frac{\partial h}{\partial p_i} > h(\mathbf{p})[1 - h(\mathbf{p})] \quad \text{for } \mathbf{0} < \mathbf{p} < \mathbf{1}. \tag{5.4}$$

(b) $$pq \frac{dh}{dp} > h(p)[1 - h(p)] \quad \text{for } 0 < p < 1. \tag{5.5}$$

(c) If $h(p_0) = p_0$ for some $0 < p_0 < 1$, then $h(p) < p$ for $0 < p < p_0$, whereas $h(p) > p$ for $p_0 < p < 1$.

(d) If the structure has no path sets or cut sets of size 1, then there exists $0 < p_0 < 1$ such that $h(p_0) = p_0$.

PROOF. (a) For any Bernoulli random variable Z with probability of success p, $\operatorname{var} Z = p(1 - p)$. Thus $\operatorname{var} \phi(\mathbf{X}) = h(\mathbf{p})[1 - h(\mathbf{p})]$. Using (5.2) and (5.3), we obtain (5.4).

(b) Equation (5.5) follows from (5.4) by taking $p_1 = \cdots = p_n = p$.

(c) Note that at p_0, $dh/dp > 1$, from (b). Thus at an intersection $h(p)$ crosses the diagonal p from below. The conclusion follows.

(d) See Exercise 6.

Example. For a 6-out-of-10 structure of like components, the reliability function is given by

$$h(p) = \binom{10}{6} p^6(1-p)^4 + \binom{10}{7} p^7(1-p)^3 + \binom{10}{8} p^8(1-p)^2 + 10p^9(1-p) + p^{10}.$$

As stated in Theorem 5.4(c), the reliability function is S-shaped. See Figure 2.5.1.

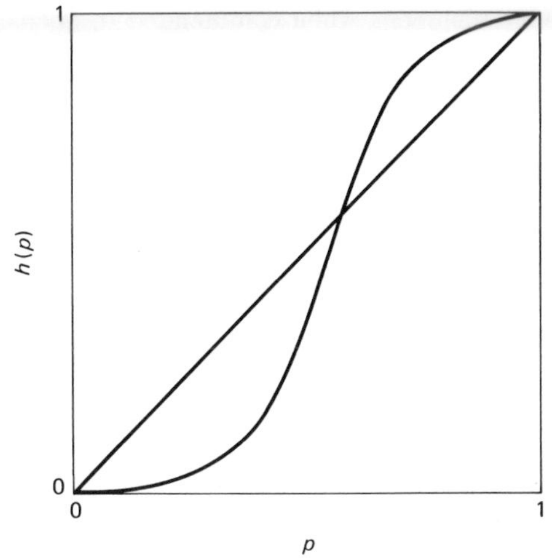

Figure 2.5.1.
S-shaped function. Reliability function for a 6-out-of-10 structure. [SOURCE: From "Multi-Component Systems and Structures and Their Reliability" by Birnbaum, Z. W., J. D. Esary, and S. C. Saunders, *Technometrics* (1961) 3: 55–77.]

6. APPLICATIONS TO RELAY CIRCUITS AND SAFETY MONITORING SYSTEMS

Relay Circuits

Moore and Shannon (1956) first developed and exploited the S-shapedness property to show how to construct relay circuits of arbitrarily high reliability from like components of any given degree of unreliability. The exact model treated may be summarized as follows.

As in Section 3 of this chapter, suppose like idealized relays are subject to two kinds of failure: failure to close and failure to open. Similarly, circuits constructed from these relays are subject to two kinds of failure: failure to close, that is, no closed path is achieved from input wire to output wire when the circuit is commanded to close, and failure to open, that is, a closed path exists from input wire to output wire even though the circuit is commanded to open. Each relay operates independently of the others, with probability p (preferably large) of closing when commanded to close (energized), and probability p' (preferably small) of closing or remaining closed when commanded to open. These probabilities do not vary with time.

Let $X_i = 1$ if relay i closes when commanded to close, and 0 otherwise. Let $\phi = 1$ if the circuit closes when commanded to close, and 0 otherwise. Then $h_\phi(p) \stackrel{\text{def}}{=} E\phi(\mathbf{X})$ is the probability the circuit closes when commanded to close.

Let $Y_i = 1$ if relay i opens when commanded to open, and 0 otherwise. Let $\psi = 1$ if the circuit opens when commanded to open, and 0 otherwise. As we saw in 3.6 of Chapter 1, structure function ψ is dual to structure function ϕ. Thus if $h_\psi(1 - p')$ is the probability that the circuit opens when commanded to open, we may write

$$h_\psi(1 - p') = 1 - h_\phi(p'). \tag{6.1}$$

The basic problem is to achieve a high value for probability $h_\phi(p)$, preferably $h_\phi(p) > p$, and a low value for probability $h_\phi(p')$, preferably $h_\phi(p') < p'$, by appropriate arrangement of a sufficient number of relays. It is clear that the relay circuit represents an improvement in reliability over the individual relay when the function $h_\phi(p)$ is S-shaped.

Moore and Shannon (1956) prove (b) and (c) of Theorem 5.4 for the probability $h(p)$ of relay network closure. Using this S-shapedness property of the $h(p)$ function, Moore and Shannon then calculate the number of individual relays and their configuration required to achieve any specified (high) reliability for the relay network, given the individual relay reliability.

Safety Systems

A second application of S-shapedness occurs in the reliability analysis of certain monitoring systems. For example, nuclear reactors often use three identical and independently functioning counters to monitor the radioactivity of the air in the ventilation system, with the aim of initiating reactor shutdown when a dangerous level of radioactivity is present. When two or more counters register a dangerous level of radioactivity, the reactor automatically shuts down.

Two kinds of error are possible: (a) the monitoring system may be incapable of detecting dangerous radioactivity due to the breakdown of two or more counters; (b) the monitoring system may trigger reactor shutdown even when the radioactivity level is not dangerous, as a result of a false alarm by two or more of the counters.

To analyze the monitoring system, let $X_i = 1$ if the ith counter is able to detect dangerous levels of radioactivity, and $X_i = 0$ otherwise; let $p = EX_i$. Let ϕ denote the structure function of a 2-out-of-3 system, and $h_\phi(p) = E\phi(\mathbf{X})$ denote the corresponding reliability function. Thus $h_\phi(\mathbf{p}) = 3p^2(1 - p) + p^3$ is the probability that the monitoring system is able to detect dangerous radioactivity.

Next let $Y_i = 0$ if monitor i gives a false alarm, and 1 otherwise, with $EY_i = 1 - p'$. Let $\psi = 0$ if the monitoring system shuts down the reactor, and 1 otherwise. As in the relay circuit case, structure function ψ is dual to structure function ϕ. Thus if $h_\psi(1 - p')$ is the probability that the monitoring system does not give a false alarm as a function of $1 - p'$, the probability that the individual monitor does not give a false alarm, then (6.1) holds, as in the relay circuit case. Since the dual of a 2-out-of-3 system is also a 2-out-of-3 system, $\psi \equiv \phi$.

When p is close to 1, we want $h_\phi(p)$ also close to 1, preferably $h_\phi(p) > p$. Similarly, when p' is close to 0, we want $h_\phi(p')$ also close to 0, preferably $h_\phi(p') < p'$. By the S-shaped character of a 2-out-of-3 structure function, we know that these inequalities will hold; that is, the 2-out-of-3 monitoring system gives performance superior to that of a single counter.

By building a 2-out-of-3 system out of modules which are themselves 2-out-of-3 systems, S-shapedness may be increased further, and system performance correspondingly improved.

In general, define the composition

$$h_n(p) = h[h_{n-1}(p)], \tag{6.2}$$

where $h(p) = h_1(p)$ is the reliability function of a coherent structure of independent components each having reliability p. If $h(p_0) = p_0$, $0 < p_0 < 1$, then, as $n \to \infty$,

$$\left.\begin{array}{lll} h_n(p) \downarrow 0 & \text{for} & 0 \le p < p_0, \\ h_n(p) \equiv p_0 & \text{for} & p = p_0, \\ h_n(p) \uparrow 1 & \text{for} & p_0 < p \le 1. \end{array}\right\} \tag{6.3}$$

The proof is left to Exercise 9.

For example, in the 2-out-of-3 case, (6.3) holds with $p_0 = \frac{1}{2}$. Thus repeated use of 2-out-of-3 systems built out of 2-out-of-3 modules steadily improves monitoring performance. Obviously, engineering as well as practical considerations must temper this theoretical conclusion.

EXERCISES

1. Let $X_1, \ldots, X_n$ be independent binary random variables, with $EX_i = p_i$, $i = 1, \ldots, n$. For any functions $f_j(\mathbf{X})$, $j = 1, 2$, and any X_i,

$$\begin{aligned}\operatorname{cov}[f_1(\mathbf{X}), f_2(\mathbf{X})] = {} & p_i \operatorname{cov}[f_1(1_i, \mathbf{X}), f_2(1_i, \mathbf{X})] \\ & + q_i \operatorname{cov}[f_1(0_i, \mathbf{X}), f_2(0_i, \mathbf{X})] \\ & + p_i q_i E[f_1(1_i, \mathbf{X}) - f_1(0_i, \mathbf{X})]E[f_2(1_i, \mathbf{X}) - f_2(0_i, \mathbf{X})].\end{aligned}$$

2. Under the hypothesis of Exercise 1 and the additional assumption that $0 < p_i < 1$ for $i = 1, \ldots, n$, a necessary and sufficient condition for

$$\operatorname{cov}[f_1(\mathbf{X}), f_2(\mathbf{X})] > 0$$

is that some variable X_i be relevant to both f_1 and f_2.

3. Prove Lemma 5.2.

*4. Prove

(a) $$\frac{dh(p)}{dp} = \sum_{i=1}^{n} E[\phi(1_i, \mathbf{X}) - \phi(0_i, \mathbf{X})],$$

(b) $$\left.\frac{dh}{dp}\right|_{p=0} = \sum_{i=1}^{n} \phi(1_i, \mathbf{0}),$$

(c) $$\left.\frac{dh}{dp}\right|_{p=0} = \text{number of path sets of size 1.}$$

5. Prove $\left.\frac{dh}{dp}\right|_{p=1}$ = number of cut sets of size 1.

6. Prove Theorem 5.4(d).

7. Show that $h_\phi(\mathbf{p}) = l_\phi(\mathbf{p})$, $0 < p_i < 1$, $i = 1, \ldots, n$, if and only if the minimal cut sets are disjoint.

*8. Consider a coherent system of n independent components. Let p_i = probability that the ith component functions, $i = 1, \ldots, n$, and $h(\mathbf{p})$ = probability that the system functions.

The information obtained by observing that the ith component is functioning or failed may be quantified as $-[p_i \log p_i + (1 - p_i) \log (1 - p_i)]$. Show that the total information obtained by observing each of the component states is greater than the information obtained by observing just the system state, that is,

$$-\sum_{i=1}^{n} [p_i \log p_i + (1 - p_i) \log (1 - p_i)] \geq -[h(\mathbf{p}) \log h(\mathbf{p}) + (1 - h(\mathbf{p})) \log (1 - h(\mathbf{p}))].$$

9. Prove (6.3).

7. NOTES AND REFERENCES

Section 1. The concept of reliability importance is due to Z. W. Birnbaum (1969).

Section 2. The theory of associated random variables is developed in Esary-Proschan-Walkup (1967). A closely related paper by Lehmann (1966) presents other concepts of dependence. Lehmann's classes of dependent random variables are restricted to the bivariate case; the class of associated random variables applies in the multivariate case.

Section 3. The application of association to the establishment of minimal cut lower bounds for system reliability is presented in Esary-Proschan (1970). The model treated in this paper is more general than the one discussed in Section 3 in that repair of failed parts is permitted. Still more general models are treated in Esary-Proschan (1968); in these models a variety of maintenance policies are considered. In addition, the failure of parts of the system may result in increased failure rates for the remaining unfailed parts, due to the added load on them.

Section 4. The results reported in Section 4 are based on Bodin (1970).

Section 5. The S-shapedness property of reliability functions is developed for relay networks in Moore-Shannon (1956). Extension to coherent systems is accomplished in Birnbaum-Esary-Saunders (1961). The methods used in Section 4 are developed in Esary-Proschan (1963a); this paper also presents further results concerning the crossing by the reliability function of various members of a family of curves.

3

PARAMETRIC FAMILIES OF DISTRIBUTIONS OF DIRECT IMPORTANCE IN RELIABILITY THEORY

In Chapters 1 and 2 we have considered systems of components almost exclusively in terms of their functioning or failure at a fixed time. In this chapter we take a different point of view and consider the life lengths of components and the corresponding life lengths of systems of components. In general, life length is random, and so we are led to a study of life distributions; we confine ourselves to those of direct importance in reliability analysis.

To understand better which life distributions are important in reliability models, we consider first a notion of aging. Aging is conveniently studied in terms of the failure rate function (defined in Section 1). In the simplest case, where no aging is present, we obtain a constant failure rate, corresponding to the exponential distribution. The exponential distribution is in several senses the most fundamental distribution in reliability theory, as we will see repeatedly throughout the book. In Section 2 we obtain probabilistic properties of the exponential distribution that not only are useful in probabilistic reliability models, but also are needed in the statistical applications in the forthcoming volume.

It is natural to consider next the Poisson process (Section 3), the stochastic process in which intervals between successive events are independently distributed according to a common exponential distribution. The Poisson process has many direct and indirect applications in reliability, especially in formulating shock models. (See Section 3 of Chapter 4 and Section 2 of Chapter 6.) We are led next to the Poisson distribution, a discrete distribution used often for spares determination.

Finally, in Section 5 we study parametric families of life distributions for which the failure rate is monotone (increasing or decreasing) over time.

These include the Weibull, the gamma, and the truncated normal distributions. In Chapter 4 we study in a more general way classes of distributions with monotone failure rates, as well as other broad classes of life distributions.

1. A NOTION OF AGING

The reliability (or survival probability) of a fresh unit corresponding to a mission of duration x is, by definition, $\bar{F}(x) \equiv 1 - F(x)$, where F is the life distribution of the unit. The corresponding *conditional reliability of a unit of age t* is

$$\bar{F}(x \mid t) = \frac{\bar{F}(t + x)}{\bar{F}(\ )} \qquad \text{if} \quad \bar{F}(t) > 0.$$

Similarly, the conditional probability of failure during the next interval of duration x of a unit of age t is

$$F(x \mid t) = \frac{F(t + x) - F(t)}{\bar{F}(t)} = 1 - \bar{F}(x \mid t).$$

Finally, we may obtain a conditional *failure rate* $r(t)$ at time t:

$$r(t) = \lim_{x \to 0} \frac{1}{x} \frac{F(t + x) - F(t)}{\bar{F}(t)},$$

so that

$$r(t) = \frac{f(t)}{\bar{F}(t)}, \tag{1.1}$$

when $f(t)$ exists and $\bar{F}(t) > 0$. [Alternate names for $r(t)$ defined in (1.1) are hazard rate, force of mortality, and intensity rate.]

Useful identities are readily obtained by integrating both sides of (1.1),

$$\int_0^x r(t)\, dt = -\log \bar{F}(x), \tag{1.2}$$

and then exponentiating,

$$\bar{F}(x) = \exp\left[-\int_0^x r(t)\, dt\right]. \tag{1.3}$$

The cumulative failure rate, $R(x) = \int_0^x r(t)\, dt$, is referred to as the *hazard* function, or simply the hazard. Equation (1.3) gives a useful theoretical representation of reliability as a function of failure rate. An alternate representation,

$$\bar{F}(x) = e^{-R(x)}, \tag{1.4}$$

gives reliability in terms of hazard.

Now consider a device which does not age stochastically; that is, its survival probability over an additional period of duration x is the same regardless of its present age t. Symbolically,

$$\bar{F}(x \mid t) = \bar{F}(x) \qquad \text{for all } x, \quad t \geq 0.$$

Equivalently,

$$\bar{F}(t + x) = \bar{F}(t)\bar{F}(x) \qquad \text{for all } x, \quad t \geq 0. \tag{1.5}$$

Equation (1.5) is classical; taking into account the requirement that $0 \leq \bar{F}(x) \leq 1$, its solution is of the form

$$\bar{F}(x) = e^{-\lambda x}, \qquad \lambda > 0, \quad x \geq 0, \tag{1.6}$$

an exponential survival probability, as shown in Theorem 2.2 below. In Section 2 we derive a number of useful properties of the exponential distribution.

Note that the failure rate of the exponential distribution $r(t) = f(t)/\bar{F}(t) = \lambda e^{-\lambda t}/e^{-\lambda t} = \lambda$ is constant. Conversely, a distribution with constant failure rate is of the form (1.6).

Suppose now that the unit ages adversely in the sense that the conditional survival probability is a decreasing function of age:

$$\bar{F}(x \mid t) = \frac{\bar{F}(t + x)}{\bar{F}(t)} \qquad \text{is decreasing in } -\infty < t < \infty \text{ for each } x \geq 0. \tag{1.7}$$

As a consequence, we obtain

$$r(t) = \lim_{x \to 0} \frac{1}{x}\left[1 - \frac{\bar{F}(t + x)}{\bar{F}(t)}\right] \qquad \text{is increasing in } t \geq 0,$$

when the density $f(t)$ exists. Conversely, $r(t)$ is increasing implies

$$\bar{F}(x \mid t) = \exp\left[-\int_t^{t+x} r(u)\, du\right]$$

is decreasing in $t \geq 0$ for each $x \geq 0$, so that (1.7) holds. Thus *when the density exists,* (1.7) *is equivalent to the failure rate* $r(t)$ *increasing.* Note that (1.7) does not require the existence of a density, whereas $r(t)$ increasing does. We are thus led to the following definition of a class of distributions corresponding to adverse aging:

1.1. Definition. F is an increasing failure rate (IFR) distribution if F satisfies (1.7).

Finally, suppose aging is beneficial in the sense that the device has *increasing* conditional survival probability as a function of age:

$$\bar{F}(x \mid t) = \frac{\bar{F}(x + t)}{\bar{F}(t)} \qquad \text{is increasing in } t \geq 0 \quad \text{for each } x \geq 0. \tag{1.8}$$

In analogy with IFR distributions, we may present Definition 1.2.

1.2. Definition. F is a decreasing failure rate (DFR) distribution if F satisfies (1.8).

Arguing as we did in the IFR case, we may show that *if the density exists, $r(t)$ decreasing is equivalent to* (1.8).

DFR distributions may arise in a variety of ways:

(a) Certain metals increase in strength as they are work-hardened.
(b) Mixtures of exponential distributions are DFR. (See Section 4 of Chapter 4.)
(c) Devices displaying "infant mortality" (for example, human beings) have a failure rate decreasing over the early part of the time axis. For some of these devices a positive probability exists of being in a failed state at time 0, that is, the device is nonfunctional before it is ever used, or—for an electronic device, say—fails upon the initial surge of current.

Another class of life distributions arising naturally in reliability situations may be constructed by assuming a failure rate initially decreasing during the infant mortality phase, next constant during the so-called "useful life" phase, and, finally, increasing during the so-called "wearout" phase. In reliability literature such failure rate functions are said to have a "bathtub" shape.

Figure 3.1.1 presents an empirical illustration of a bathtub-shaped

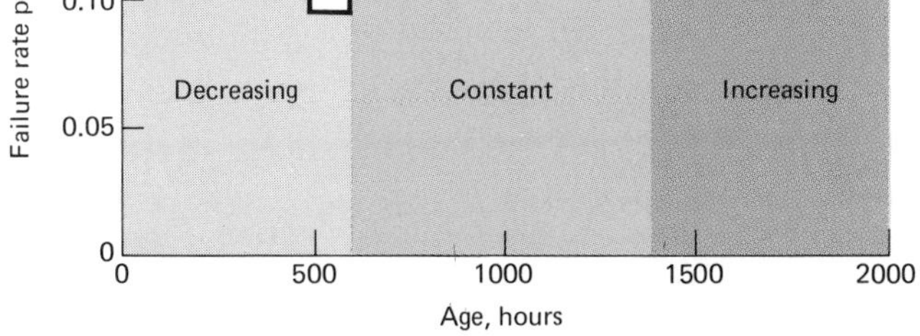

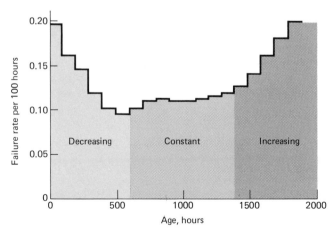

Figure 3.1.1.
Failure history of engine accessory.

failure rate (Kamins, 1962). The ordinate represents the empirical failure rate per 100 hours for a hot-gas generating system used for starting the engines of a particular commercial airliner. During the first 600 hours of operation the observed failure rate decreases by about half. From 600 hours of age to 1400 hours of age, the failure rate remains very nearly constant. Finally, after 1400 hours of operation, the failure rate seems to increase steadily, reflecting wearout.

Other notions of aging and wearout will be discussed in Chapter 4.

2. THE EXPONENTIAL DISTRIBUTION

We have seen in Section 1 that the exponential life distribution provides a good description of the life length of a unit which does not age with time. In this section we develop properties of the exponential distribution helpful in solving both predictive and inferential reliability problems.

As mentioned in Section 1, the *exponential distribution* $G_\lambda(t)$ is given by

$$G_\lambda(t) = 1 - e^{-\lambda t} \qquad \text{for} \quad t \geq 0, \tag{2.1}$$

where λ is a fixed strictly positive parameter. Its survival probability $\bar{G}_\lambda(t) \equiv 1 - G_\lambda(t)$ is given by

$$\bar{G}_\lambda(t) = e^{-\lambda t} \qquad \text{for} \quad t \geq 0, \tag{2.2}$$

and its density $g_\lambda(t) \equiv G_\lambda'(t)$ by

$$g_\lambda(t) = \lambda e^{-\lambda t} \qquad \text{for} \quad t \geq 0. \tag{2.3}$$

Often we will delete the subscript λ.

A property of the exponential distribution which makes it especially important in reliability theory and application is that *the remaining life of a used component is independent of its initial age* (*the "memoryless" property*). This is the property obtained in showing that the functional equation (1.5) is satisfied by the exponential. More precisely, we may summarize the result in the following theorem.

2.1. Theorem. Let T have exponential distribution $G(t) = 1 - e^{-\lambda t}$ for $t \geq 0$. Then, the conditional survival probability $P[T > t + x \mid T > t] = e^{-\lambda x}$ for $x \geq 0$, independent of t.

This property tells us that a used exponential component is essentially "as good as new."

We also pointed out in Section 1 that the converse holds; we state this formally in Theorem 2.2.

2.2. Theorem. Let the survival probability $\bar{G}$ of a nondegenerate nonnegative random variable satisfy (2.4).

$$\bar{G}(t + x) = \bar{G}(t)\bar{G}(x) \qquad \text{for all} \quad t \geq 0, \quad x \geq 0. \tag{2.4}$$

Then G is an exponential distribution for some $\lambda \geq 0$, as given in (2.1).

PROOF. Let $c > 0$ and m and n be positive integers. From (2.4) it follows that

$$\bar{G}(nc) = [\bar{G}(c)]^n \quad \text{and} \quad \bar{G}(c) = [\bar{G}(c/m)]^m.$$

We claim $0 < \bar{G}(1) < 1$. If $\bar{G}(1) = 1$, then $\bar{G}(n) = [\bar{G}(1)]^n = 1$, which contradicts $\bar{G}(+\infty) = 0$. If $\bar{G}(1) = 0$, then $\bar{G}(1/m) = 0$ and by right continuity, $\bar{G}(0) = 0$, another contradiction.

Since $0 < \bar{G}(1) < 1$, write $\bar{G}(1) = e^{-\lambda}$ where $0 < \lambda < \infty$. It follows from the above that $\bar{G}(1/m) = e^{-\lambda/m}$. Hence $\bar{G}(n/m) = e^{-\lambda n/m}$, that is, $\bar{G}(y) = e^{-\lambda y}$ for positive rational y. By right continuity it follows that $\bar{G}(y) = e^{-\lambda y}$ for all $y > 0$. ||

Theorems 2.1 and 2.2 together state that the functional equation (2.4) *characterizes* the exponential survival probability, that is, is both necessary and sufficient for it. More intuitively, only the exponential distribution possesses the memoryless property. This property has important practical and theoretical consequences. Assuming an exponential life distribution, it follows that:

(a) Since a used component is as good as new (stochastically), there is no advantage in following a policy of planned replacement of used components known to be still functioning.
(b) In statistical estimation of mean life, percentiles, reliability, and so on, data may be collected consisting only of the number of hours of observed life and of the number of observed failures; the ages of components under observation are irrelevant.

We will develop the ramifications of (a) and (b) in greater detail in Chapter 6.

For the exponential distribution we compute the failure rate function as

$$r(t) = \frac{\lambda e^{-\lambda t}}{e^{-\lambda t}} = \lambda \qquad \text{for all} \quad t \geq 0.$$

Note that the failure rate function of the exponential distribution is given by the parameter λ, constant, as t ranges over $[0, \infty)$. Conversely, suppose the failure rate function $r(t)$ is constant, say, λ. Symbolically, for $t \geq 0$:

$$\frac{f(t)}{\overline{F}(t)} = \lambda.$$

Integrating, we obtain

$$-\log \overline{F}(t) = \lambda t,$$

or

$$\overline{F}(t) = e^{-\lambda t}.$$

Thus we obtain a second characterization of the exponential distribution.

2.3. Theorem. The exponential distribution is the only life distribution with constant failure rate function.

Next we present a result of direct use in generating exponential random variables.

2.4. Theorem. Let U be a uniform random variable on $[0, 1]$. Then $Y = -\log U$ has exponential distribution $1 - e^{-t}$.

PROOF. For $t \geq 0$, $P[Y \leq t] = P[-\log U \leq t] = P[U \geq e^{-t}] = 1 - e^{-t}$. ||

Thus one way to generate an exponential random variable is first to generate a uniform random variable U by any one of various standard procedures and then to form $-\log U$. Generation of exponential random variables is often required in carrying out Monte Carlo studies of statistical tests or estimates.

Next we obtain the moments and related quantities. For $s > -1$, the sth moment μ_s of the exponential is given by

$$\mu_s = \int_0^\infty t^s \lambda e^{-\lambda t}\, dt = \frac{1}{\lambda^s} \int_0^\infty x^s e^{-x}\, dx,$$

so that

$$\mu_s = \frac{1}{\lambda^s} \Gamma(s + 1). \tag{2.5}$$

In particular, for integer values $n = 1, 2, \ldots,$

$$\mu_n = \frac{n!}{\lambda^n}. \tag{2.6}$$

[A direct verification of (2.6) is left for Exercise 1.] Thus *the mean of the exponential distribution is the reciprocal of the constant failure rate.* (This relationship holds only for the exponential distribution.) Another easy consequence of (2.6) is the *coefficient of variation of the exponential distribution is 1.* The proof is left for Exercise 2.

The *moment generating function* (MGF) $M(s)$ of a distribution F is defined by

$$M(s) = \int_{-\infty}^{\infty} e^{sx}\, dF(x), \tag{2.7}$$

assuming that the integral converges in a neighborhood of the origin. It is easy to see that finite integer moments of the distribution may be obtained from the MGF as follows:

$$M^{(n)}(0) = \int_0^{\infty} x^n e^{sx}\, dF(x)|_{s=0} = \int_0^{\infty} x^n\, dF(x),$$

so that

$$\mu_n = M^{(n)}(0) \qquad \text{for} \quad n = 1, 2, \ldots. \tag{2.8}$$

For the exponential distribution $1 - e^{-\lambda t}$, the MGF is given by

$$M(s) = \int_0^{\infty} e^{sx} \lambda e^{-\lambda x}\, dx = \frac{\lambda}{\lambda - s} \qquad \text{for} \quad s < \lambda. \tag{2.9}$$

Thus, using (2.8) and (2.9), we may verify formula (2.6) giving integer moments for the exponential distribution.

Note that the MGF of a density $f(x)$ is essentially the same as its Laplace transform; for the MGF we compute Ee^{sX}, while for the Laplace transform we compute Ee^{-sX}.

Next we study order statistics and spacings between order statistics from an exponential distribution. Let $Y_1, \ldots, Y_n$ be a sample of size n (that is, a set of n independent, identically distributed observations) from the exponential distribution $G_\lambda(t) = 1 - e^{-\lambda t}, t \geq 0$. Then $Y_{1:n} \leq \cdots \leq Y_{n:n}$, the corresponding ordered values, are called the *order statistics* from G_λ. Note that the order statistics may be interpreted as successive times of failure of the components of a system, so that the kth order statistic may be considered as the time of failure of a $(n - k + 1)$-out-of-n system.

The kth *spacing* D_k between order statistics $Y_{1:n}, Y_{2:n}, \ldots, Y_{n:n}$ is given by $D_k = Y_{k:n} - Y_{k-1:n}$ for $k = 1, 2, \ldots, n$, where $Y_{0:n} \equiv 0$. For the exponential distribution, the memoryless property makes it particularly simple to compute the joint distribution of the spacings and their means and variances.

2.5. Theorem. Let $D_1, \ldots, D_n$ be successive spacings between order statistics from the exponential distribution $G_\lambda(t) = 1 - e^{-\lambda t}$. Then

(a) $P[D_k \leq t] = G_{(n-k+1)\lambda}(t) = 1 - e^{-(n-k+1)\lambda t}$ for $k = 1, \ldots, n$;
(b) $D_1, \ldots, D_n$ are mutually independent;
(c) $ED_k = 1/(n - k + 1)\lambda$ and $\text{var}(D_k) = 1/[(n - k + 1)\lambda]^2$ for $k = 1, \ldots, n$.

PROOF. (a) First note that $P[D_1 > t] = e^{-n\lambda t}$, so that (a) holds for $k = 1$. Next, note that by the memoryless property of the exponential distribution (Theorem 2.1), D_2 has the same marginal distribution as does the first spacing in a sample of size $n - 1$ from G_λ; that is, $P[D_2 > t] = e^{-(n-1)\lambda t}$, so that (a) holds for $k = 2$. By repeated applications of this argument we may show that (a) holds for $k = 1, 2, \ldots, n$.

(b) The joint likelihood of observing the n spacings $d_1, \ldots, d_n$ is given by

$$L(d_1, \ldots, d_n) = n! \prod_{i=1}^{n} \lambda e^{-\lambda(d_1 + \cdots + d_i)}$$

$$= \prod_{i=1}^{n} (n - i + 1)\lambda e^{-(n-i+1)\lambda d_i}.$$

Note that the ith factor $(n - i + 1)\lambda e^{-(n-i+1)\lambda d_i}$ represents the marginal density of D_i obtained in (a) above. Since the joint density of $D_1, \ldots, D_n$ factors into the product of the individual densities of the D_i, then $D_1, \ldots, D_n$ are mutually independent.

(c) Part (c) follows from (a) and (2.6). ‖

2.6. Corollary. Under the hypothesis of Theorem 2.5, the *normalized spacings* $nD_1, (n - 1)D_2, \ldots, D_n$ from the exponential distribution $G_\lambda(t) = 1 - e^{-\lambda t}$ are independently distributed with common exponential distribution G_λ.

Since the order statistic $Y_{k:n} = D_1 + D_2 + \cdots + D_k$, we have the following corollary.

2.7. Corollary. The order statistics $Y_{1:n} \leq Y_{2:n} \leq \cdots \leq Y_{n:n}$ from $G_\lambda(t) = 1 - e^{-\lambda t}$ have expected values

$$EY_{k:n} = \frac{1}{\lambda}\left(\frac{1}{n} + \frac{1}{n-1} + \cdots + \frac{1}{n-k+1}\right), \tag{2.10a}$$

and variances

$$\mathrm{var}(Y_{k:n}) = \frac{1}{\lambda^2}\left(\frac{1}{n^2} + \frac{1}{(n-1)^2} + \cdots + \frac{1}{(n-k+1)^2}\right), \tag{2.10b}$$

$k = 1, 2, \ldots, n$.

PROOF. Equation (2.10a) follows immediately from Theorem 2.5(c). Equation (2.10b) requires, in addition, the independence of the D_i, $i = 1, \ldots, n$, obtained in Theorem 2.5(b). ‖

In life testing from an exponential distribution, a statistic that plays a central role is the total time on test. Assume n items are placed on test at time 0 and that successive failures are observed at times $Y_{1:n} \leq$

$Y_{2:n} \leq \cdots \leq Y_{n:n}$. Let $Y_{i:n} < t \leq Y_{i+1:n}$. Then the *total time on test*, $\tau(t)$, *during* $[0, t]$ is given by

$$\tau(t) = nY_{1:n} + (n-1)(Y_{2:n} - Y_{1:n}) + \cdots \\ + (n-i+1)(Y_{i:n} - Y_{i-1:n}) + (n-i)(t - Y_{i:n}). \qquad (2.11)$$

Note that $nY_{1:n}$ represents the total test time observed between 0 and $Y_{1:n}$, $(n-1)(Y_{2:n} - Y_{1:n})$ represents the total test time observed between $Y_{1:n}$ and $Y_{2:n}, \ldots$, and $(n-i)(t - Y_{i:n})$ represents the total test time observed between $Y_{i:n}$ and t. It follows that $\tau(t)$ represents the total test time observed between 0 and t.

To obtain the distribution of $\tau(Y_{i:n})$, we first need to establish the following theorem.

2.8. Theorem. Let $Y_1, \ldots, Y_n$ be independently and identically distributed with exponential density $\lambda e^{-\lambda t}$, $t > 0$. Then $Z = Y_1 + \cdots + Y_n$ has density

$$g_{\lambda,n}(z) = \frac{\lambda^n z^{n-1}}{(n-1)!} e^{-\lambda z}, \quad z \geq 0. \qquad (2.12)$$

PROOF. First note that (2.12) holds for $n = 1$. Assume (2.12) holds for $n = k$. Then for $z \geq 0$,

$$g_{\lambda,k+1}(z) = \int_0^z g_{\lambda,k}(x)\lambda e^{-\lambda(z-x)}\, dx,$$

where the integrand represents the joint probability density that $Y_1 + \cdots + Y_k = x$ and $Y_{k+1} = z - x$, and the range $[0, z]$ of integration corresponds to the set of mutually exclusive and exhaustive outcomes for $Y_1 + \cdots + Y_k$. Hence

$$g_{\lambda,k+1}(z) = \int_0^z \frac{\lambda^k x^{k-1}}{(k-1)!} e^{-\lambda x}\lambda e^{-\lambda(z-x)}\, dx = \lambda^{k+1}e^{-\lambda z} \int_0^z \frac{x^{k-1}}{(k-1)!}\, dx$$
$$= \frac{\lambda^{k+1}z^k}{k!} e^{-\lambda z}.$$

Thus (2.12) holds for $n = k + 1$. By induction it follows that the theorem holds for $n = 1, 2, \ldots$. ||

One of the pleasant properties of the exponential life distribution is that it is preserved under the formation of series systems. Specifically, we show the following theorem.

2.9. Theorem. Let Y be the life length of a series system of n independent components. Let Y_i, the life length of component i, have exponential distribution $G_{\lambda_i}(t) = 1 - e^{-\lambda_i t}$, $i = 1, \ldots, n$. Then Y has exponential distribution $G_\lambda(t) = 1 - e^{-\lambda t}$, where $\lambda = \sum_1^n \lambda_i$.

PROOF. $P[Y > t] = P[Y_1 > t, \ldots, Y_n > t]$ since the system is series. Thus $P[Y > t] = \prod_1^n P[Y_i > t]$, by the independence of components. The conclusion is immediate. ||

Remark. Theorem 2.9 represents the theoretical basis for the "parts count" method of system reliability analysis often used in practice. Under this method the reliability analyst counts the number n_i of parts of type i present in the system, multiplies by λ_i, the failure rate for that type of part, and sums over all part types appearing in the system. Specifically, he calculates

$$\lambda = \sum_1^k n_i \lambda_i,$$

where k denotes the number of different part types present.

Example. Bazovsky (1961, pp. 90–91) presents a simple example of the parts count method of reliability analysis. An electronic circuit consists of 10 silicon diodes, 4 silicon transistors, 20 composition resistors, and 10 ceramic capacitors, all connected in series. Assume that the wiring (printed circuit) and the solder connections are perfectly reliable, and that the components are independently exponentially distributed with failure rates (per million hours):

Silicon diode: $\lambda_d = 2$,
Silicon transistor: $\lambda_t = 10$,
Composition resistor: $\lambda_r = 1$,
Ceramic capacitor: $\lambda_c = 2$.

Then by Theorem 2.9, system life is exponentially distributed with failure rate given by: $\lambda = 10 \times 2 + 4 \times 10 + 20 \times 1 + 10 \times 2 = 100$.

From Theorem 2.9 we see that the parts count method yields the correct system reliability when the components are independently distributed with constant failure rates. In Esary, Marshall, and Proschan (1971) it is shown that this condition is also a necessary condition.

EXERCISES

1. Verify by integration by parts that the nth moment μ_n of the exponential distribution $1 - e^{-\lambda t}$ is given by

$$\mu_n = \frac{n!}{\lambda^n}.$$

2. Prove that for the exponential distribution, the coefficient of variation $\sigma/\mu = 1$, where σ is the standard deviation.

3. Use (2.8) and (2.9) to obtain integer moments of the exponential distribution.
4. Let $X_{1:n} \leq X_{2:n} \leq \cdots \leq X_{n:n}$ be order statistics from a distribution F of a positive random variable. Prove

(a) $$P[X_{i:n} \leq x] = \sum_{j=i}^{n} \binom{n}{j} [F(x)]^j [\bar{F}(x)]^{n-j}$$
$$\equiv \frac{n!}{(i-1)!\,(n-i)!} \int_0^{F(x)} t^{i-1}(1-t)^{n-i}\,dt.$$

(b) $$EX_{i:n} = \int_0^{\infty} \sum_{j=0}^{i-1} \binom{n}{j} [F(x)]^j [\bar{F}(x)]^{n-j}\,dx.$$

Note that in the exponential case, the expected values of order statistics are much more conveniently obtained by the arguments leading to Corollary 2.7 than by the use of formula (b) just above.

5. Let $G_{\lambda,n}(z)$ be the gamma distribution with integer shape parameter n:
$$G_{\lambda,n}(t) = \int_0^t \frac{\lambda^n z^{n-1}}{(n-1)!} e^{-\lambda z}\,dz.$$
Prove that
$$\bar{G}_{\lambda,n}(t) = \sum_{i=0}^{n-1} \frac{(\lambda t)^i}{i!} e^{-\lambda t}.$$

6. By direct argument, show that the total time on test $\tau(t)$, $Y_{i:n} < t \leq Y_{i+1:n}$, is given by $\tau(t) = Y_{1:n} + \cdots + Y_{i:n} + (n-i)t$. Verify that this is equivalent to (2.11).
7. Suppose the n components of a parallel system have independent exponential life lengths with respective failure rates $\lambda_1, \ldots, \lambda_n$. Compute system reliability and mean system life.

3. THE POISSON PROCESS

Suppose we operate a "socket" for an indefinite length of time; when the unit fails, we replace it immediately by a new unit. Assume all units are stochastically independent, with common exponential life distribution $G_\lambda(t) = 1 - e^{-\lambda t}$. Let $N(t)$ denote the number of failed units by time t. Then $\{N(t); t \geq 0\}$ is an example of a Poisson process. A general definition is now given.

3.1. Definition. Suppose events are occurring successively in time, with the intervals between successive events independently and identically distributed according to an exponential distribution $G_\lambda(t) = 1 - e^{-\lambda t}$. Denote the number of events during $[0, t]$ by $N(t)$. Then the stochastic process $\{N(t); t \geq 0\}$ is called a *Poisson process with mean rate (or intensity)* λ.

In the terminology of Section 3, Chapter 6, a Poisson process is a renewal process for which the underlying distribution is exponential. Actually, the Poisson process can be derived under several alternative assumptions. To present a second definition of the Poisson process, recall that a stochastic process $\{X(t); 0 \leq t < \infty\}$ has independent increments if for all $t_0 < t_1 < \cdots < t_n$, $n = 2, 3, \ldots$, $X(t_1) - X(t_0), \ldots, X(t_n) - X(t_{n-1})$ are independent random variables. If, in addition, $X(t_2 + h) - X(t_1 + h)$ has the same distribution as $X(t_2) - X(t_1)$ for all $t_1, t_2, h > 0$, then the process is said to have *stationary independent increments*.

3.2. Second Definition of Poisson Process. An integer valued process $\{N(t); t \geq 0\}$ is a Poisson process with mean rate λ if:

(a) $\{N(t); t \geq 0\}$ has stationary independent increments.
(b) For time points s and t, $s < t$, the number $N(t) - N(s)$ of counts in the interval $[s, t]$ has a Poisson distribution with mean $\lambda(t - s)$:

$$P[N(t) - N(s) = k] = \frac{e^{-\lambda(t-s)}[\lambda(t - s)]^k}{k!}. \tag{3.1}$$

See Section 4 for a discussion of the Poisson distribution, its properties, and some of its applications in reliability.

Next we prove the equivalence of the two definitions of a Poisson process.

3.3. Theorem. Definitions 3.1 and 3.2 of a Poisson process are equivalent.

PROOF. Let $\{N(t); t \geq 0\}$ satisfy the conditions of Definition 3.1. By Theorem 2.1, the increments $N(t_1) - N(t_0), \ldots, N(t_n) - N(t_{n-1})$ are stationary and independent, so that (a) of Definition 3.2 holds.

To demonstrate (b) of Definition 3.2, it suffices to show that

$$P[N(t) \leq k] = \sum_{i=0}^{k} e^{-\lambda t} \frac{(\lambda t)^i}{i!}, \tag{3.2}$$

since the increments are stationary. Letting $Y_1, Y_2, \ldots$, be independently and identically distributed according to exponential distribution $G_\lambda(t) = 1 - e^{-\lambda t}$, we have

$$P[N(t) \leq k] = P[Y_1 + \cdots + Y_{k+1} > t].$$

By Theorem 2.8, the last probability is given by

$$\int_t^\infty \frac{\lambda^{k+1} z^k}{k!} e^{-\lambda z}\, dz.$$

By Exercise 5 of Section 2, this integral may be expressed as

$$\sum_{i=0}^{k} \frac{(\lambda t)^i}{i!} e^{-\lambda t}.$$

Thus condition (b) of Definition 3.2 holds.

The proof that the conditions of Definition 3.2 imply those of Definition 3.1 is left to Exercise 1. ||

A third definition of the Poisson process may be developed using an axiomatic approach.

3.4. Third Definition of Poisson Process. As in Definition 3.1, let $N(t)$ denote the number of events that have occurred during $[0, t]$. Suppose that whatever the number of events during [0, t],

(a) The probability of an event during $[t, t + h]$ is $\lambda h + o(h)$.
(b) The probability of more than one event during $[t, t + h]$ is $o(h)$.

Then the stochastic process $\{N(t); t \geq 0\}$ is called a *Poisson process with mean rate* λ.

The third definition of the Poisson process is equivalent to each of the first two definitions of the Poisson process; the proof is left to Exercise 2.

3.5. Examples of the Poisson Process Arising in Life Testing. The Poisson process arises quite naturally in reliability and life testing situations when the underlying life distribution is exponential. The following are commonly occurring examples:

(a) *Maintained Unit.* A unit is put into operation at time 0. Each time failure occurs, the failed unit is replaced by a fresh unit of the same type. Assume life lengths are independently and identically distributed according to an exponential distribution $G_\lambda(t) = 1 - e^{-\lambda t}$. If $N(t)$ denotes the number of failures observed during $[0, t]$, then $\{N(t); t \geq 0\}$ is a Poisson process with mean rate λ.

(b) *Sampling with Replacement.* A sample of n units is randomly selected from a population having life distribution $G_\lambda(t) = 1 - e^{-\lambda t}$. The failure rate λ is unknown and is to be estimated from data collected during the life testing experiment. The n units are put on test at time 0; each time a failure occurs, it is replaced by a new, randomly selected unit. If $N(t)$ denotes the number of failures observed during $[0, t]$, then $\{N(t); t \geq 0\}$ is a Poisson process with mean rate $n\lambda$.

(c) *Sampling without Replacement.* As in (b), it is desired to estimate the unknown failure rate λ of an exponential life distribution. However, now failed units are *not* replaced; the experiment ends in one of the following ways:

Case 1. At a specified time t_0.
Case 2. At the rth failure, where r is specified in advance, $1 \le r \le n$.
Case 3. At the earlier of time t_0 and the time of the rth failure, where t_0 and r $(1 \le r \le n)$ are specified in advance.

Let $\tau(t)$ represent the total time on test during $[0, t]$, as defined by (2.11), and let τ_0 represent the total time on test accumulated when the experiment ends. Let $N^*(\tau)$ represent the number of failures observed when the total time on test (not elapsed time) has reached τ. Then $\{N^*(\tau); 0 \le \tau \le \tau_0\}$ is stochastically equivalent to a Poisson process with mean rate λ, conditioned on the fact that the process stops in case 1 at $\tau(t_0)$, in case 2 at the rth failure, and in case 3 at the earlier of $\tau(t_0)$ and the rth failure.

To verify the last assertion, note that the total time on test elapsed between the $(i - 1)$st and the ith failure is $(n - i + 1)(Y_{i:n} - Y_{i-1:n})$, where $Y_{1:n} \le Y_{2:n} \le \cdots \le Y_{n:n}$ are the order statistics in a sample of size n. But by Corollary 2.6, $nY_{1:n}$, $(n - 1)(Y_{2:n} - Y_{1:n}), \ldots, Y_{n:n} - Y_{n-1:n}$ are independently and identically distributed with exponential distribution $G_\lambda(t) = 1 - e^{-\lambda t}$. It follows from Definition 3.1 that $\{N^*(\tau); 0 \le \tau \le \tau_0\}$ is a Poisson process with mean rate λ, conditioned as specified above.

We have seen from Definition 3.1 that in a Poisson process the intervals Y_i between the $i - 1$st and ith events, $i = 1, 2, \ldots$, are independently and identically distributed with exponential distribution $G_\lambda(t) = 1 - e^{-\lambda t}$. Let $S_k = \sum_{i=1}^{k} Y_i$ denote the time of the kth event, with $S_0 \equiv 0$ by definition. Then we may show the following theorem.

3.6. Theorem. In a Poisson process, given that $S_{k+1} = s$, then the conditional joint distribution of $S_1, S_2, \ldots, S_k$ is that of the order statistics in a sample of k from the uniform distribution on $[0, s]$. Symbolically, if $f(s_1, \ldots, s_k \mid s)$ is the conditional joint density of $S_1, \ldots, S_k$, given that $S_{k+1} = s$, we claim

$$f(s_1, \ldots, s_k \mid s) = \frac{k!}{s^k} \quad \text{for} \quad 0 \le s_1 \le \cdots \le s_k \le s. \tag{3.3}$$

PROOF. Assuming a mean rate of λ in the Poisson process, the conditional joint density may be expressed as follows:

$$f(s_1, \ldots, s_k \mid s) = \frac{\lambda e^{-\lambda s_1} \lambda e^{-\lambda(s_2 - s_1)} \cdots \lambda e^{-\lambda(s - s_k)}}{\dfrac{\lambda^{k+1} s^k}{k!} e^{-\lambda s}}$$

$$\text{for} \quad 0 \le s_1 \le \cdots \le s_k \le s.$$

The denominator, a gamma density of order $k + 1$, represents the density of S_{k+1} by Theorem 2.8. The numerator expresses directly the joint density of $k + 1$ independent observations from the exponential G. Simplifying, we obtain (3.3). ||

Remark. Note that given that $S_{k+1} = s$ the set $\{S_1, \ldots, S_k\}$, when unordered, are distributed independently and uniformly on $[0, s]$.

A closely related result is given in Theorem 3.7.

3.7. Theorem. Given that k events have occurred during $[0, t_0]$ in a Poisson process, then their successive times, $S_1, \ldots, S_k$, of occurrence are jointly distributed as the order statistics in a sample of size k from the uniform distribution on $[0, t_0]$. Symbolically, if $f(s_1, \ldots, s_k \mid k)$ represent the conditional joint density of $S_1, \ldots, S_k$ given that $N(t_0) = k$, then we claim

$$f(s_1, \ldots, s_k \mid k) = \frac{k!}{t_0^k} \quad \text{for} \quad 0 \leq s_1 \leq \cdots \leq s_k \leq t_0. \tag{3.4}$$

PROOF. Assuming the mean rate of the Poisson process is λ, the conditional joint density may be expressed as follows:

$$f(s_1, \ldots, s_k \mid k) = \frac{\lambda e^{-\lambda s_1} \cdot \lambda e^{-\lambda(s_2 - s_1)} \cdots \lambda e^{-\lambda(s_k - s_{k-1})} \cdot e^{-\lambda(t_0 - s_k)}}{\dfrac{(\lambda t_0)^k}{k} e^{-\lambda t_0}}$$

$$\text{for} \quad 0 \leq s_1 \leq \cdots \leq s_k \leq t_0.$$

The denominator represents the probability of the conditioning event $N(t_0) = k$, by Definition 3.2(b). The numerator represents the joint probability density of k successive exponential observations of respective lengths $s_1, s_2 - s_1, \ldots, s_k - s_{k-1}$, followed by survival over the remaining partial interval of length $t_0 - s_k$. Upon simplifying, we obtain (3.4). ||

Remark. Given that k events have occurred during $[0, t_0]$, then their successive times, $S_1, \ldots, S_k$, *when unordered* are uniformly and independently distributed on $[0, t_0]$.

Theorems 3.6 and 3.7 and Corollary 3.8 below lead to useful tests of the exponentiality of the underlying distribution, as will be shown in the forthcoming volume on reliability inference.

Using the notions concerning total time on test developed in Example 3.5(c), we may immediately obtain Corollary 3.8.

3.8. Corollary. Assume sampling without replacement, as described in Example 3.5(c). Let $\tau(Y_{i:n})$ represent the total time on test observed by the time $Y_{i:n}$ of the ith failure.

(a) Given that $\tau(Y_{k+1:n}) = \tau$, the conditional joint distribution of $\tau(Y_{1:n}), \ldots, \tau(Y_{k:n})$ is that of the order statistics in a sample of k from the uniform distribution on $[0, \tau]$.

(b) Given that k failures have occurred during $[0, t_0]$, the conditional joint distribution of $\tau(Y_{1:n}), \ldots, \tau(Y_{k:n})$ is that of the order statistics in a sample of k from the uniform distribution on $[0, \tau(t_0)]$.

Next consider the *superposition* of independent Poisson processes, that is, the process resulting from pooling together the time points of events occurring in each of the separate Poisson processes. In Figure 3.3.1 the time points of events of Process 1 are shown on line 1, and of Process 2 on line 2; the superposed process is shown below.

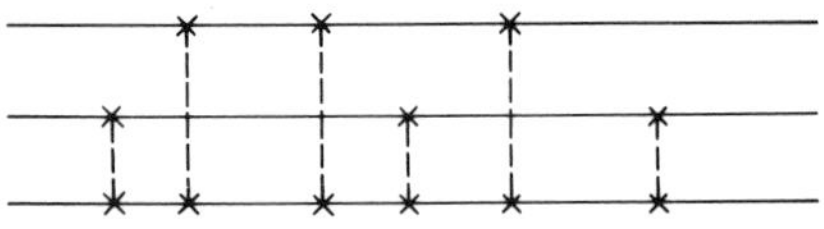

Figure 3.3.1.
Superposition of two processes.

We prove Theorem 3.9.

3.9. Theorem. The superposition of two independent Poisson processes with mean rates λ_1 and λ_2 is itself a Poisson process with mean rate $\lambda = \lambda_1 + \lambda_2$.

PROOF. Let $N(t)$ be the number of events observed during $[0, t]$ in the pooled process. Then $\{N(t); t \geq 0\}$ has stationary independent increments since each process being superposed does. Thus (a) of Definition 3.2 holds.

Also, for time points s and t, $s < t$, $N(t) - N(s)$ is the sum of two independent Poisson random variables with respective means $\lambda_1(t - s)$ and $\lambda_2(t - s)$. By Theorem 4.2 below, $N(t) - N(s)$ is itself a Poisson random variable with mean $\lambda_1(t - s) + \lambda_2(t - s) \equiv \lambda(t - s)$. Hence (b) of Definition 3.2 holds. ‖

An important example of the superposition of independent Poisson processes arises as follows. Consider a series system with component i having life distribution $G_{\lambda_i}(t) = 1 - e^{-\lambda_i t}$, $i = 1, \ldots, n$. When a component fails, it is replaced immediately by a similar component having the same distribution. All components are assumed stochastically independent. By Definition 3.1, the sequence of failures of component i and its replacements determines a Poisson process with mean rate λ_i. Since the system fails each time any one of the n components comprising it fails, the sequence of *system* failures is obtained as a superposition of the n individual component Poisson processes. Thus by Theorem 3.9, the sequence of system failures is governed by a Poisson process with mean rate $\lambda = \sum_{i=1}^{n} \lambda_i$.

EXERCISES

1. Prove that the conditions of Definition 3.2 imply those of Definition 3.1.
2. Prove that Definition 3.4 of the Poisson process is equivalent to Definition 3.1 or Definition 3.2.
3. A sample of n items is randomly selected and tested until failure, yielding successive life lengths $Y_{1:n} \le Y_{2:n} \le \cdots \le Y_{n:n}$. If the underlying life distribution is the exponential $G_\lambda(t) = 1 - e^{-\lambda t}$, and $\tau(Y_{r:n}) = nY_{1:n} + (n-1)(Y_{2:n} - Y_{1:n}) + \cdots + (n-r+1)(Y_{r:n} - Y_{r-1:n})$ is the total time on test until the rth failure, prove that $2\lambda\tau(Y_{r:n})$ has a χ^2-distribution with $2r$ degrees of freedom for $1 \le r \le n$.
4. Under the same assumptions as in Exercise 3 above, show that
$$\sum_{i=1}^{r-1} \frac{\tau(Y_{i:n})}{\tau(Y_{r:n})}$$
is asymptotically normally distributed with mean $(r-1)/2$ and variance $(r-1)/12$, as $r \to \infty$ (and, of course, $n \to \infty$).
5. Let the number $N(t)$ of shocks during $[0, t]$ be governed by a Poisson process with rate λ. Let the probability that a given shock is fatal be q, independent of all other events. Then the sequence of *fatal* shocks constitutes a Poisson process with rate $q\lambda$.
6. Let the number $N_i(t)$ of shocks during $[0, t]$ be governed by a Poisson process with rate λ_i, $i = 1, \ldots, n$, with the n processes mutually independent. Then the total number $N(t) = \sum_{i=1}^{n} N_i(t)$ of shocks of all kinds is governed by a Poisson process with rate $\lambda = \sum_{i=1}^{n} \lambda_i$.

4. THE POISSON DISTRIBUTION

In Definition 3.2 of the Poisson process, the Poisson distribution was seen to govern the number of events during a specified time interval. In this section we present some useful properties of the Poisson distribution and their applications in reliability.

4.1. Definition. An integer-valued random variable N is said to have a Poisson distribution with parameter m if

$$P[N = i] = \frac{m^i}{i!} e^{-m} \qquad \text{for} \quad i = 0, 1, 2, \ldots. \tag{4.1}$$

For the Poisson distribution, the moment generating function (MGF) is given by

$$M(t) = \sum_{i=0}^{\infty} P[N = i]e^{it} = \sum_{i=0}^{\infty} \frac{(me^t)^i}{i!} e^{-m},$$

or

$$M(t) = e^{m(e^t - 1)}. \tag{4.2}$$

From (4.2) we obtain the first two moments μ_1 and μ_2 and the variance σ^2 of the Poisson as follows:

$$\begin{aligned} \mu_1 &= M'(t)|_{t=0} = e^{me^t - m}(me^t)|_{t=0} = m, \\ \mu_2 &= M''(t)|_{t=0} = e^{me^t - m}(me^t)^2 + e^{me^t - m}(me^t) \\ &= m^2 + m. \end{aligned}$$

Thus $\sigma^2 = \mu_2 - {\mu_1}^2 = m$. Note that both the mean and the variance of the Poisson are equal to its parameter m. For future reference, write:

$$\mu_1 = m = \sigma^2. \tag{4.3}$$

Another basic result may be obtained using the MGF.

4.2. Theorem. Let N_1 and N_2 be independent Poisson random variables with means m_1 and m_2 respectively. Then $N = N_1 + N_2$ is a Poisson random variable with mean $m = m_1 + m_2$.

PROOF. The MGF $M_N(t)$ of N is given by

$$M_N(t) = M_{N_1}(t) \cdot M_{N_2}(t),$$

where $M_{N_i}(t)$ is the MGF of N_i. Thus

$$M_N(t) = e^{m_1e^t - m_1} \cdot e^{m_2e^t - m_2} = e^{(m_1 + m_2)e^t - m_1 + m_2)},$$

which, by (4.2), is the MGF of a Poisson random variable with mean $m = m_1 + m_2$. ||

An application may be made to spares provisioning.

4.3. The Spare Parts Problem. Consider a system of k component positions using parts of a single type, having life distribution $G_\lambda(t) = 1 - e^{-\lambda t}$, with all life lengths stochastically independent. When a part fails it is immediately replaced by a spare, if available. Component position j is required to operate for t_j hours during the mission, $j = 1, \ldots, k$. (The t_j may differ since it may not be necessary for all component positions to be in use throughout the mission.) Determine the number n of spares to provide to achieve assurance $\geq \alpha$ that the system will operate throughout the mission without shutdown due to shortage of spares.

Let $N_j(t_j)$ denote the number of failures (or replacements) occurring in the jth component position, assuming spares are available when needed. Then n will be the smallest integer r satisfying

$$P[N_1(t_1) + \cdots + N_k(t_k) \leq r] \geq \alpha. \tag{4.4}$$

First, note that $N_j(t_j)$ is a Poisson random variable with mean λt_j, by Definition 3.2. Thus by Theorem 4.2 above, $N_1(t_1) + \cdots + N_k(t_k)$ is also a Poisson random variable with mean $m = \lambda(t_1 + \cdots + t_k)$. Note that the expected total number m of failures occurs as the product of the failure rate λ and the total exposure time $(t_1 + \cdots + t_k)$. Hence

$$P[N_1(t_1) + \cdots + N_k(t_k) \leq r] = \sum_{i=0}^{r} \frac{m^i}{i!} e^{-m}. \tag{4.5}$$

From (4.4) and (4.5) we conclude that the solution n is the smallest integer r satisfying

$$\sum_{i=0}^{r} \frac{m^i}{i!} e^{-m} \geq \alpha. \tag{4.6}$$

We may obtain n from the tables of Molina (1942), Pearson and Hartley (1956), or General Electric Co. (1962).

Remark. An easy generalization is possible to cover the case in which component position failure rates are not necessarily alike, even though all units are functionally alike and a single common supply of spares is maintained. For example, different component positions may be operating at different temperatures. Assume then a failure rate of λ_j for the jth component position, $j = 1, \ldots, k$. We proceed as above except that now $N_1(t_1) + \cdots + N_k(t_k)$ is a Poisson random variable with mean $m = \lambda_1 t_1 + \cdots + \lambda_k t_k$.

Example. A silicon diode having exponential life length is used in a number of component positions in an electronic circuit. In the first five positions, its failure rate is .000,002; in the next three positions, its failure rate is .000,001; finally, in the last four positions, its failure rate is .000,003. A supply depot is required to provide spares to keep 100 such circuits in operation during a period containing 1,000 operating hours. How many spares should the depot carry so that the probability of shortage is $\leq .05$?

The total number N of silicon diode failures experienced during the 1,000 operating hours is a Poisson random variable with expected value $(5 \times .000{,}002 + 3 \times .000{,}001 + 4 \times .000{,}003) \times 100 \times 1{,}000 = 2.5$. From Molina's Tables (1942), we find $P[N > 5] = .0420$, while $P[N > 4] = .1088$. Thus the smallest number of spare silicon diodes to stock so as to achieve an assurance of at least .95 of no shortage is 5.

4.4. Binomial Moments of Poisson. As a final result, we obtain the binomial moment B_k of the Poisson distribution, $k = 1, 2, \ldots$, for use in Chapter 6. For a discrete random variable N, the kth *binomial moment* B_k is defined as

$$B_k = E\binom{N}{k}, \qquad k = 1, 2, \ldots.$$

Thus for the Poisson random variable with mean m, we have

$$B_k = \sum_{i=k}^{\infty} \binom{i}{k} \frac{m^i}{i!} e^{-m} = \frac{e^{-m} m^k}{k!} \sum_{i=k}^{\infty} \frac{m^{i-k}}{(i-k)!}$$

Thus

$$B_k = \frac{m^k}{k!} \qquad \text{for} \quad k = 1, 2, \ldots. \tag{4.7}$$

EXERCISES

1. In the spare parts problem of 4.3, suppose $k = 3$, $t_1 = 1000$, $t_2 = 2000$, $t_3 = 1500$, $\lambda = .001$, and $\alpha = .95$. Find n.

2. Suppose that a separate supply of spares is maintained for each of the three component positions of Exercise 1. Find numbers n_1, n_2, n_3 required to provide assurance $\geq .95$ of no shortage in any of the three positions. Show that $n_1 + n_2 + n_3$ is greater than the solution n of Exercise 1.

3. In the model of 4.3, show in general that a single common pool of spares for all positions (using like parts, of course) requires fewer spares for a given assurance than does a separate supply of spares for each position.

4. The variance to mean ratio is a measure of a discrete distribution's departure from the Poisson. (For the Poisson, recall var $N/EN = 1$.) Show that var $N/EN \geq 1$ when N is a negative binomial random variable; that is,

$$P[N = k] = \binom{r + k - 1}{k} p^r q^k \qquad \text{for} \quad k = 0, 1, 2, \ldots,$$

where $q = 1 - p$ (var N means variance of N).

5. Assume var $N_\alpha/EN_\alpha > 1$ where α is a parameter indexing the distribution of N_α. Let G be a mixing distribution on α. Show that var $N/EN \geq 1$ when

$$P[N = k] = \int_0^\infty P[N_\alpha = k] \, dG(\alpha);$$

that is, mixing preserves the variance to mean inequality.

5. PARAMETRIC FAMILIES OF LIFE DISTRIBUTIONS WITH MONOTONE FAILURE RATE

In Section 1 we introduced several broad classes of life distributions based on notions of aging. In Section 2 we presented the properties of the exponential distribution, for which the failure rate is constant. In this section we present some parametric families of life distributions popular in reliability applications which represent special cases of the classes introduced in Section 1; in each case the failure rate is monotone.

5.1. Weibull Distribution.[1] This distribution is given by

$$F_\alpha(t) = 1 - e^{-(\lambda t)^\alpha} \qquad \text{for} \quad t \geq 0, \quad \text{where } \lambda, \alpha > 0. \tag{5.1}$$

The Weibull distribution is an extreme value distribution in that it is the limiting distribution of the minimum of independent random variables. (Extreme value theory is discussed in detail in Chapter 8.)

The failure rate $r(t)$ takes a particularly simple form:

$$r(t) = \frac{f(t)}{\bar{F}(t)} = \alpha\lambda(\lambda t)^{\alpha-1} \qquad \text{for} \quad t > 0. \tag{5.2}$$

Thus the Weibull distribution F_α is IFR for $\alpha \geq 1$ and DFR for $0 < \alpha \leq 1$; for $\alpha = 1$, $F_\alpha(t) = 1 - e^{-\lambda t}$, the exponential distribution, which is both DFR and IFR. The failure rate function r of the Weibull is plotted in Figure 3.5.1. Note that $r(t) \uparrow \infty$ for $\alpha > 1$ and $\downarrow 0$ for $\alpha < 1$ as $t \to \infty$. The parameter α is called the *shape parameter*; as α increases, the failure rate function rises more steeply and the probability density becomes more peaked. The parameter λ is a *scale parameter*; that is, the distribution depends on λ and t only through their product λt.

The Weibull distribution has been used to describe fatigue failure (Weibull, 1939), vacuum tube failure (Kao, 1958), and ball bearing failure (Lieblein and Zelen, 1956). It is perhaps the most popular parametric family of failure distributions at the present time.

5.2. Gamma Density. This is given by

$$g_{\lambda,\alpha}(t) = \frac{\lambda^\alpha t^{\alpha-1}}{\Gamma(\alpha)} e^{-\lambda t} \qquad \text{for} \quad t \geq 0, \quad \text{where } \lambda, \alpha > 0. \tag{5.3}$$

[1] *Notation.* In presenting a distribution F, it will be understood that $F(t) \equiv 0\ (1)$ for t to the left (right) of the domain specified. Thus (5.1) implicitly states that $F_\alpha(t) \equiv 0$ for $t < 0$. Similarly, if a probability density is explicitly described on the domain $[a, b]$, it will be implicitly understood to be 0 elsewhere.

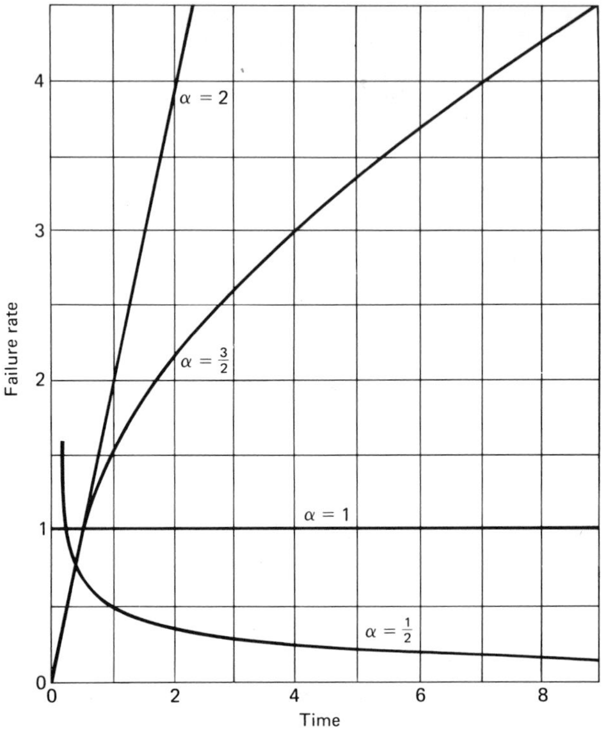

Figure 3.5.1.
Failure rate curves of the Weibull distribution for $\lambda = 1$.

For general α, the distribution function $G_{\lambda,\alpha}$ may be written as

$$G_{\lambda,\alpha}(t) = \int_0^t g_{\lambda,\alpha}(x)\, dx.$$

When α is a positive integer, $G_{\lambda,\alpha}(t)$ may be written in closed form as

$$G_{\lambda,\alpha}(t) = 1 - \sum_{i=0}^{\alpha-1} \frac{(\lambda t)^i}{i!} e^{-\lambda t} \qquad \text{for} \quad t \geq 0, \tag{5.4}$$

as may be shown directly by integration by parts. (See Exercise 5 of Section 2.) The gamma distribution $G_{\lambda,\alpha}$ with integer parameter α is the distribution of the sum of α independent exponential random variables, each with failure rate λ, as shown in Theorem 2.8.

To study the monotonicity properties of the failure rate of the gamma, write

$$[r(t)]^{-1} = \frac{\bar{G}_{\lambda,\alpha}(t)}{g_{\lambda,\alpha}(t)} = \int_t^\infty \left(\frac{x}{t}\right)^{\alpha-1} e^{-\lambda(x-t)}\, dx.$$

With the change of variable $u = x - t$, we obtain

$$[r(t)]^{-1} = \int_0^\infty \left(1 + \frac{u}{t}\right)^{\alpha - 1} e^{-\lambda u}\, du.$$

Hence $[r(t)]^{-1}$ is increasing for $0 < \alpha \leq 1$, decreasing for $\alpha \geq 1$. We thus come to the following conclusion.

5.3. Gamma Distribution. The distribution $G_{\lambda,\alpha}$ is DFR for $0 < \alpha \leq 1$ and IFR for $\alpha \geq 1$. For $\alpha = 1$, $G_{\lambda,\alpha} = 1 - e^{-\lambda t}$, an exponential distribution.

The failure rate of the gamma is plotted in Figure 3.5.2 for various values of α. Note that $r(t) \uparrow \lambda$ for $\alpha > 1$ and $\downarrow \lambda$ for $\alpha < 1$ as $t \to \infty$. See Exercise 2.

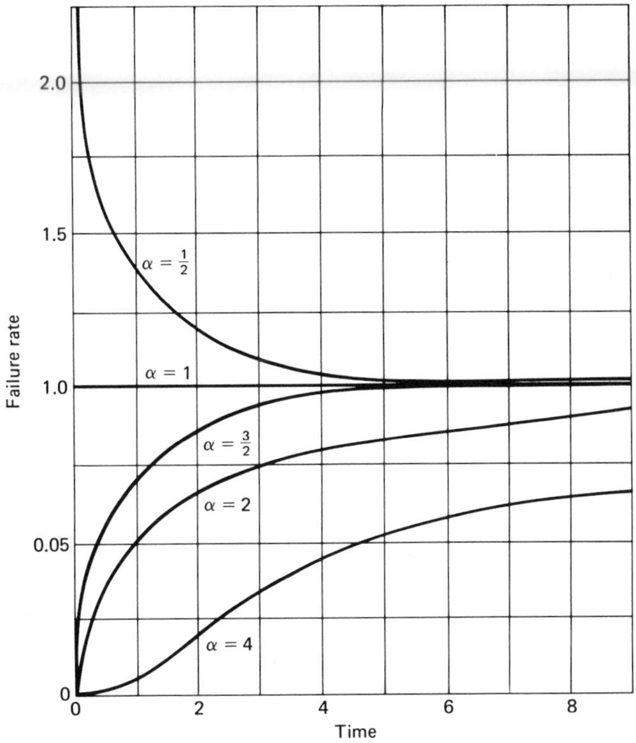

Figure 3.5.2.
Failure rate curves of the gamma distribution for $\lambda = 1$.

As with the Weibull family, as the shape parameter α increases the probability density becomes more peaked. Also the parameter λ is a scale parameter.

5.4. Truncated Normal Density. This density is given by

$$f(t) = \frac{1}{a\sigma\sqrt{2\pi}} e^{-(t-\mu)^2/2\sigma^2} \quad \text{for} \quad 0 \leq t < \infty, \tag{5.5}$$

where $\sigma > 0$, $-\infty < \mu < \infty$, $a = \int_0^\infty (1/\sigma\sqrt{2\pi})e^{-(t-\mu)^2/2\sigma^2}\,dt$. The introduction of a insures that $\int_0^\infty f(t)\,dt = 1$, so that f is the density of a (nonnegative) life length. Note that for $\mu \gg 3\sigma$, the value of a is very close to 1, and for most practical purposes may be omitted, so that (5.5) reduces to the usual normal density.

Davis (1952), after examining failure data for a wide variety of items, has shown empirically that items manufactured and tested under close control may be fitted nicely with truncated normal life distributions of the form (5.5).

In order to study the monotonicity properties of the failure rate $r(t)$ of (5.5), it is useful to consider the class of Pólya frequency functions of order 2 (PF_2), which are of considerable independent interest (Karlin, 1968).

5.5. Definition. A function $h(x)$, $-\infty < x < \infty$, is PF_2 if

(a) $h(x) \geq 0$ for $-\infty < x < \infty$, and

(b) $$\begin{vmatrix} h(x_1 - y_1) & h(x_1 - y_2) \\ h(x_2 - y_1) & h(x_2 - y_2) \end{vmatrix} \geq 0 \tag{5.6}$$

for all $-\infty < x_1 < x_2 < \infty$ and $-\infty < y_1 < y_2 < \infty$, or equivalently,

(b′) $\log h(x)$ is concave on $(-\infty, \infty)$, or equivalently,

(b″) for fixed $\Delta > 0$, $h(x + \Delta)/h(x)$ is decreasing in x for $a \leq x \leq b$, where $a = \inf_{h(y)>0} y$, $b = \sup_{h(y)>0} y$.

The proof of the equivalence of (b), (b′), and (b″) is left to Exercise 12.

Note that a Pólya frequency function f of order 2 is not necessarily a probability frequency function, in that $\int_{-\infty}^{\infty} f(x)\,dx$ need not be 1, nor even finite. Totally positive functions, a generalization of Pólya frequency functions, will be discussed in Section 4 of Chapter 4.

Next we show that an IFR survival probability is PF_2.

5.6. Theorem. Let F be a distribution function. Then F IFR is equivalent to $\bar{F}$ PF_2.

PROOF. By Definition 1.1, F IFR is equivalent to $\bar{F}(t + \Delta)/\bar{F}(t)$ decreasing in $-\infty < t < \infty$ for each $\Delta \geq 0$. Thus by Definition 5.5(b″), F IFR is equivalent to $\bar{F}$ PF_2. ‖

Because of its logarithmic concavity, an IFR survival probability displays certain smoothness properties, as shown in Lemma 5.7.

5.7. Lemma. If F is IFR and $F(z) < 1$, then F is absolutely continuous on $(-\infty, z)$; that is, F has a probability density on $(-\infty, z)$.

PROOF. Let $R(z) \equiv -\log \bar{F}(z)$ denote the hazard function. [See (1.4).] Let $\varepsilon > 0$, and $\alpha_1 < \beta_1 < \alpha_2 < \beta_2 < \cdots < \alpha_m < \beta_m \leq z$ satisfying $\sum_{i=1}^m (\beta_i - \alpha_i) < \varepsilon/r^+(z)$, where $r^+(z) = \lim_{\delta \downarrow 0} [R(z + \delta) - R(z)]/\delta$ exists finitely since $R \equiv -\log \bar{F}$ is convex. Then $\sum_{i=1}^m |R(\beta_i) - R(\alpha_i)| = \sum_{i=1}^m \{[R(\beta_i) - R(\alpha_i)]/(\beta_i - \alpha_i)\}(\beta_i - \alpha_i) \leq r^+(z) \sum_{i=1}^m (\beta_i - \alpha_i) \leq \varepsilon$. Thus R is absolutely continuous on $(-\infty, z)$, and the result follows. ||

Note, however, that an IFR distribution may have a jump at the right-hand endpoint of its interval of support, if finite.

5.8. Lemma. If f is a PF_2 density on $[0, \infty)$, then the corresponding distribution $F(t) = \int_0^t f(x)\, dx$ is IFR.

PROOF. To show $r(t) = f(t)/\bar{F}(t)$ increasing in t, it suffices to show that for $t_1 < t_2$,

$$0 \geq \begin{vmatrix} f(t_1) & f(t_2) \\ \bar{F}(t_1) & \bar{F}(t_2) \end{vmatrix} = \int_0^\infty \begin{vmatrix} f(t_1) & f(t_2) \\ f(t_1 + x) & f(t_2 + x) \end{vmatrix} dx.$$

Since f is PF_2, the determinant is ≤ 0. Hence F is IFR. ||

An analogous result is shown in Lemma 5.9.

5.9. Lemma. If f is a density on $[0, \infty)$ for which $\log f(x)$ is convex on $[0, \infty)$, then the corresponding distribution function F is DFR.

The proof is left to Exercise 13.

Returning to the study of the monotonicity of the failure rate of the truncated normal density f given in (5.5), we note that

$$\log f(t) = -\log(a\sigma\sqrt{2\pi}) - \frac{(t - \mu)^2}{2\sigma^2}, \qquad t \geq 0,$$

a concave function on $[0, \infty)$. Thus the density is PF_2, and by Lemma 5.8, the distribution is IFR. The failure rate function of the normal distribution is plotted in Figure 3.5.3.

Similarly, Lemmas 5.8 and 5.9 can be used to show that the gamma distribution with density (5.3) is IFR for $\alpha \geq 1$ and DFR for $\alpha \leq 1$. See Exercise 10.

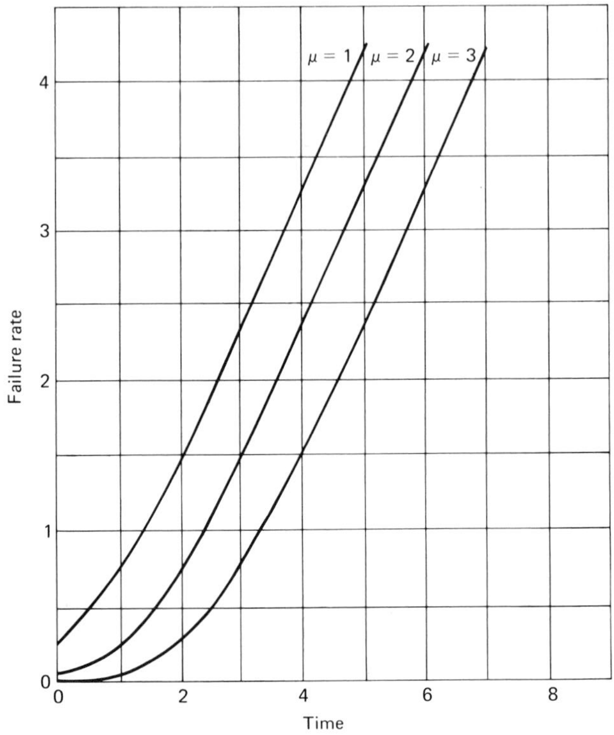

Figure 3.5.3.
Failure rate curves of the normal distribution for $\sigma = 1$.

EXERCISES

1. Prove by the method of moment generating functions that if $Y_1, \ldots, Y_n$ are independent random variables each with exponential distribution $1 - e^{-\lambda x}$, then $P[Y_1 + \cdots + Y_n \le t] = 1 - \sum_{i=0}^{n-1} [(\lambda t)^i/i!]e^{-\lambda t}$, a gamma distribution with shape parameter n (Theorem 2.8).

2. Prove that the failure rate $r(t)$ for the gamma distribution $G_{\lambda,\alpha} \uparrow \lambda$ for $\alpha > 1$ as $t \to \infty$ and $\downarrow \lambda$ for $\alpha < 1$ as $t \to \infty$.

3. Prove that if X_1 has gamma distribution G_{λ,α_1}, X_2 has gamma distribution G_{λ,α_2}, and X_1 and X_2 are independent, then $X_1 + X_2$ has gamma distribution $G_{\lambda,\alpha_1+\alpha_2}$.

4. Compute the mean and variance of the Weibull distribution $F_\alpha(t) = 1 - e^{-(\lambda t)^\alpha}$.

5. Compute the mean and variance of the gamma distribution $G_{\lambda,\alpha}(t) = \int_0^t [(\lambda^\alpha t^{\alpha-1})/\Gamma(\alpha)]e^{-\lambda t}\,dt$.

6. Show that a PF_2 density is unimodal.

7. Verify that the following densities are PF_2:

   (a) Exponential: $f(x) = \lambda e^{-\lambda x}$, $x \geq 0$;
   (b) Gamma: $g_{\lambda,\alpha}(x) = [(\lambda^\alpha x^{\alpha-1})/\Gamma(\alpha)]e^{-\lambda x}$, $x \geq 0$ for $\alpha \geq 1$;
   (c) Weibull: $f_\alpha(x) = \alpha\lambda(\lambda x)^{\alpha-1}e^{-(\lambda x)^\alpha}$, $x \geq 0$, $\alpha \geq 1$;
   (d) Truncated normal: $f(x) = (1/a\sqrt{2\pi}\sigma)e^{-(x-\mu)^2/2\sigma^2}$, $x \geq 0$, for each $-\infty < \mu < \infty$ and $\sigma > 0$;
   (e) Normal: $f(x) = (1/\sqrt{2\pi}\sigma)e^{-(x-\mu)^2/2\sigma^2}$, $-\infty < x < \infty$.
   (f) Laplace: $f(x) = \frac{1}{2}e^{-|x|}$, $-\infty < x < \infty$.

8. Show that the following densities are *not* PF_2:

   (a) Weibull with shape parameter <1: $f(x) = \alpha\lambda(\lambda x)^{\alpha-1}e^{-(\lambda x)^\alpha}$, $x \geq 0$, $\lambda > 0$, $0 < \alpha < 1$.
   (b) Cauchy: $f(x) = 1/\pi(1 + x^2)$, $-\infty < x < \infty$.

9. Give an example of an IFR distribution whose density is *not* PF_2.

10. Using Lemmas 5.8 and 5.9, prove that the gamma distribution with density (5.3) is IFR for $\alpha \geq 1$ and DFR for $\alpha \leq 1$.

11. Show that if f is the density of a DFR distribution, then $f(x) > 0$ for $x \geq 0$.

*12. Prove that (b), (b′), and (b″) in Definition 5.5 are equivalent.

13. Prove Lemma 5.9.

6. NOTES AND REFERENCES

Section 1. Barlow, Marshall, and Proschan (1963) present a systematic treatment of properties of distributions with monotone failure rate. Bryson and Siddiqui (1969) discuss a variety of notions of aging.

Section 2. Many of the basic results of this section are developed in Sukhatme (1936, 1937), Epstein and Sobel (1953, 1954), and Rényi (1953).

Section 3. Feller (1968) develops the axiomatic approach to the Poisson process given in 3.4. Other equivalent definitions are developed in Parzen

(1962). For a discussion of superposition of processes, see Cox and Lewis (1966).

Section 4. A useful and comprehensive discussion of the Poisson distribution and its properties is given in Haight (1967).

Section 5. The most complete and authoritative treatment of Pólya frequency functions appears in Karlin (1968).

4

CLASSES OF LIFE DISTRIBUTIONS BASED ON NOTIONS OF AGING

1. INTRODUCTION

In Section 1 of Chapter 3 we considered a notion of aging based on the behavior of the failure rate function. In particular, we defined increasing failure rate (IFR) and decreasing failure rate (DFR), and in Section 5 of Chapter 3 we gave examples of well-known families of life distributions having either increasing or decreasing failure rate.

In this chapter we study systems of components with monotone failure rate. In Section 2 we show that a coherent system of independent IFR components need not have an IFR life distribution; however, the system failure rate does *increase on the average*. In Section 3 we show that life distributions with failure rates increasing on the average also arise in a natural way under shock models in which damage accumulates until a critical threshold is exceeded. In Section 4 we study the effects of other reliability operations, such as forming the sum of independent life lengths (that is, convolving the corresponding life distributions) and forming mixtures of life distributions. We determine for each class of life distributions whether a given reliability operation yields a life distribution within the same class. In Section 5 we set up a partial ordering of distributions within a class; so that, roughly speaking, within the IFR class, distributions are ordered according to the degree of their IFR-ness, and similarly for the other classes of life distributions. In Section 6 we develop bounds on reliability assuming some partial a priori knowledge, such as the underlying distribution IFR with known mean. Finally, in Section 7 we obtain comparisons for the mean life of series systems and for parallel systems under various assumptions on component distributions.

Looking ahead, in Chapter 6 we will develop additional classes of life distributions, appropriate in replacement models. We will show how these classes of life distributions may arise quite naturally from shock models. For these classes we will perform an analysis similar to the analysis in this chapter for the IFR and DFR classes.

In Chapter 2 we were concerned with system reliability at a single *instant* in time. In this chapter, we are concerned with system reliability for the *interval* $[0, t]$. This is the probability that the system functions successfully throughout the interval $[0, t]$. The connection between the two concepts can be made in terms of random indicator processes. Let $T_1, T_2, \ldots, T_n$ be component lifetimes (not necessarily independent) in a coherent system. Let

$$X_i(t) = \begin{cases} 1 & \text{if } T_i > t, \\ 0 & \text{otherwise,} \end{cases}$$

be the random indicator process for component i. Let T be the system life length with distribution F. Let

$$X(t) = \begin{cases} 1 & \text{if } T > t, \\ 0 & \text{otherwise.} \end{cases}$$

Then

$$X(t) = \phi[X_1(t), \ldots, X_n(t)],$$

and

$$\begin{aligned} \bar{F}(t) &= P[X(s) = 1 \quad \text{for} \quad 0 \le s \le t] \\ &= P[X(t) = 1] = EX(t) \end{aligned}$$

since ϕ is coordinatewise nondecreasing. If components are *statistically independent*, then

$$\bar{F}(t) = h_\phi[\bar{F}_1(t), \bar{F}_2(t), \ldots, \bar{F}_n(t)]$$

by (1.2) of Chapter 2. If components are *not* statistically independent, then $\bar{F}(t)$ depends on more than the component marginal distributions at time t.

In this chapter, we are basically interested in the connection between properties of *independent* component life distributions, F_i $(i = 1, 2, \ldots, n)$, and properties of the system life distribution, F. We consider models for *dependent* component life distributions in Chapter 5.

2. LIFE DISTRIBUTIONS OF COHERENT SYSTEMS

In this section we study the class of distributions describing the life of coherent systems of independent IFR components. We show that this class is closed under the formation of coherent systems; that is, any coherent system of independent components whose life distributions belong to this class has a life distribution which is also a member of the class.

At first glance, it would seem reasonable to suppose that if each component of a coherent system has increasing failure rate, then the system itself would have increasing failure rate. After all, each time a component fails, the system becomes structurally weaker (because of the monotonicity of the structure function of a coherent system). In addition, as time passes the failure rate of each component increases. However, the following example shows that it is *not* true that a coherent system of independent IFR components is necessarily IFR.

2.1. Example. Let F be the life distribution of a parallel system of two independent components having respective life distributions $F_1(t) = 1 - e^{-\lambda_1 t}$, $F_2(t) = 1 - e^{-\lambda_2 t}$.
Then

$$\bar{F}(t) = 1 - (1 - e^{-\lambda_1 t})(1 - e^{-\lambda_2 t}), \tag{2.1}$$

so that

$$r(t) = \frac{\lambda_1 e^{-\lambda_1 t} + \lambda_2 e^{-\lambda_2 t} - (\lambda_1 + \lambda_2)e^{-(\lambda_1+\lambda_2)t}}{e^{-\lambda_1 t} + e^{-\lambda_2 t} - e^{-(\lambda_1+\lambda_2)t}}. \tag{2.2}$$

It is easy to verify that $r(t)$ is increasing on $[0, t_0)$ and decreasing on (t_0, ∞), where t_0 depends on λ_1, λ_2. (See Exercise 1.) Figure 4.2.1 shows this behavior for various combinations of λ_1, λ_2, normalized so that $\lambda_1 + \lambda_2 = 1$.

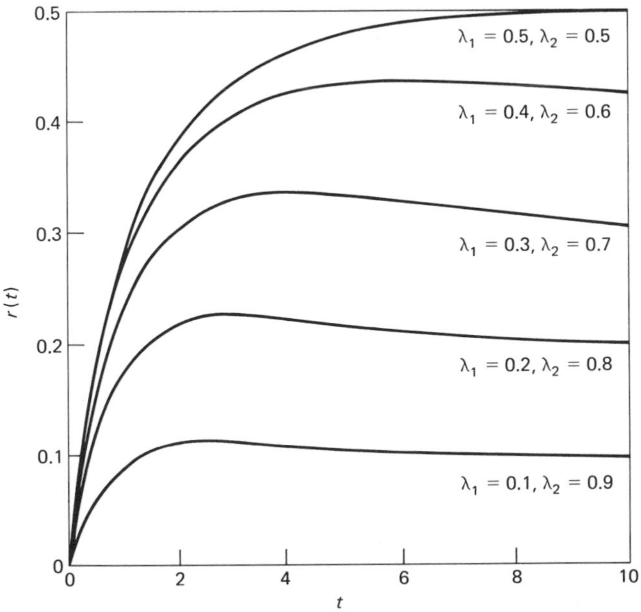

Figure 4.2.1.
Representative shapes of the system failure rate when the system consists of two exponential components in parallel with failure rates λ_1, λ_2 such that $\lambda_1 + \lambda_2 = 1$.

Since coherent systems of independent components need not be IFR, we are motivated to find the smallest class of distributions describing life lengths of coherent systems of IFR components. We will show in this section that this is the class of distributions with increasing failure rate average (IFRA), defined precisely in the following definition.

2.2. Definition. A distribution F has *increasing failure rate average* (F is IFRA) if $-(1/t)\log \bar{F}(t)$ is increasing in $t \geq 0$. Recall from (1.2) of Chapter 3 that $-\log \bar{F}(t)$ represents the cumulative failure rate $\int_0^t r(u)\,du$ when the failure rate $r(u)$ exists. Note however that in general the failure rate need not exist.

Similarly, F has *decreasing failure rate average* (F is DFRA) if $-(1/t)\log \bar{F}(t)$ is decreasing in $t \geq 0$.

Remark. It is obvious that an IFRA distribution F is characterized by $\bar{F}^{1/t}(t)\downarrow$ on $[0, \infty)$, while a DFRA distribution F is characterized by $\bar{F}^{1/t}(t)\uparrow$ on $[0, \infty)$. Hence F is IFRA (DFRA) if and only if $\bar{F}(\alpha t) \geq (\leq) \bar{F}^{\alpha}(t)$ for all $0 < \alpha < 1$ and $t \geq 0$.

We will prove that a coherent system of independent IFRA components itself has an IFRA life distribution. To prove this *IFRA Closure Theorem* we will need the following lemma.

2.3. Lemma. Let $0 \leq \alpha \leq 1$, $0 \leq \lambda \leq 1$, and $0 \leq x \leq y$. Then

$$\lambda^{\alpha} y^{\alpha} + (1 - \lambda^{\alpha}) x^{\alpha} - [\lambda y + (1 - \lambda) x]^{\alpha} \geq 0.$$

PROOF. Since $f(x) = x^{\alpha}$ is concave in $x \geq 0$ for $0 \leq \alpha \leq 1$, then $f(u_1 + \delta) - f(u_1) \geq f(u_2 + \delta) - f(u_2)$, $u_1 \leq u_2$. Take $\delta = \lambda(y - x)$, $u_1 = \lambda x$, and $u_2 = x$ to complete the proof. ||

Most practical systems are coherent. (See Definition 2.1 of Chapter 1.) However, for purposes of mathematical proof, we are also interested in the broader class of *monotonic* structures.

2.4. Definition. A structure function $\phi(x_1, \ldots, x_n)$ is *monotonic* if ϕ is increasing in each argument.

Note that the only difference between a coherent structure and a monotonic structure is that a monotonic structure function ϕ may have irrelevant components. In carrying out inductive proofs by pivotal decomposition (see Lemma 1.7 of Chapter 1 and Lemma 1.1 of Chapter 2), we sometimes encounter monotonic structures, rather than coherent structures.

We also need to prove the following theorem.

2.5. Theorem. Let $h(\mathbf{p})$ be the reliability function of a monotonic system. Then

$$h(\mathbf{p}^\alpha) \geq h^\alpha(\mathbf{p}) \qquad \text{for} \quad 0 < \alpha \leq 1, \tag{2.3}$$

where $\mathbf{p}^\alpha = (p_1{}^\alpha, \ldots, p_n{}^\alpha)$.

PROOF. We use induction on the number n of components in the system.

Let $n = 1$. Then either $h(p) \equiv p$, $h(p) \equiv 0$, or $h(p) \equiv 1$. In each case, (2.3) holds.

Assume (2.3) holds for all monotonic systems of $n - 1$ components. Then for a monotonic system of n components, we have

$$h(\mathbf{p}^\alpha) = p_n{}^\alpha h(1_n, \mathbf{p}^\alpha) + (1 - p_n{}^\alpha)h(0_n, \mathbf{p}^\alpha),$$

by the pivotal decomposition formula given in Lemma 1.1 of Chapter 2. Since $h(1_n, \mathbf{p}^\alpha)$ and $h(0_n, \mathbf{p}^\alpha)$ are each reliability functions of monotonic systems of $n - 1$ components, it follows by (2.3) that

$$h(1_n, \mathbf{p}^\alpha) \geq h^\alpha(1_n, \mathbf{p})$$

and

$$h(0_n, \mathbf{p}^\alpha) \geq h^\alpha(0_n, \mathbf{p}).$$

Hence

$$h(\mathbf{p}^\alpha) \geq p_n{}^\alpha h^\alpha(1_n, \mathbf{p}) + (1 - p_n{}^\alpha)h^\alpha(0_n, \mathbf{p}).$$

Next apply Lemma 2.3, choosing $\lambda = p_n$, $y = h(1_n, \mathbf{p})$, and $x = h(0_n, \mathbf{p})$. We obtain

$$p_n{}^\alpha h^\alpha(1_n, \mathbf{p}) + (1 - p_n{}^\alpha)h^\alpha(0_n, \mathbf{p}) \geq [p_n h(1_n, \mathbf{p}) + (1 - p_n)h(0_n, \mathbf{p})]^\alpha.$$

Applying the pivotal decomposition formula (Lemma 1.1 of Chapter 2) again, we conclude that

$$h(\mathbf{p}^\alpha) \geq h^\alpha(\mathbf{p}). \ \|$$

We can now prove the IFRA Closure Theorem.

2.6. Theorem. Suppose each of the independent components of a coherent system has an IFRA life distribution. Then the system itself has an IFRA life distribution.

PROOF. Let F denote the system life distribution, while F_i denotes the life distribution of the ith component, $i = 1, \ldots, n$. Then, for $0 < \alpha \leq 1$, $\bar{F}(\alpha t) = h[\bar{F}_1(\alpha t), \bar{F}_2(\alpha t), \ldots, \bar{F}_n(\alpha t)]$. Since F_i is IFRA, then $\bar{F}_i(\alpha t) \geq \bar{F}_i{}^\alpha(t)$, $i = 1, \ldots, n$. Since h is increasing in each argument (Theorem 1.2 of Chapter 2), it follows that $\bar{F}(\alpha t) \geq h[\bar{F}_1{}^\alpha(t), \ldots, \bar{F}_n{}^\alpha(t)]$. But by Theorem 2.5,

$$h[\bar{F}_1{}^\alpha(t), \ldots, \bar{F}_n{}^\alpha(t)] \geq h^\alpha[\bar{F}_1(t), \ldots, \bar{F}_n(t)].$$

Combining the last two inequalities, and using the identity

$$h[\overline{F}_1(t), \ldots, \overline{F}_n(t)] = \overline{F}(t),$$

we conclude that

$$\overline{F}(\alpha t) \geq \overline{F}^{\alpha}(t).$$

Thus system life distribution F is IFRA. ||

As a special case of Theorem 2.6, it follows that a coherent system of independent IFR components has an IFRA life distribution, as stated at the beginning of the chapter.

As a further specialization of Theorem 2.6, we see that a coherent system of independent exponential components is IFRA. Is it possible that for each IFRA life distribution F, we can find a corresponding coherent system of exponential components whose life distribution F^* is "close" to F? Putting it more precisely, is the class of IFRA distributions the smallest class containing the exponential distributions which is obtained from the formation of coherent systems and limits in distribution? We will obtain an affirmative answer in Theorem 2.13 below. First we need some notation and preliminary results.

2.7. Definition. Let A be a class of life distributions. Then A^{CS} denotes the class of life distributions of coherent systems of independent components with life distributions in A. We say A^{CS} is the *closure* of A under the formation of coherent systems. Symbolically,

$$F \in A^{CS} \qquad \text{if} \quad \overline{F}(t) \equiv h(\overline{F}_1(t), \ldots, \overline{F}_n(t)),$$

where h is the reliability function of a coherent system and $F_1, \ldots, F_n \in A$.

Note that $A \subset A^{CS}$ since a single component constitutes a coherent system. Also $A \subset B$ implies $A^{CS} \subset B^{CS}$, and $(A^{CS})^{CS} = A^{CS}$.

2.8. Definition. Let A be a class of distributions. Then A^{LD} is the class obtained by taking limits in distribution of sequences of members of A.

Recall that F is the *limit in distribution* of $F_1, F_2, \ldots$ (written $F \stackrel{L}{=} \lim_{n\to\infty} F_n$) if $F(t) = \lim_{n\to\infty} F_n(t)$ for all continuity points t of F.

2.9. Example. Let $F_n(x) = x^n$ for $0 \leq x \leq 1$. Clearly,

$$\lim_{n\to\infty} F_n(x) = \begin{cases} 0 & \text{for} \quad x < 1, \\ 1 & \text{for} \quad x \geq 1, \end{cases}$$

a distribution in $\{F_1, F_2, \ldots\}^{LD}$, but not in $\{F_1, F_2, \ldots\}$.

Notation. Let {IFRA} denote the class of IFRA distributions, {exp} the class of exponential distributions including the distributions degenerate at 0 and ∞, and {deg} the class of degenerate distributions.

2.10. Lemma. $\{\text{IFRA}\} \subset \{\exp, \deg\}^{CS,LD}$; that is, every IFRA distribution can be obtained as a limit in distribution of a sequence of coherent systems of components which have either exponential or degenerate distributions.

PROOF. The only IFRA distribution placing mass at $0\,(\infty)$ is the distribution degenerate at $0\,(\infty)$. Such a distribution is in both $\{\exp\}$ and $\{\deg\}$, and so is in $\{\exp, \deg\}^{CS}$. Thus it suffices to consider only IFRA distributions F for which (a) $\bar{F}(0) = 1$, and (b) given $\varepsilon > 0$, there exists t_ε such that $\bar{F}(t_\varepsilon) \le \varepsilon$. We prove the desired result by showing that (i) an IFRA survival probability $\bar{F}$ satisfying (a) and (b) can be approximated uniformly from below by a survival probability $\bar{H}_n$ of the form

$$\bar{H}_n(t) = \begin{cases} \exp[-(\lambda_1 + \lambda_2 + \cdots + \lambda_i)t] & \text{if } t_{i-1} \le t < t_i, \quad i = 1, \ldots, n, \\ 0 & \text{if } t_n \le t, \end{cases} \tag{2.4}$$

where $0 = t_0 \le t_1 < \cdots < t_n < \infty$ and $0 \le \lambda_i < \infty$, and (ii) piecewise exponential survival probabilities of the form (2.4) are in the class $\{\exp, \deg\}^{CS}$.

The following construction indicates how (i) may be verified. (See Figure 4.2.2.) It is convenient and harmless to substitute the restrictions $0 < t_1 \le t_2 \le \cdots \le t_n < \infty$ and $0 < \sum_{i=1}^{k} \lambda_i \le \sum_{i=1}^{k+1} \lambda_i < \infty$, $k = 1, \ldots, n-1$, for those of (2.4).

Given $\varepsilon > 0$, let $\delta = \log(1+\varepsilon)$. We determine t_i by $-\log \bar{F}(t_i^-) \le i\delta \le -\log \bar{F}(t_i^+)$. From (a), $t_1 > 0$; also $t_1 \le t_2 \le \cdots$ since $-\log \bar{F}(t)$ is increasing. From (b), we can find an n such that $t_\varepsilon \le t_n < \infty$. It follows from (2.4) that $0 \le \bar{F}(t) - \bar{H}_n(t) < \varepsilon$ [since $\bar{H}_n(t) = 0$ and $\bar{F}(t) < \varepsilon$] for $t \ge t_n$.

Next define $\sum_{i=1}^{k} \lambda_i = k\delta/t_k$. Note that $\lambda_1 > 0$ since $t_1 \le t_n < \infty$, $\sum_{i=1}^{k} \lambda_i \le \sum_{i=1}^{k+1} \lambda_i$ since $-\log \bar{F}(t)/t$ is increasing, and $\sum_{i=1}^{n} \lambda_i < \infty$ since $0 < t_1 \le t_n$. From (2.4), $0 \le -\log \bar{H}(t) - [-\log \bar{F}(t)] \le \delta$ for $0 \le t < t_n$. Thus $0 \le \bar{F}(t) - \bar{H}(t) \le \varepsilon$ on $[0, t_n)$.

To show (ii) we consider the coherent system $\phi_n(e_1, \ldots, e_n, d_1, \ldots, d_n) = e_1(e_2 \amalg d_1)(e_3 \amalg d_2) \cdots (e_n \amalg d_{n-1})d_n$, where e_i is the performance indicator of a component with exponential survival probability $\bar{G}_i(t) = e^{-\lambda_i t}$, and d_i is the performance indicator of a component with survival probability $\bar{D}_i$ degenerate at t_i [that is, $\bar{D}_i(t) = 1$ for $t < t_i$ and $\bar{D}_i(t) = 0$ for $t \ge t_i$]. (See Figure 4.2.3.) The survival probability of the system is

$$\bar{H}_n(t) = \bar{G}_1(t)[\bar{G}_2(t) \amalg \bar{D}_1(t)][\bar{G}_3(t) \amalg \bar{D}_2(t)] \cdots [\bar{G}_n(t) \amalg \bar{D}_{n-1}(t)]\bar{D}_n(t),$$

which reduces to (2.4). ‖

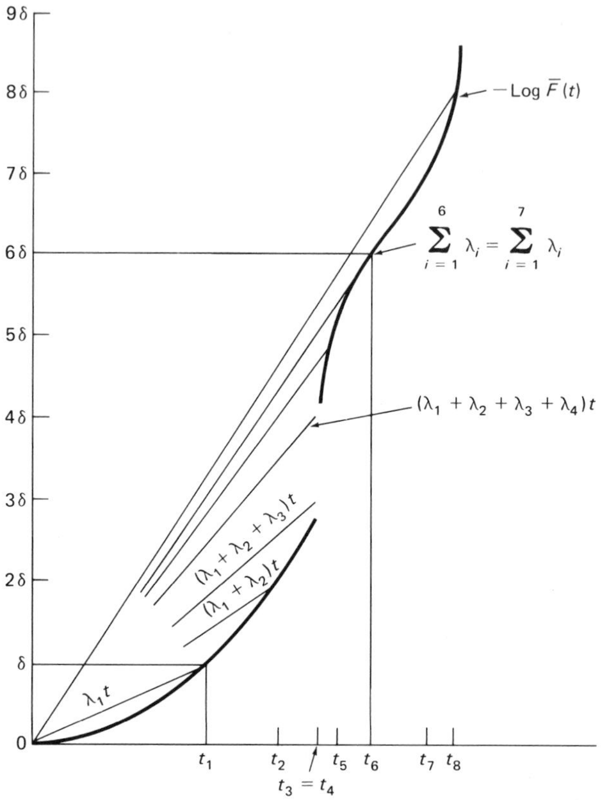

Figure 4.2.2.
Piecewise exponential approximation to an IFRA survival probability $\bar{F}$.

2.11. Lemma. $\{\text{deg}\} \subset \{\exp\}^{CS,LD}$.
The proof is left to Exercise 4.

2.12. Lemma. For any class of distributions A, $A^{LD,CS} \subset A^{CS,LD}$.
The proof is left to Exercise 5.

2.13. Theorem. $\{\text{IFRA}\} = \{\exp\}^{CS,LD}$.

PROOF. Since $\{\exp\} \subset \{\text{IFRA}\}$, then $\{\exp\}^{CS} \subset \{\text{IFRA}\}^{CS} = \{\text{IFRA}\}$ by Theorem 2.6. Thus $\{\exp\}^{CS,LD} \subset \{\text{IFRA}\}^{LD} = \{\text{IFRA}\}$. (See Exercise 9.)

To show the reverse inclusion, note that $\{\text{IFRA}\} \subset \{\exp, \text{deg}\}^{CS,LD}$ by Lemma 2.10. But $\{\text{deg}\} \subset \{\exp\}^{CS,LD}$ by Lemma 2.11. Thus $\{\text{deg}, \exp\} \subset \{\exp\}^{CS,LD}$, so that $\{\text{deg}, \exp\}^{CS} \subset \{\exp\}^{CS,LD,CS} \subset \{\exp\}^{CS,CS,LD}$ by Lemma

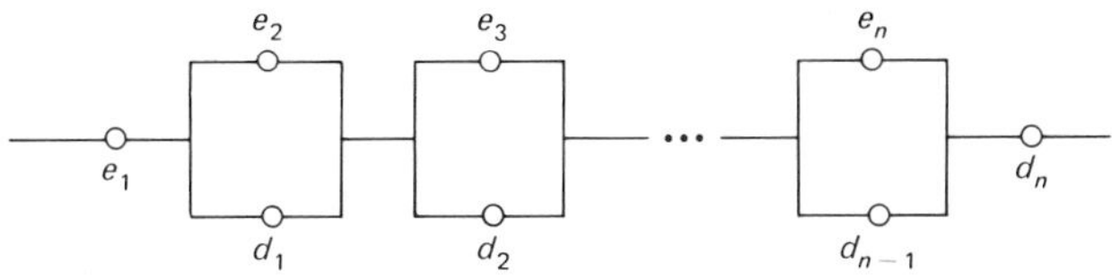

Figure 4.2.3.
$\phi_n(e_1, \ldots, e_n, d_1, \ldots, d_n) = e_1(e_2 \amalg d_1)(e_3 \amalg d_2) \cdots (e_n \amalg d_{n-1})\, d_n.$

2.12. Hence $\{\deg, \exp\}^{CS,LD} \subset \{\exp\}^{CS,CS,LD,LD} \equiv \{\exp\}^{CS,LD}$. It follows that $\{\text{IFRA}\} \subset \{\exp\}^{CS,LD}$. ‖

2.14. Summary Result. $\{\text{IFRA}\}^{CS} = \{\text{IFRA}\} = \{\exp\}^{CS,LD}$.

This is a combination of Theorem 2.6 and Theorem 2.13. It states that the IFRA class of distributions is the smallest class containing the exponential distributions, closed under formation of coherent systems and taking limits in distribution.

We have shown above how IFRA life distributions arise in a very natural way when coherent systems of independent IFR distributions are formed. Next we present a few basic characterizations of IFRA distributions. We will find quite useful the concept of a star-shaped function.

2.15. Definition. A function $g(x)$ defined on $[0, \infty)$ such that $(1/x)g(x)$ is increasing on $[0, \infty)$ is called *star-shaped*.

2.16. Equivalent Definition. A function $g(x)$ defined on $[0, \infty)$ and satisfying

$$g(\alpha x) \leq \alpha g(x) \qquad \text{for} \quad 0 \leq \alpha \leq 1, \quad x \geq 0, \tag{2.5}$$

is called *star-shaped*.

The proof of the equivalence is left for Exercise 3.

From Definition 2.2 of an IFRA distribution and the definition of a hazard function given in (1.3) of Chapter 3 and the discussion there, we immediately conclude the following.

2.17. Theorem. Let F be an IFRA distribution. Then its hazard function, $-\log \bar{F}(t)$, is a star-shaped function.

A property that we will find useful in the shock model derivation of IFRA distributions (Section 3) and in establishing bounds for IFRA distributions (Section 6) is presented next.

2.18. Theorem. A distribution F is IFRA (DFRA) if and only if for each $\lambda > 0$, $\bar{F}(t) - e^{-\lambda t}$ has at most one change of sign, and if one change of sign actually occurs, it occurs from $+$ to $-$ (from $-$ to $+$).

PROOF. Let F be IFRA. Then by Theorem 2.17, $-\log \bar{F}(t)$ is star-shaped. On the other hand, $-\log(e^{-\lambda t})$ is linear. Both functions pass through the origin. Thus $-\log \bar{F}(t)$ crosses $-\log(e^{-\lambda t})$ at most once, and if a crossing does occur, $-\log \bar{F}(t)$ crosses $-\log (e^{-\lambda t})$ from below. It follows that $\bar{F}(t)$ crosses $e^{-\lambda t}$ at most once, and if it crosses, it does so from above.
The proof of the converse is left to Exercise 12.

For F DFRA the proof is similar. ||

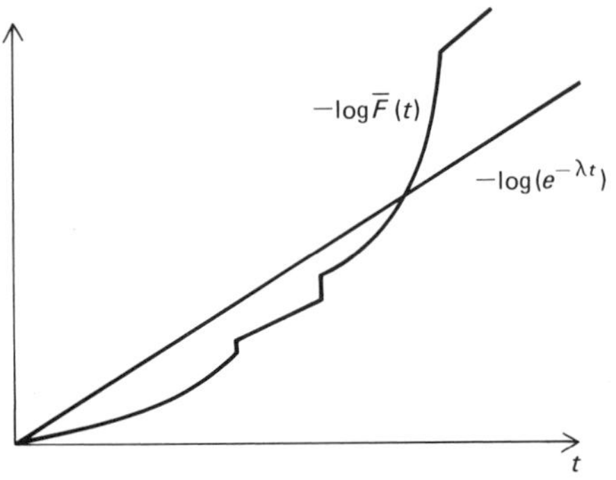

Figure 4.2.4.
IFRA hazard function crosses exponential hazard function at most once, and from below.

2.19. Remark. It is easy to verify that a convex function passing through the origin is star-shaped. (The proof is left to Exercise 2.) This observation applied to the hazard function immediately shows that an IFR distribution is IFRA. (See Exercise 11.) Of course, the converse need not be true, as we have seen in Example 2.1.

In Section 5 we show how to obtain a partial ordering of distributions in the IFRA class according to the degree of star-shapedness.

EXERCISES

1. Prove that the failure rate $r(t)$ of a parallel system of two independent exponential components given in (2.2) is increasing on $[0, t_0)$ and decreasing on (t_0, ∞), where t_0 depends on λ_1, λ_2. $r(t)$ is increasing on $[0, \infty)$ if and only if $\lambda_1 = \lambda_2$.
2. Let $g(x)$ be convex on $[0, \infty)$, with $g(0) \leq 0$. Then $g(x)$ is star-shaped. Does the conclusion still hold if the assumption that $g(0) \leq 0$ is dropped?
3. Show that Definition 2.15 and Definition 2.16 of star-shaped functions are equivalent.

4. Show that $\{\text{deg}\} \subset \{\text{exp}\}^{CS,LD}$ (Lemma 2.11).
5. For any class of distributions A, prove that $A^{LD,CS} \subset A^{CS,LD}$. Give an example to show that the reverse inclusion need not hold.
6. Give an example of an IFRA distribution F with a discontinuity at each positive integer.

*7. Show that if a coherent system of independent components has a constant failure rate when each component has some constant failure rate, then the system is series.

*8. Show that if a coherent system of independent components has an IFR life distribution whenever the components have IFR life distributions, then the system is series.

9. Show that $\{\text{IFRA}\}^{LD} = \{\text{IFRA}\}$.
10. Show that the hazard function of an exponential distribution is linear, of an IFR distribution is convex, and of an IFRA distribution is star-shaped.
11. Prove that if F is IFR, then F is IFRA. Construct a counterexample to show that the converse is not true.
12. Let life distribution F satisfy the property that for each $\lambda > 0$, $\bar{F}(t) - e^{-\lambda t}$ has at most one change of sign, and if one change of sign does occur, it goes from $+$ to $-$. Then F is an IFRA distribution. (See Theorem 2.18.)

3. DISTRIBUTIONS WITH INCREASING FAILURE RATE AVERAGE ARISING FROM SHOCK MODELS

In Section 2 we showed how an IFRA distribution provides a natural description of coherent system life when components are independent IFR. In this section we show that the IFRA distribution also arises naturally in a quite different setting, namely, when shocks occur according to a Poisson process in time, each independently causing random damage to a device, the damages accumulating until a critical threshold is exceeded, at which time the device fails. This time of failure will be shown to be governed by an IFRA distribution. The fact that the IFRA class of distributions furnishes an appropriate description of life length in two such widely diverse models would seem to establish the IFRA class as basic in reliability theory.

3.1. Cumulative Damage Model. A device is subject to shocks occurring randomly in time according to a Poisson process with intensity λ. The ith shock causes a random amount X_i of damage, where $X_1, X_2, \ldots$ are independently distributed with common distribution F. The device fails when the total accumulated damage exceeds a specified capacity or threshold x.

Let $\bar{H}_F(t)$ denote the probability that the device survives $[0, t]$. Then

$$\bar{H}_F(t) = \sum_{k=0}^{\infty} e^{-\lambda t} \frac{(\lambda t)^k}{k!} F^{(k)}(x) \qquad \text{for} \quad 0 \le t < \infty. \tag{3.1}$$

Note that $e^{-\lambda t}[(\lambda t)^k/k!]$ represents the Poisson probability that the device experiences exactly k shocks in $[0, t]$ (see Definition 3.2, Chapter 3, of the Poisson process), while the k-fold convolution $F^{(k)}(x)$ represents the probability that the total damage accumulated over the k shocks does not exceed the threshold x; for $k = 0$, we define $F^{(k)}(x) = 1$ for $x \ge 0$, and 0 otherwise. Summing over the mutually exclusive and exhaustive outcomes $k = 0, 1, 2, \ldots$ yields the survival probability $\bar{H}_F(t)$ for the device.

The main result of this section, presented formally in Theorem 3.8 below, states that for *any* damage distribution F, the survival probability $\bar{H}_F(t)$ is IFRA. To prove this, we need several preliminary results in shock model theory of independent interest. To prove these results, we review some basic concepts and tools from the theory of total positivity.

3.2. Definition. Let A and B be subsets of the real line. A function $K(x, y)$ on $A \times B$ is said to be *totally positive of order n* (TP_n) if

$$\begin{vmatrix} K(x_1, y_1) & \cdots & K(x_1, y_r) \\ \vdots & \ddots & \vdots \\ K(x_r, y_1) & \cdots & K(x_r, y_r) \end{vmatrix} \ge 0 \tag{3.2}$$

for all $x_1 < \cdots < x_r$ in A, $y_1 < \cdots < y_r$ in B, $r = 1, 2, \ldots, n$.

A function which is totally positive of all finite orders is said to be *totally positive* (TP).

Note that the Pólya frequency function $h(x)$ of order 2 (PF_2) defined in 5.5 of Chapter 3 is a TP_2 function $K(x, y) = h(x - y)$, where x and y range over the entire real line.

3.3. Example. The function $K_1(x, y) = e^{xy}$ is totally positive (TP) in $x, y \in (-\infty, \infty)$ (Karlin, 1968, p. 15), so that $K_2(r, t) = t^r$ is TP in $t \in (0, \infty)$ and $r \in (-\infty, \infty)$. It follows by Exercise 3 below that $K_3(k, t) = e^{-\lambda t}[(\lambda t)^k/k!]$ is TP in $t \in (0, \infty)$ and $k \in \{0, 1, 2, \ldots\}$.

3.4. Definition. Let a function $f(t)$ be defined on I, where I is an ordered set of the real line. Let

$$S(f) = \sup S[f(t_1), f(t_2), \ldots, f(t_m)], \tag{3.3}$$

where the supremum is extended over all sets $t_1 < t_2 < \cdots < t_m (t_i \in I)$, m is arbitrary but finite, and $S(x_1, \ldots, x_m)$ is the *number of sign changes* of the sequence $x_1, \ldots, x_m$, zero terms being discarded.

A basic property of a totally positive function of finite or infinite order is its variation diminishing property, stated in Theorem 3.5.

3.5. Theorem. Let $K(x, y)$ be TP_r on $A \times B$. Let f be a bounded and Borel measurable function on B. Let the transformation

$$g(x) = \int_B K(x, y) f(y)\, dy \tag{3.4}$$

be finite for each x in A. Then

$$S(g) \le S(f) \qquad \text{provided} \quad S(f) \le r - 1. \tag{3.5}$$

Moreover, if $S(g) = S(f) \le r - 1$, then f and g exhibit the same sequence of signs when their respective arguments traverse the domain of definition from left to right. See Karlin (1968), Chapter 5, for a proof.

Now consider a somewhat more general shock model than that of 3.1. As before, shocks occur randomly in time in accordance with a Poisson process with intensity λ. However, now the probability $\bar{P}_k$ of surviving k shocks is assumed to be a deterministic function of k alone, and not of any random damages. The resulting survival probability $\bar{H}(t)$ for the period $[0, t]$ is now given by

$$\bar{H}(t) = \sum_{k=0}^{\infty} \bar{P}_k \frac{(\lambda t)^k}{k!} e^{-\lambda t} \qquad \text{for} \quad 0 \le t < \infty. \tag{3.6}$$

It is reasonable to assume that $\bar{P}_0 = 1$, and $\bar{P}_k$ is decreasing for $k = 0, 1, 2, \ldots$.

A result of general interest is given in the following theorem.

3.6. Theorem. Let $\bar{H}(t)$ be given by (3.6), where $1 = \bar{P}_0 \ge \bar{P}_1 \ge \cdots$, and

$$\bar{P}_k^{1/k} \qquad \text{is decreasing in} \quad k = 1, 2, \ldots. \tag{3.7}$$

Then H is an IFRA distribution, that is,

$$\bar{H}^{1/t}(t) \qquad \text{is decreasing in} \quad t > 0. \tag{3.8}$$

PROOF. Since $\bar{P}_k^{1/k}$ is decreasing in k, then $\bar{P}_k^{1/k} - \zeta$ $(0 \le \zeta \le 1)$ has at most one sign change, and from + to − if one occurs. Also by Example 3.3, the Poisson frequency function $e^{-\lambda t}[(\lambda t)^k/k!]$ is TP_2 in $t \in (0, \infty)$ and $k \in \{0, 1, 2, \ldots\}$. This means that

$$\bar{H}(t) - e^{-(1-\zeta)\lambda t} = \sum_{k=0}^{\infty} (\bar{P}_k - \zeta^k) e^{-\lambda t} \frac{(\lambda t)^k}{k!}$$

has the same sign change property in t, by the variation diminishing property of Theorem 3.5.

Since $\bar{H}(t) = e^{-\lambda t} + \sum_{k=1}^{\infty} \bar{P}_k e^{-\lambda t}[(\lambda t)^k/k!]$, it follows that $\bar{H}(t) \geq e^{-\lambda t}$ for $t \geq 0$, and hence trivially that for $\theta \geq \lambda$, $\bar{H}(t) \geq e^{-\theta t}$ for all $t \geq 0$.

We may now conclude that for all $\theta > 0$, $\bar{H}(t) - e^{-\theta t}$ has at most one sign change in t, and from + to −, if one occurs. By Theorem 2.18, H is IFRA. ‖

Theorem 3.6 states that if the probability $\bar{P}_k$ of surviving k shocks possesses the discrete IFRA property of (3.7), then the corresponding induced survival probability $\bar{H}(t)$ over time possesses the continuous IFRA property (3.8). In a similar way, we may show that the discrete IFR property in P_k is transformed into the continuous IFR property in $H(t)$. See Exercise 7 for the details.

Now we return to the cumulative damage model of 3.1. In this model the probability $\bar{P}_k$ of surviving k shocks is given by

$$\bar{P}_k = F^{(k)}(x) \qquad \text{for} \quad k = 0, 1, 2, \ldots. \tag{3.9}$$

This fact, together with Theorem 3.6, motivates the following lemma, of interest in its own right.

3.7. Lemma. Every distribution F such that $F(z) = 0$ for $z < 0$ satisfies

$$[F^{(k)}(x)]^{1/k} \qquad \text{is decreasing in} \quad k = 1, 2, \ldots. \tag{3.10}$$

PROOF. $F^{(2)}(x) = \int_0^x F(x - z)\, dF(z) \leq \int_0^x F(x)\, dF(z) = [F(x)]^2$, so that $F(x) \geq [F^{(2)}(x)]^{1/2}$.

Now suppose that for all x, $[F^{(k-1)}(x)]^{1/(k-1)} \geq [F^{(k)}(x)]^{1/k}$, that is, $F^{(k-1)}(x) \geq [F^{(k)}(x)]^{(k-1)/k}$. Then

$$\begin{aligned}
[F^{(k)}(x)]^{k+1} &= F^{(k)}(x)\left[\int F^{(k-1)}(x - z)\, dF(z)\right]^k \\
&\geq F^{(k)}(x)\left\{\int [F^{(k)}(x - z)]^{(k-1)/k}\, dF(z)\right\}^k \\
&= \left\{\int [F^{(k)}(x)]^{1/k}[F^{(k)}(x - z)]^{1-(1/k)}\, dF(z)\right\}^k \\
&\geq \left\{\int F^{(k)}(x - z)\, dF(z)\right\}^k = [F^{(k+1)}(x)]^k.
\end{aligned}$$

Equation (3.10) follows by induction. ‖

We are now able to prove the following fundamental result.

3.8. Theorem. Let $\bar{H}_F(t) = \sum_{k=0}^{\infty} [(\lambda t)^k/k!]e^{-\lambda t}F^{(k)}(x)$ represent the survival probability in the cumulative damage model 3.1, where (nonnegative) damage follows arbitrary distribution F. Then H_F is an IFRA life distribution.

PROOF. By Lemma 3.7, $\bar{P}_k \equiv F^{(k)}(x)$ satisfies (3.7). By Theorem 3.6, it follows that H_F is IFRA. ‖

The cumulative damage model of 3.1 may be generalized to cover other practical shock and damage situations; in these extensions the resulting life distribution remains IFRA.

3.9. Successive Shocks Cause Greater Damage. A realistic extension of the cumulative damage model of 3.1 may be formulated in which, as before, shocks occur in time according to a Poisson process with rate λ. However, now successive shocks become increasingly effective in causing damage or wear, even though they are independent. This means that $F_i(z)$, the distribution function for the amount of damage caused by the ith shock, is decreasing in $i = 1, 2, \ldots$ for each z. The probability $\bar{P}_k$ of surviving k shocks is given by

$$\bar{P}_0 = 1 \quad \text{and} \quad \bar{P}_k = F_1 * F_2 * \cdots * F_k(z), \quad k = 1, 2, \ldots, \tag{3.11}$$

where $*$ denotes convolution, as usual.

The following generalization of Lemma 3.7 will be used.

3.10. Lemma. Let distribution function F_i satisfy $F_i(z) = 0$ for $z < 0$, $i = 1, 2, \ldots$, and let $F_i(z)$ be decreasing in i for all z. Then

$$[F_1 * F_2 * \cdots * F_k(x)]^{1/k} \quad \text{is decreasing in} \quad k = 1, 2, \ldots. \tag{3.12}$$

PROOF. The inductive proof of Lemma 3.7 requires little modification to apply here. We simply add the step that

$$\int F_1 * \cdots * F_k(x - z)\, dF_k(z) \geq \int F_1 * \cdots * F_k(x - z)\, dF_{k+1}(z). \quad \|$$

We may now prove that the life distribution $H(t)$ of a device subject to shocks and damage as described in the model of 3.9 is IFRA.

3.11. Theorem. Let $F_i(z) = 0$ for $z < 0$, and $F_i(z)$ is decreasing in $i = 1, 2, \ldots$; then

$$\bar{H}(t) = \sum_{k=0}^{\infty} e^{-\lambda t} \frac{(\lambda t)^k}{k!} F_1 * \cdots * F_k(x) \tag{3.13}$$

is IFRA.

PROOF. By Lemma 3.10, $\bar{P}_k \equiv F_1 * \cdots * F_k(x)$ satisfies (3.7). By Theorem 3.6, H is IFRA. ‖

In our last model, generalizing both of the previous models, we assume successive damages are neither independent nor identically distributed. A primary reason for this is that an accumulation of damage may result in a loss of resistance to further damage. In this case the magnitudes of successive damages are dependent, so that (3.11) does not apply. On the other hand, it may be reasonable to assume the following model.

3.12. General Cumulative Damage Model. Shocks occur in time according to a Poisson process with rate λ. The kth shock causes damage X_k satisfying

$$P[X_k > u \mid X_1, \ldots, X_{k-1}] \quad \text{depends on} \quad X_1, \ldots, X_{k-1} \quad \text{only via } Z_{k-1} = X_1 + \cdots + X_{k-1}; \tag{3.14}$$

$$P[X_k > u \mid Z_{k-1} = z] \quad \text{is increasing in} \quad z \geq 0; \tag{3.15}$$

$$P[X_k > u \mid Z_{k-1} = z] \leq P[X_{k+1} > u \mid Z_k = z], \quad z \geq 0, \quad k = 1, 2, \ldots, \quad \text{where} \quad Z_0 = 0. \tag{3.16}$$

Condition (3.15) states that an accumulation of damage lowers resistance to further damage. Condition (3.16) says that for any given accumulation of damage, later shocks are apt to be more severe.

Under these conditions, we may identify $\bar{P}_k$ as:

$$\bar{P}_0 = 1 \quad \text{and} \quad \bar{P}_k = P[X_1 + \cdots + X_k \leq x], \quad k = 1, 2, \ldots. \tag{3.17}$$

Of course, (3.9) and (3.11) are special cases.

As before, we need to show that $\bar{P}_k^{1/k}$ is decreasing in $k = 1, 2, \ldots$.

3.13. Lemma. Let $X_1, X_2, \ldots$ be nonnegative random variables with a joint distribution satisfying (3.14), (3.15), and (3.16). Then

$$P^{1/k}[X_1 + \cdots + X_k \leq x] \quad \text{is decreasing in} \quad k = 1, 2, \ldots. \tag{3.18}$$

PROOF. Let

$$F^{[k]}(x) = P[X_1 + \cdots + X_k \leq x] \equiv \int_0^x P[X_k \leq x - z \mid Z_{k-1} = z]\, dF^{[k-1]}(z).$$

Using (3.15), then (3.16) with $z = 0$ and $k = 1$, we obtain

$$F^{[2]}(x) = \int_0^x P[X_2 \leq x - z \mid X_1 = z]\, dF^{[1]}(z) \leq \int_0^x P[X_2 \leq x - z \mid X_1 = 0]\, dF^{[1]}(z) \leq \int_0^x P[X_1 \leq x - z]\, dF^{[1]}(z) \leq [F^{[1]}(x)]^2.$$

To complete the induction, suppose that $\{F^{[k]}(z)\}^{1/k} \le \{F^{[k-1]}(z)\}^{1/(k-1)}$. Then for $z \le x$,

$$\begin{aligned} F^{[k]}(z) &= \{F^{[k]}(z)\}^{1/k}\{F^{[k]}(z)\}^{(k-1)/k} \\ &\le \{F^{[k]}(z)\}^{1/k}F^{[k-1]}(z). \end{aligned}$$

Using this and (3.15), and then using (3.16), we obtain

$$\begin{aligned} \{F^{[k+1]}(x)\}^k &= \left\{\int_0^x P[X_{k+1} \le x - z \mid Z_k = z]\, dF^{[k]}(z)\right\}^k \\ &\le \left\{\int_0^x P[X_{k+1} \le x - z \mid Z_k = z]\, dF^{[k-1]}(z)\right\}^k F^{[k]}(x) \\ &\le \left\{\int_0^x P[X_k \le x - z \mid Z_{k-1} = z]\, dF^{[k-1]}(z)\right\}^k F^{[k]}(x) \\ &= \{F^{[k]}(x)\}^{k+1}. \quad \| \end{aligned}$$

Using Lemma 3.13 and Theorem 3.6, we immediately obtain the following theorem.

3.14. Theorem. Let $X_1, X_2, \ldots$ be nonnegative random variables satisfying (3.14), (3.15), and (3.16). Then

$$\bar{H}(t) = \sum_{k=0}^{\infty} e^{-\lambda t}\frac{(\lambda t)^k}{k!} P[X_1 + \cdots + X_k \le x] \tag{3.19}$$

is IFRA.

We wish to emphasize that the IFRA property has been obtained in Theorem 3.14 as an implication of a natural physical model. The only hypothesis imposed upon the damage magnitudes $X_1, X_2, \ldots$ is given in (3.14), (3.15), and (3.16), which in the original special model 3.1 reduces to the requirement that the random damage be nonnegative.

EXERCISES

1. Let $K(x) = \sum_{k=0}^{\infty} e^{-\lambda t}[(\lambda t)^k/k!]F^{(k)}(x)$ for $0 \le x < \infty$ and fixed $t > 0$. Then $K(x)$ is a distribution function (compound Poisson distribution).
2. Show that $\bar{H}(t)$ given in (3.6) is a decreasing function of λ.
3. Let $K(x, y)$ be TP_r for $x \in A$, $y \in B$, and $f(x) \ge 0$, $x \in A$, $g(y) \ge 0$, $y \in B$. Then $L(x, y) = f(x)K(x, y)g(y)$ is TP_r for $x \in A$, $y \in B$.
4. Let $K(x, y)$ be TP_r for $x \in A$, $y \in B$, and $f(x)$ be increasing on A, $g(y)$ be increasing on B. Then $L(x, y) = K(f(x), g(y))$ is TP_r for $x \in A$, $y \in B$.

5. Let $f(x) = 0$ for $x \leq 0$ and PF_2 for $-\infty < x < \infty$. Then $K(n, x) = f^{(n)}(x)$ is TP_2 for $n = 1, 2, \ldots$ and $-\infty < x < \infty$.

6. Let $K(x, y)$ and $L(x, y)$ each be TP_2 for $x \in A$, $y \in B$. Then $M(x, y) = K(x, y)L(x, y)$ is TP_2 for $x \in A$, $y \in B$.

*7. Let $\bar{H}(t) = \sum_{k=0}^{\infty} \bar{P}_k e^{-\lambda t}[(\lambda t)^k/k!]$, where $1 = \bar{P}_0 \geq \bar{P}_1 \geq \cdots$. If $\bar{P}_{k+1}/\bar{P}_k \downarrow$ in $k = 0, 1, \ldots$ (that is, P_k is discrete IFR), then H is IFR.

8. Show by counterexample that Lemma 3.7 is not true if $F(x) > 0$ for $x < 0$. In fact show that $[F^{(k)}(x)]^{1/k}$ can be strictly increasing in $k = 1, 2, \ldots$ when F has support on the whole line.

9. Let distribution function F satisfy $F(z) = 0$ for $z < 0$. Then $[F^{(n_k)}(x)]^{1/k} \downarrow$ in $k = 1, 2, \ldots$, where n_k is a convex, increasing, integer-valued function, with $n_1 = 1$.

10. Let $\bar{H}(t)$ be given by (3.6), representing the survival probability of a device subject to shocks. Compute the corresponding density $h(t)$ and failure rate $r(t) = h(t)/\bar{H}(t)$. Show that $r(t) \leq \lambda$ for $t \geq 0$.

11. Let $\bar{H}(t)$ be given by (3.6). Show that the jth moment μ_j is given by

$$\mu_j = \frac{j!}{\lambda^j} \sum_{k=0}^{\infty} \binom{j + k - 1}{k} \bar{P}_k \qquad \text{for } j = 1, 2, \ldots.$$

12. Show that the following functions are TP_2:

(a) $f(x, y) = xy$ for $x \geq 0$, $y \geq 0$.
(b) $f(x, y) = 1$ if $x \geq y$, 0 if $x < y$, for real x, y.
(c) $f(x, y) = \begin{cases} g(x) & \text{for } x \geq y, \\ h(y) & \text{for } x < y, \end{cases}$
for real x, y where g and h are nonnegative functions.
(d) $f(x, y) = e^{xy}$ for real x, y.

If the domain in (a) is extended to negative x, y, is $f(x, y)$ still TP_2?

13. Find F satisfying $F(z) = 0$ for $z < 0$ and a value x for which $[F^{(k)}(x)]^{1/k} \equiv$ constant for $k = 1, 2, \ldots$.

4. PRESERVATION OF LIFE DISTRIBUTION CLASSES UNDER RELIABILITY OPERATIONS

In Section 2 we noted that the formation of a coherent system of independent IFR components does not necessarily yield an IFR system life. However, the formation of a coherent system of independent IFRA components *does* yield an IFRA system life. This raises the general question, under what reliability operations is a given class of life distributions preserved?

In this section we will consider the reliability operations of

(a) Formation of coherent systems,
(b) Addition of life lengths,
(c) Mixture of distributions,

applied to the classes of life distributions introduced thus far:

(i) IFR
(ii) IFRA
(iii) DFR
(iv) DFRA.

Formation of Coherent Systems

To complete our analysis of the formation of coherent systems, we note that a parallel system of unlike exponential components is neither DFR nor DFRA. Remembering that exponential life distributions are both DFR and DFRA, we conclude that the DFR (DFRA) class is *not* closed under the formation of coherent systems.

Addition of Life Lengths

Next we consider the addition of life lengths. Note that when a failed component is replaced by a spare, the total life accumulated is obtained by the addition of the two life lengths. Thus operation (b) is central in the study of maintenance policies. To express the distribution of the sum of independent life lengths X_1 and X_2, suppose X_1 has distribution F_1, X_2 has distribution F_2, and F is the distribution of $X_1 + X_2$. Then

$$F(t) = \int_0^t F_2(t - x)\, dF_1(x). \tag{4.1}$$

[Equation (4.1) corresponds to the fact that if component 1 fails at any time x preceding t, while component 2 fails during the interval of time $t - x$ remaining, then the sum $X_1 + X_2$ of life lengths does not exceed t.] We say that F is the *convolution* of F_1 and F_2, written $F = F_1 * F_2$. For general distributions on the whole axis, we have the following definition.

4.1. Definition. Let F_1 and F_2 be distributions not necessarily confined to the positive half of the axis. Then

$$F(t) = \int_{-\infty}^{\infty} F_1(t - x)\, dF_2(x) \tag{4.2}$$

is the *convolution* of F_1 and F_2. If X_1 has distribution F_1 and X_2 has distribution F_2, with X_1 and X_2 independent, then F is the distribution of $X_1 + X_2$.

Next we prove that the sum of independent IFR random variables is itself an IFR random variable.

4.2. Theorem. If F_1 and F_2 are IFR, then their convolution $F(t) = \int_0^t F_2(t - x)\, dF_1(x)$ is IFR.

PROOF. First assume F_1 has density f_1, F_2 has density f_2. For $t_1 < t_2$, $u_1 < u_2$, write

$$D = \begin{vmatrix} \bar{F}(t_1 - u_1) & \bar{F}(t_1 - u_2) \\ \bar{F}(t_2 - u_1) & \bar{F}(t_2 - u_2) \end{vmatrix}$$

$$= \begin{vmatrix} \displaystyle\int \bar{F}_1(t_1 - s) f_2(s - u_1)\, ds & \displaystyle\int \bar{F}_1(t_1 - s) f_2(s - u_2)\, ds \\ \displaystyle\int \bar{F}_1(t_2 - s) f_2(s - u_1)\, ds & \displaystyle\int \bar{F}_1(t_2 - s) f_2(s - u_2)\, ds \end{vmatrix}$$

$$= \iint_{s_1 < s_2} \begin{vmatrix} \bar{F}_1(t_1 - s_1) & \bar{F}_1(t_1 - s_2) \\ \bar{F}_1(t_2 - s_1) & \bar{F}_1(t_2 - s_2) \end{vmatrix} \begin{vmatrix} f_2(s_1 - u_1) & f_2(s_1 - u_2) \\ f_2(s_2 - u_1) & f_2(s_2 - u_2) \end{vmatrix} ds_2\, ds_1,$$

by the Basic Composition Formula[1] (Karlin, 1968, p. 17). Throughout, the domain of integration is $(-\infty, \infty)$. Integrating the inner integral by parts, we obtain

$$D = \iint_{s_1 < s_2} \begin{vmatrix} \bar{F}_1(t_1 - s_1) & f_1(t_1 - s_2) \\ \bar{F}_1(t_2 - s_1) & f_1(t_2 - s_2) \end{vmatrix} \begin{vmatrix} f_2(s_1 - u_1) & f_2(s_1 - u_2) \\ \bar{F}_2(s_2 - u_1) & \bar{F}_2(s_2 - u_2) \end{vmatrix} ds_2\, ds_1.$$

The sign of the first determinant is the same as that of

$$\frac{f_1(t_2 - s_2}{\bar{F}_1(t_2 - s_2)} \frac{\bar{F}_1(t_2 - s_2)}{\bar{F}_1(t_2 - s_1)} - \frac{f_1(t_1 - s_2)}{\bar{F}_1(t_1 - s_2)} \frac{\bar{F}_1(t_1 - s_2)}{\bar{F}_1(t_1 - s_1)},$$

[1] *Basic Composition Formula.* Let $w(x, z) = \int u(x, y) v(y, z)\, d\sigma(y)$ converge absolutely, where $d\sigma(y)$ is a sigma-finite measure. Then

$$\begin{vmatrix} w(x_1, z_1) & \cdots & w(x_1, z_n) \\ \vdots & \ddots & \vdots \\ w(x_n, z_1) & \cdots & w(x_n, z_n) \end{vmatrix} = \int \cdots \int_{y_1 < \cdots < y_n} \begin{vmatrix} u(x_1, y_1) & \cdots & u(x_1, y_n) \\ \vdots & \ddots & \vdots \\ u(x_n, y_1) & \cdots & u(x_n, y_n) \end{vmatrix}$$

$$\times \begin{vmatrix} v(y_1, z_1) & \cdots & v(y_1, z_n) \\ \vdots & \ddots & \vdots \\ v(y_n, z_1) & \cdots & v(y_n, z_n) \end{vmatrix} d\sigma(y_1) \cdots d\sigma(y_n).$$

assuming nonzero denominators. But

$$\frac{f_1(t_2 - s_2)}{\overline{F}_1(t_2 - s_2)} \geq \frac{f_1(t_1 - s_2)}{\overline{F}_1(t_1 - s_2)}$$

by hypothesis, while

$$\frac{\overline{F}_1(t_2 - s_2)}{\overline{F}_1(t_2 - s_1)} \geq \frac{\overline{F}_1(t_1 - s_2)}{\overline{F}_1(t_1 - s_1)}$$

by Theorem 5.6 of Chapter 3. Thus the first determinant is nonnegative. A similar argument holds for the second determinant, so that $D \geq 0$. But by Theorem 5.6 of Chapter 3, this implies F is IFR.

If F_1 or F_2 do not have a density at the right-hand endpoint(s), the theorem may be proved in a similar fashion using limiting arguments. ||

It is easy to see that convolution does *not* preserve DFR or DFRA. As an example, let $F_1 \equiv F_2 \equiv G_\alpha$, the gamma distribution with shape parameter $\frac{1}{2} < \alpha < 1$, whose density is defined in (5.3) of Chapter 3. Then as shown in 5.3 of Chapter 3, F_1 and F_2 are DFR. However, as stated in Exercise 3 of Section 5, Chapter 3, $F_1 * F_2 \equiv G_{2\alpha}$, a gamma distribution with shape parameter >1. Thus $F_1 * F_2$ is *not* DFR or DFRA.

The question as to whether convolution preserves IFRA remains unanswered at this time. No proof has been given and, alternately, no counterexample has been found. In a systematic numerical attempt to find a possible counterexample, all pairs of distributions placing mass (in multiples of .1) on the points 0, 1, 2, 3 were convolved. Every one of the 1988 convolutions of pairs of IFRA distributions turned out to be IFRA also. No counterexamples were uncovered.

Mixture of Distributions

Mixtures of distributions arise naturally in a number of reliability situations. For example, suppose a manufacturer produces 60 percent of a certain product in Assembly Line 1 and 40 percent in Assembly Line 2. Because of differences in machines, personnel, and so on, the life length of a unit produced in Assembly Line 1 has distribution F_1, whereas the life length of a unit produced in Assembly Line 2 has distribution $F_2 \not\equiv F_1$. After production, units from both assembly lines flow into a common shipping room, so that outgoing lots consist of a random mixture of the output of the two lines. It is clear that a unit selected at random from a lot would have life distribution $F \equiv .6F_1 + .4F_2$, a mixture of the two underlying distributions.

More generally, the distributions being mixed may be uncountably infinite in number. For example, suppose an important quality characteristic of the product being manufactured depends on the amount α of impurity present in the raw material; specifically, the probability distribution of the quality characteristic is F_α. Suppose α itself is random with distribution

$G(\alpha)$. Then the resultant distribution F of the quality characteristics is given by

$$F(x) = \int_{-\infty}^{\infty} F_\alpha(x)\, dG(\alpha). \tag{4.3}$$

We are thus led to the following definition.

4.3. Definition. Let $\{F_\alpha\}$ be a set of probability distributions, where the index α is governed by the distribution G. Then the *mixture* F *of* F_α *according to* G is given by (4.3).

We will establish preservation properties of mixtures by using their hazard transforms, defined as follows:

4.4. Definition. The *hazard transform of the mixture* $F(x) = \int F_\alpha(x)\, dG(\alpha)$ is

$$\eta(\mathbf{u}) = -\log \int e^{-u_\alpha}\, dG(\alpha), \tag{4.4}$$

where the vector $\mathbf{u}$ has elements u_α, $0 \le u_\alpha \le \infty$, $-\infty < \alpha < \infty$.

It follows immediately that the hazard function $R(t)$ of the mixture is given by

$$R(t) = \eta(\mathbf{R}(t)) \equiv -\log \int e^{-R_\alpha(t)}\, dG(\alpha), \qquad 0 \le t < \infty, \tag{4.5}$$

where $R_\alpha(t)$ is the hazard function of F_α and boldface $\mathbf{R}(t)$ is a vector.

To study preservation properties of mixtures, we will find it useful to establish an interesting geometric property of the hazard transform.

4.5. Theorem. The hazard transform of a mixture is concave; that is, if η is the hazard transform (4.4) of the mixture given by (4.3), then

$$\eta[a\mathbf{u} + (1 - a)\mathbf{v}] \ge a\eta(\mathbf{u}) + (1 - a)\eta(\mathbf{v}), \tag{4.6}$$

$$0 \le a \le 1, \quad 0 \le u_\alpha \le \infty, \quad 0 \le v_\alpha \le \infty, \quad -\infty < \alpha < \infty.$$

PROOF. By the Hölder inequality,[2]

$$\int e^{-au_\alpha} e^{-(1-a)v_\alpha}\, dG(\alpha)$$

$$\le \left\{\int e^{-u_\alpha}\, d\ (\alpha)\right\}^a \left\{\int e^{-v_\alpha}\, dG(\alpha)\right\}^{1-a}, \qquad 0 \le a \le 1.$$

From (4.4), the desired conclusion (4.6) follows. ‖

We will need the following lemma.

[2] *Hölder's Inequality.* Let $\int |f(x)|^p\, d\mu(x) < \infty$, $\int |g(x)|^q\, d\mu(x) < \infty$, $p > 1$, and $1/p + 1/q = 1$. Then $\int |f(x)g(x)|\, d\mu(x) \le [\int |f(x)|^p\, d\mu(x)]^{1/p}[\int |g(x)|^q\, d\mu(x)]^{1/q}$. See Munroe (1953), p. 238. The special case $p = 2$ yields the following inequality.

Schwarz's Inequality. Let $\int |f(x)|^2\, d\mu(x) < \infty$, $\int |g(x)|^2\, d\mu(x) < \infty$. Then $\{\int |f(x)g(x)|\, d\mu(x)\}^2 \le \int |f(x)|^2\, d\mu(x) \int |g(x)|^2\, d\mu(x)$.

4.6. Lemma. Let $h(\mathbf{u})$ be concave (convex) and increasing in each argument. Let each $u_\alpha(t)$ be concave (convex). Then $g_{\mathbf{u}}(t) \equiv h(\mathbf{u}(t))$ is concave (convex) in t.

PROOF. $g_{\mathbf{u}}[at + (1 - a)t'] \equiv h[\mathbf{u}(at + (1 - a)t')]$. But $u_\alpha[at + (1 - a)t'] \geq (\leq)\, au_\alpha(t) + (1 - a)u_\alpha(t')$ since u_α is concave (convex). Since h is increasing, then $h[\mathbf{u}(at + (1 - a)t')] \geq (\leq)\, h[a\mathbf{u}(t) + (1 - a)\mathbf{u}(t')]$. Since h is concave (convex), $h[a\mathbf{u}(t) + (1 - a)\mathbf{u}(t')] \geq (\leq)\, ah(\mathbf{u}(t)) + (1 - a)h(\mathbf{u}(t')) \equiv ag_{\mathbf{u}}(t) + (1 - a)g_{\mathbf{u}}(t')$ so that $g_{\mathbf{u}}(t)$ is concave (convex) in t. ||

The one-dimensional version of Lemma 4.6 is stated as problem 114 in Hardy, Littlewood, and Pólya (1952), p. 97.

Using Theorem 4.5 and Lemma 4.6, we may now show that a mixture of DFR distributions is DFR and similarly a mixture of DFRA distributions is DFRA.

4.7. Theorem. Let $F(t)$ be the mixture given by (4.3).

(a) If each F_α is DFR, then F is DFR.
(b) If each F_α is DFRA, then F is DFRA.

PROOF. (a) The result follows from Theorem 4.5 and Lemma 4.6.

(b) The hazard function $R(t)$ of the mixture satisfies

$$R(at) = \eta(\mathbf{R}(at)) \geq \eta(a\mathbf{R}(t)) \qquad \text{for} \quad 0 \leq a \leq 1,$$

since each F_α is DFRA and η is increasing. But choosing $\mathbf{v} = \mathbf{0}$ in (4.6), we see that

$$\eta(a\mathbf{R}(t)) \geq a\eta(\mathbf{R}(t)) \qquad \text{for} \quad 0 \leq a \leq 1.$$

Thus

$$R(at) \geq aR(t) \qquad \text{for} \quad 0 \leq a \leq 1,$$

and hence F is DFRA. ||

In particular, note that a mixture of exponential distributions, each of which is DFR, is itself DFR. As an application of this observation to reliability situations, suppose a given electronic component has a constant failure rate corresponding to a particular environment. If life data for this component are collected from various dissimilar environments, the inadvertent pooling of the unlike data will display an apparent decreasing failure rate. A real life occurrence of this phenomenon is described in Proschan (1963).

To complete our discussion of mixtures of distributions, we conclude immediately from our discussion just above that mixtures of IFR (IFRA) are not necessarily IFR (IFRA), since exponential distributions are IFR (IFRA), and their mixture is not IFR (IFRA), but rather strictly DFR (assuming the distributions are not all identical).

We may summarize the results obtained in this section concerning preservation of life distribution classes under various reliability operations.

Table 4.1. Preservation of Life Distribution Classes under Reliability Operations

Life Distribution Class	RELIABILITY OPERATION		
	Formation of Coherent Systems	*Addition of Life Lengths (Convolution of Distributions)*	*Mixture of Distributions*
IFR	Not preserved	Preserved	Not preserved
IFRA	Preserved	?	Not preserved
DFR	Not preserved	Not preserved	Preserved
DFRA	Not preserved	Not preserved	Preserved

EXERCISES

1. Let F_1 and F_2 be log concave life distribution functions. Then the convolution $F(t) = \int_0^t F_1(t - x)\, dF_2(x)$ is log concave.

2. Give an example of an IFR distribution with finite support $[0, b]$ having a jump at b.

3. Show that the formation of series systems preserves each of the classes:

   (a) IFR,
   (b) IFRA,
   (c) DFR,
   (d) DFRA.

4. All distinct coherent structures of order 3 (up to permutations of components) are listed in Chapter 1, Section 2. Assume the components are independent with respective IFR failure distributions F_1, F_2, and F_3. Which structures have IFR distributions, if

   (a) $F_1 \equiv F_2 \equiv F_3$,
   (b) $F_2 \equiv F_3$,
   (c) $F_1 \not\equiv F_2 \not\equiv F_3$?

   Support your assertions.

5. Let $\overline{F}_\alpha(t) = e^{-\alpha t}$, an exponential survival probability. Let failure rate α be random with gamma density $g(\alpha) = [\beta^\gamma \alpha^{\gamma-1}/\Gamma(\gamma)]e^{-\beta\alpha}$, $0 \leq \alpha < \infty$, where $\beta, \gamma > 0$.

(a) Show that the mixture F of F_α according to G given in (4.3) has survival probability $\bar{F}(t) = (1 + t/\beta)^{-\gamma}$.

(b) Show explicitly that the failure rate is decreasing, in accordance with Theorem 4.7(a).

*6. Show that the harmonic mean of two IFR survival probabilities is an IFR survival probability.

7. Let $f(x, y)$ be TP_r for $x \in A$, $y \in B$, and $g(y, z)$ be TP_r for $y \in B$, $z \in C$. Then $h(x, z) = \int_B f(x, y)g(y, z)\,dy$ is TP_r for $x \in A$, $z \in C$. (Use the Basic Composition Formula.)

8. Use the result of Exercise 7 to show that the following functions are TP_2:

(a) $f(x, y) = \int_{-\infty}^{x} g(u, y)\,du$, where $g(u, y)$ is TP_2 for real u, y.

(b) $f(x, y) = \int_{x}^{\infty} g(u, y)\,du$, where $g(u, y)$ is TP_2 for real u, y.

(c) $f(r, s) = \int_0^\infty x^r x^s g(x)\,dx$ for real r, s such that the integral is finite, and $g(x) \geq 0$ for $x \geq 0$. It follows that if $\mu_r = \int_0^\infty x^r g(x)\,dx$ is the rth moment of a nonnegative random variable, assumed finite for $r \geq 0$, then μ_{r+s} is TP_2 for $r \geq 0$, $s \geq 0$.

(d) Let the Laplace transform $f^*(s) = \int_0^\infty e^{-xs} f(x)\,dx$ exist for $0 \leq s < \infty$. Then $f^*(s + t)$ is TP_2 in $s \geq 0$, $t \geq 0$.

9. Let f, g be nonnegative functions. Then the Schwarz inequality states:

$$\begin{vmatrix} \int f(x)f(x)\,d\mu(x) & \int f(x)g(x)\,d\mu(x) \\ \int g(x)f(x)\,d\mu(x) & \int g(x)g(x)\,d\mu(x) \end{vmatrix} \geq 0.$$

Use the Basic Composition Theorem to give a direct proof.

5. PARTIAL ORDERINGS OF LIFE DISTRIBUTIONS

Convex and Star-shaped Orderings

An examination of the Weibull family of life distributions reveals an ordering with respect to the shape parameter α. Thus for $0 < \alpha_1 < \alpha_2$,

$$F_{\alpha_1}^{-1}F_{\alpha_2}(t) = \lambda^{(\alpha_2/\alpha_1)-1}t^{\alpha_2/\alpha_1}, \qquad t \geq 0,$$

a convex function. A similar ordering of the gamma family of life distributions G_α exists with respect to the shape parameter α.

The ordering suggested above may be motivated from another point of view. Let $G(t) = 1 - e^{-\lambda t}$, $t \geq 0$, an exponential distribution, and F be IFR. Then

$$G^{-1}F(t) = -\frac{1}{\lambda}\log \bar{F}(t)$$

is convex, since the first derivative, a positive multiple of the failure rate of F, is increasing. This motivates the following definition.

5.1. Definition. Let F and G be continuous distributions, G be strictly increasing on its support, an interval, and $F(0) = G(0) = 0$. Then F is *convex with respect to* G (written $F \underset{c}{<} G$) if $G^{-1}F(x)$ is a convex function in x on the support[3] of F, assumed an interval.

5.2. Convex Ordering Is a Partial Ordering. Note that $F \underset{c}{<} G$ is equivalent to $F(\alpha x) \underset{c}{<} G(\beta x)$ for all $\alpha > 0$, $\beta > 0$. Thus in discussing convex ordering, we may group into equivalence classes distributions that differ only by a positive scale factor: that is, $\{F(\theta x); \theta > 0\}$ constitutes an equivalence class of distributions. Hence when we write F in this section we are designating the family $\{F(\theta x): \theta > 0\}$.

(a) Let $F \underset{c}{<} G$ and $G \underset{c}{<} H$. Then $F \underset{c}{<} H$.

PROOF. By hypothesis, $G^{-1}F$ is convex, $H^{-1}G$ is convex, and $H^{-1}G$ is increasing. Thus $H^{-1}GG^{-1}F$ is convex by Lemma 4.6. It follows that $H^{-1}F$ is convex. ||

Thus *convex ordering is transitive.*

(b) $F \underset{c}{<} F$ for each distribution F.

PROOF. $F^{-1}F(x) = x$ (up to a scale factor), a linear and hence a convex function.

Thus *convex ordering is reflexive.*

(c) $F \underset{c}{<} G$ and $G \underset{c}{<} F$ imply that $F = G$ (up to a scale factor).

Thus *convex ordering is antisymmetric.*

It follows that convex ordering is a partial ordering of the scale equivalent classes of distribution. See Hewitt and Stromberg (1965), p. 8.

A second type of ordering is defined as follows.

5.3. Definition. Let F and G be continuous distributions, G be strictly increasing on its support, and $F(0) = G(0) = 0$. Then F is *star-shaped with respect to* G (written $F \underset{*}{<} G$) if $G^{-1}F(x)$ is star-shaped [that is, $(1/x)G^{-1}F(x)$ is increasing for $x \geq 0$].

As in 5.2 above, we group into equivalence classes distributions that differ only by a positive scale factor. We may show that *star-shaped ordering is a partial ordering* of the scale equivalent classes of distributions. The details are left to Exercise 1.

The following results are immediate.

[3] A point x is in the *support* of a distribution F if for every $\varepsilon > 0$, $F(x + \varepsilon) - F(x - \varepsilon) > 0$.

5.4. Results. (a) $F \underset{c}{<} G$ implies $F \underset{*}{<} G$.

(b) The relationship $F \underset{c}{<} G$ is unaffected by a translation transformation of either F or G, assuming the random variables remain nonnegative.

(c) The relationship $F \underset{*}{<} G$ may be destroyed by a translation transformation of either F or G, assuming the random variables remain nonnegative.

(d) Let $G(x) = 1 - e^{-\lambda x}$, F be a continuous distribution function, with $F(0) = 0$. Then $F \underset{*}{<} G$ is equivalent to F IFR.

(e) Let $G(x) = 1 - e^{-\lambda x}$, F be a continuous distribution function, with $F(0) = 0$. Then $F \underset{c}{<} G$ is equivalent to F IFRA.

See Exercise 2.

5.5. Single Crossing Property. Let $F \underset{*}{<} G$. (a) Then $\bar{F}(x)$ crosses $\bar{G}(\theta x)$ at most once, and from above, as x increases from 0 to ∞, for each $\theta > 0$.

(b) If, in addition, F and G have the same mean, then a single crossing does occur, and F has smaller variance than G.

If we take G to be the exponential distribution, then F must be IFRA by 5.4(e), as we have seen in Theorem 2.18.

Preservation of Partial Ordering under Formation of k-out-of-n Systems of Like Components

Next we study the effects on a partial ordering of forming a k-out-of-n system of like components or, equivalently, taking order statistics. A useful identity is given in the following lemma.

5.6. Lemma. Let F_{jn} be the distribution of the jth order statistic in a sample of n from F. Then

$$F_{jn}(x) = B_{jn}[F(x)] \qquad \text{for} \quad 0 < x < \infty, \tag{5.1}$$

where

$$B_{jn}(p) = \frac{n!}{(j-1)!\,(n-j)!} \int_0^p u^{j-1}(1-u)^{n-j}\,du \qquad \text{for} \quad 0 \le p \le 1. \tag{5.2}$$

PROOF.

$$F_{jn}(x) = P[X_{j:n} \le x] = \sum_{i=j}^{n} \binom{n}{i} [F(x)]^i [\bar{F}(x)]^{n-i}$$

$$= \frac{n!}{(j-1)!\,(n-j)!} \int_0^{F(x)} u^{j-1}(1-u)^{n-j}\,du,$$

is a well-known formula (Mood, 1950, p. 235). Thus by (5.2), the conclusion follows. ||

We use Lemma 5.6 to obtain Theorem 5.7.

5.7. Theorem. Let $F \underset{c}{<} G$ $(F \underset{*}{<} G)$. Then $F_{jn} \underset{c}{<} G_{jn}$ $(F_{jn} \underset{*}{<} G_{jn})$, where G_{jn} is the distribution of the jth order statistic in a sample of n from G.

PROOF. $G_{jn}^{-1}F_{jn}(x) = (B_{jn}G)^{-1}B_{jn}F(x) = G^{-1}F(x)$, so that convexity (star-shapedness) is preserved. ‖

From Theorem 5.7 we obtain Theorem 5.8 immediately.

5.8. Theorem. Let F be IFR (IFRA). Then F_{jn} is IFR (IFRA).

PROOF. Let F be IFR and $G(t) = 1 - e^{-t}$. Then $F_{jn} \underset{c}{<} G_{jn}$ by Theorem 5.7. Next we will prove $G_{jn} \underset{c}{<} G$, or, equivalently, G_{jn} is IFR.

From (5.1) we have

$$\frac{\bar{G}_{jn}(t)}{g_{jn}(t)} = \frac{B_{n-j+1}[\bar{G}(t)]}{B'_{n-j+1}[\bar{G}(t)] \cdot g(t)} = \frac{1}{p}\int_0^p \left(\frac{x}{p}\right)^{n-j}\left(\frac{1-x}{1-p}\right)^{j-1} dx,$$

where $p = \bar{G}(t) = e^{-t}$. Letting $u = x/p$, we have

$$\frac{\bar{G}_{jn}(t)}{g_{jn}(t)} = \int_0^1 u^{n-j}\left(\frac{1-up}{1-p}\right)^{j-1} du.$$

Since $(1 - up)/(1 - p)$ is increasing in p, $0 \le p \le 1$, and therefore decreasing in t, $0 \le t < \infty$, we conclude that $\bar{G}_{jn}(t)/g_{jn}(t)$ is decreasing, and so G_{jn} is IFR.

It follows that $F_{jn} \underset{c}{<} G$, or, equivalently, F_{jn} is IFR.

Next let F be IFRA. Then $F_{jn} \underset{*}{<} G_{jn}$ by Theorem 5.7. Also $G_{jn} \underset{*}{<} G$ since $G_{jn} \underset{c}{<} G$. Hence $F_{jn} \underset{*}{<} G$, and so F_{jn} is IFRA. ‖

Remark. An alternate method of proving that G_{jn} is IFR is to note that successive spacings are independent exponential random variables, and then use the fact that the convolution of IFR distributions is IFR.

EXERCISES

1. Show that star-shaped ordering is a partial ordering.
2. Verify results 5.4(a) through (e).
3. Prove properties 5.5(a) and (b).
4. Let $F \underset{c}{<} G$ with respective densities f and g. Then show (a) $[f(t)/gG^{-1}F(t)] \uparrow$, and (b) $[fF^{-1}(t)/gG^{-1}(t)] \uparrow$. ($gG^{-1}$ and fF^{-1} denote functional composition.)
5. Let F and G be distributions such that $F(0) = 0 = G(0)$ and $G^{-1}F(x + y) \ge G^{-1}F(x) + G^{-1}F(y)$ for all $x, y \ge 0$. (We say F is superadditive with respect to G and write $F \underset{su}{<} G$.) Show

(a) $F \underset{\text{su}}{\leq} G$ is a partial ordering of scale equivalent classes of distributions.
(b) $F \underset{*}{\leq} G \Rightarrow F \underset{\text{su}}{\leq} G$.
(c) $F \underset{\text{su}}{\leq} G \Rightarrow F_{jn} \underset{\text{su}}{\leq} G_{jn}$.

6. Assume a structure with reliability function $h(p)$, with each component life independently distributed according to distribution F having density f. Define $u_h(p) = ph'(p)/h(p)$. Then

(a) $u_h(\overline{F}(t)) = r_h(t)/r(t)$, where $r(t) = f(t)/\overline{F}(t) =$ component failure rate at time t, and $r_h(t) =$ system failure rate at time t.
(b) $u_h(p)$ is a decreasing function of p if and only if $r_h(t)/r(t)$ is an increasing function of t.
(c) If $r(t)$ is an increasing function of t and $u_h(p)$ is a decreasing function of p, then $r_h(t)$ is an increasing function of t.

7. Show that for a k-out-of-n structure with reliability function $h(p)$, $u_h(p)$ is decreasing in p.
8. Using Exercises 6 and 7, show that the life distribution of a k-out-of-n system of independent components each having an IFR distribution itself has an IFR distribution.
9.ˆ Assume $h(p) = f(g(p))$ with $g'(p) \geq 0$, $u_f(p)$ decreasing, and $u_g(p)$ decreasing. Then $u_h(p)$ is decreasing. From this derive a result concerning k-out-of-n systems having like l-out-of-n modules.

6. RELIABILITY BOUNDS

In this section we present bounds on reliability and other parameters, assuming one moment or one percentile known and the underlying distribution is IFR, IFRA, DFR, or DFRA. Such bounds are useful in reliability applications, since in a typical situation the only facts known a priori may be, for example, that the component has an increasing failure rate due to wear, and that its mean life is μ, say.

We will present only a few basic bounds among the many available in the literature. See Barlow and Marshall, I and II (1964), Barlow and Marshall (1965), Barlow (1965), and Barlow and Marshall (1967) for additional bounds on a variety of parameters under various assumptions.

Reliability Bounds Based on One Known Quantile

Our first bound is based on the fact that an IFRA (DFRA) survival probability crosses an exponential survival probability with the same pth quantile once, and from above (below).

6.1. Theorem. Let F be IFRA (DFRA) with pth quantile ξ_p [that is, $F(\xi_p) = p$]. Then

$$\bar{F}(t) \begin{cases} \geq (\leq) e^{-\alpha t} & \text{for } 0 \leq t \leq \xi_p, \\ \leq (\geq) e^{-\alpha t} & \text{for } t \geq \xi_p, \end{cases} \tag{6.1}$$

where $\alpha = -(1/\xi_p)\log(1 - p)$.

PROOF. First note that the exponential survival probability $e^{-\alpha t}$ has the same pth quantile ξ_p as does F. Thus *at least* one crossing of $e^{-\alpha t}$ by $\bar{F}(t)$ must occur, namely, at $t = \xi_p$.

By Theorem 2.18, *at most* one crossing can occur, and in the direction indicated by (6.1). ||

Example. A contractor is required to produce a component having .95 reliability corresponding to a mission length of 1000 hours; that is, $\bar{F}(1000) = .95$. The component is made of parts having IFR life distributions; part life lengths are mutually independent. Thus by Theorem 2.6, component life length F is IFRA. Assuming the contractor just meets the requirement, what is a conservative prediction of component reliability $\bar{F}(900)$ corresponding to a mission of 900 hours?

Using Theorem 6.1, we identify $t = 900$, $p = .05$, $\xi_p = 1000$. We compute $\alpha = -(1/\xi_p)\log(1 - p) = -(1/1000)\log(1 - .05)$. Noting that $t = 900 < 1000 = \xi_p$, we have from (6.1),

$$\bar{F}(t) \geq e^{-\alpha t},$$

that is,

$$\bar{F}(900) \geq e^{(1/1000)(\log 1 - .05)900} = e^{.9 \log(1 - .05)} \cong e^{-.045} \cong .955.$$

Thus, under the assumptions described above, the contractor may conservatively claim a component reliability of at least .955 corresponding to a mission length of 900 hours. [When the known reliability is high and $t < \xi_p$ are not too different, a handy approximate lower bound is $1 - pt/\xi_p$.]

Reliability Bounds Based on One Known Moment

Next we present bounds on reliability based on a single known moment. In the most practical case, the known moment would be the first moment. We will find the following notation convenient.

Notation. $X \overset{\text{st}}{=} Y$ means that X and Y are identically distributed. $X \overset{\text{st}}{\geq} Y (X \overset{\text{st}}{\leq} Y)$ means that $P[X > x] \geq (\leq) P[Y > x]$ for each x.

6.2. Theorem. If $F \underset{c}{<} G$, $F(0) = 0 = G(0)$, F and G are continuous, and $\int_0^\infty t^r\,dF(t) = \mu_r = \int_0^\infty t^r\,dG(t)$ for a fixed value of $r \geq 1$, then

$$\bar{F}(t) \geq \begin{cases} \bar{G}(t) & \text{for} \quad t < \mu_r^{1/r}, \\ 0 & \text{for} \quad t \geq \mu_r^{1/r}. \end{cases} \tag{6.2}$$

The bound is sharp.

PROOF. Let X (Y) have distribution F (G) and X^r (Y^r) have distribution F_r (G_r). First we show that $G_r^{-1}F_r(x)$ is convex in $x \geq 0$. Note that

$$G_r^{-1}F_r(x) = [G^{-1}F(x^{1/r})]^r.$$

Assuming differentiability of $G^{-1}F$,

$$\frac{d}{dx}\,G_r^{-1}F_r(x) = \left[\frac{G^{-1}F(x^{1/r})}{x^{1/r}}\right]^{r-1}(G^{-1}F)'(x^{1/r}).$$

The first factor is increasing in $x \geq 0$ since $G^{-1}F$ is star-shaped and $r \geq 1$. The second factor is increasing in x since $G^{-1}F$ is convex. Thus $(d/dx) \times G_r^{-1}F_r(x)$ is increasing in $x \geq 0$, and so $G_r^{-1}F_r$ is convex. Since convex differentiable functions are dense in the class of convex functions, the same conclusion holds even when $G^{-1}F$ is not differentiable.

Now let $X_1^r, \ldots, X_n^r(Y_1^r, \ldots, Y_n^r)$ be a random sample from $F_r(G_r)$. Since $G_r^{-1}F_r$ is convex, then

$$G_r^{-1}F_r\left(\frac{1}{n}\sum_{i=1}^n X_i^r\right) \leq \frac{1}{n}\sum_{i=1}^n {}_r^{-1}F_r(X_i^r)$$

by Jensen's inequality. It follows that

$$F_r\left(\frac{1}{n}\sum_{i=1}^n X_i^r\right) \leq G_r\left[\frac{1}{n}\sum_{i=1}^n G_r^{-1}F_r(X_i^r)\right] \overset{\text{st}}{=} G_r\left(\frac{1}{n}\sum_{i=1}^n Y_i^r\right),$$

since $G_r^{-1}F_r(X_i^r) \overset{\text{st}}{=} Y_i^r$ (see Exercise 4), and $X_1, \ldots, X_n$ are independent.

Let $n \to \infty$. By the strong law of large numbers,

$$F_r(\mu_r) \leq G_r(\mu_r).$$

By the first equation of the proof, it follows that

$$F(\mu_r^{1/r}) \leq G(\mu_r^{1/r}).$$

Since $F \underset{c}{<} G$, then $\bar{F}$ crosses $\bar{G}$ exactly once, and from above, by property 5.5. Thus

$$\bar{F}(t) \geq \bar{G}(t) \qquad \text{for} \quad t \leq \mu_r^{1/r}.$$

To verify that the bound is sharp for $t < \mu_r^{1/r}$ (that is, no larger lower bound exists), choose $F \equiv G$. For $t \geq \mu_r^{1/r}$, see Exercise 5. ‖

Choosing G exponential in Theorem 6.2, we obtain the following useful special case.

6.3. Corollary. Let F be a continuous IFR distribution, $\mu_r = \int_0^\infty t^r\, dF(t)$, and $\lambda_r = \mu_r/\Gamma(r+1)$, $r \geq 1$. Then

$$\bar{F}(t) \geq \begin{cases} \exp\left[\dfrac{-t}{\lambda_r^{1/r}}\right] & \text{for } t < \mu_r^{1/r}, \\ 0 & \text{for } t \geq \mu_r^{1/r}. \end{cases} \tag{6.3}$$

The bound is sharp.

PROOF. Observe that $\int_0^\infty (t^r/\lambda_r^{1/r}) \exp[-t/\lambda_r^{1/r}]\, dt = \Gamma(r+1)\lambda_r = \mu_r$ and that $F \underset{c}{<} G$, where $G(t) = 1 - \exp[-t/\lambda_r^{1/r}]$. Thus Theorem 6.2 applies, yielding Corollary 6.3. ‖

Next we will show how the nontrivial lower bound of Corollary 6.3 and its domain of applicability vary as a function of r. To do so, we need a lemma useful in its own right.

6.4. Lemma. Let $F \underset{*}{<} G$, $\int_0^\infty x^s\, dF(x) = \int_0^\infty x^s\, dG(x)$ for a fixed $s > 0$, and ψ be increasing. Then

$$\int_0^\infty \psi(x) x^{s-1} \bar{F}(x)\, dx \leq \int_0^\infty \psi(x) x^{s-1} \bar{G}(x)\, dx.$$

PROOF. $F \underset{*}{<} G$ implies that $\bar{F}$ crosses $\bar{G}$ at most once, and from above, if at all. $\int_0^\infty x^s\, dF(x) = \int_0^\infty x^s\, dG(x)$ implies $\bar{F}$ crosses $\bar{G}$ at least once. Hence $\bar{F}$ must cross $\bar{G}$ exactly once. Let $0 < t_0 < \infty$ be a solution of $\bar{F}(x) = \bar{G}(x)$. Then $x^{s-1}\bar{F}(x)$ crosses $x^{s-1}\bar{G}(x)$ at t_0, and from above. It follows that

$$\int_0^\infty \psi(x) x^{s-1} \bar{F}(x)\, dx - \int_0^\infty \psi(x) x^{s-1} \bar{G}(x)\, dx$$

$$= \int_0^\infty [\psi(x) - \psi(t_0)][x^{s-1}\bar{F}(x) - x^{s-1}\bar{G}(x)]\, dx,$$

since $\int_0^\infty x^{s-1}\bar{F}(x)\, dx = (1/s)\int_0^\infty x^s\, dF(x) = (1/s)\int_0^\infty x^s\, dG(x) = \int_0^\infty x^{s-1}\bar{G}(x)\, dx$. Since $\psi(x) - \psi(t_0)$ and $x^{s-1}\bar{F}(x) - x^{s-1}\bar{G}(x)$ are opposite in sign for all x, the desired conclusion follows. ‖

6.5. Corollary. Let F be IFRA (DFRA), $\mu_r = \int_0^\infty x^r\, dF(x)$, and $\lambda_r = \mu_r/\Gamma(r+1)$. Then $\lambda_r^{1/r}$ is decreasing (increasing) in $r \geq 0$.

PROOF. Let $0 < r < s$. Note that by Exercise 11 below,

$$\mu_r = r\int_0^\infty x^{r-1}\bar{F}(x)\, dx = r\int_0^\infty x^{r-1} \exp(-x/\lambda_r^{1/r})\, dx.$$

Applying Lemma 6.4 with $\psi(x) = x^{s-r}$, we obtain

$$\lambda_s = \frac{\mu_s}{\Gamma(s+1)} = \frac{1}{\Gamma(s+1)} \int_0^\infty sx^{s-1}\overline{F}(x)\,dx$$

$$\le (\ge) \int_0^\infty \frac{sx^{s-1}}{\Gamma(s+1)} \exp\left(-\frac{x}{\lambda_r^{1/r}}\right) dx = \lambda_r^{s/r}. \quad \|$$

6.6. Theorem. (a) For fixed $t \ge 0$, the nontrivial lower bound $\exp(-t/\lambda_r^{1/r})$ of Corollary 6.3 is decreasing in $r > 0$.

(b) The domain $[0, \mu_r^{1/r}]$ of applicability of this bound is increasing in $r > 0$.

PROOF. (a) By Corollary 6.5, for $r < s$, $t \ge 0$, we have $\exp(-t/\lambda_s^{1/s}) \le \exp(-t/\lambda_r^{1/r})$, and so (a) holds.

(b) See Exercise 6. ‖

The special case $r = 1$ in Corollary 6.3 is important in reliability applications.

6.7. Bound Based on First Moment. Let F be IFR with mean μ_1. Then

$$\overline{F}(t) \ge \begin{cases} e^{-t/\mu_1} & \text{for} \quad t < \mu_1, \\ 0 & \text{for} \quad t \ge \mu_1. \end{cases} \tag{6.4}$$

The bound is sharp.

6.8. Bounds on System Reliability. In the early stages of system design, it is often necessary to predict system reliability from quite minimal knowledge—say, of system structure, of component mean lives, and that each component is IFR, with all components independent. Using (6.4) we may make a conservative prediction of system reliability.

Specifically, let h be the reliability function of a coherent system of n independent components. Let component i have (unknown) life distribution F_i, IFR, with mean μ_i (known), $i = 1, \ldots, n$. Then a lower bound on system reliability $h(\overline{F}_1(t), \ldots, \overline{F}_n(t))$ for a mission of duration t is given by

$$h(\overline{F}_1(t), \ldots, \overline{F}_n(t)) \ge h(e^{-t/\mu_1}, \ldots, e^{-t/\mu_n}) \qquad \text{for} \quad t < \min(\mu_1, \ldots, \mu_n). \tag{6.5}$$

Actually, we may make a conservative reliability prediction even when the exact system structure and its associated reliability function are not yet known. Furthermore, the components need not be independent, but merely associated. (See Section 3 of Chapter 2.)

Specifically, assume (a) that we have a coherent system with (unknown) structure function ϕ, and (b) that the components are associated, with IFR marginal distributions with (known) means $\mu_1, \ldots, \mu_n$, respectively. Then we may make a conservative reliability prediction for system reliability $\bar{F}(t)$:

$$\bar{F}(t) \geq \exp\left[-t \sum_{i=1}^{n} \frac{1}{\mu_i}\right] \quad \text{for} \quad t < \min(\mu_1, \ldots, \mu_n), \tag{6.6}$$

since

$$\bar{F}(t) = P\left[\phi(X_1(t), \ldots, X_n(t)) = 1\right] \geq \prod_{i=1}^{n} P\left[X_i(t) = 1\right]$$
$$\geq \prod_{i=1}^{n} e^{-t/\mu_i} = \exp\left[-t \sum_{i=1}^{n} \frac{1}{\mu_i}\right].$$

The first inequality is a consequence of Theorem 3.2 of Chapter 2, while the second follows from (6.4). Bound (6.6), resulting from the application of two successive inequalities, may turn out to be quite conservative in many applications.

Remark. In practice in the early stages of system design, it is often assumed that all components

(a) Are connected in series.
(b) Are independent.
(c) Have exponential life distributions with known mean lives $\mu_1, \mu_2, \ldots, \mu_n$.

Then (6.6) yields a reliability estimate which is conservative if in fact all components

(a) Are part of a coherent system.
(b) Are associated.
(c) Have IFR marginal life distributions with known mean lives $\mu_1, \mu_2, \ldots, \mu_n$.

Bazovsky's Example Continued. In the example following Theorem 2.11, Chapter 3, suppose the electronic circuit consists of 10 diodes, 4 transistors, 20 resistors, and 10 capacitors, as before. Now, however, we no longer assume a series system will be used, but rather only that the system will be coherent (we may be in the very early stages of design). Further, it may not be correct to assume that components have mutually independent lives, but only that they are associated. Finally, component lives may have IFR rather than exponential distributions, with mean lives given by:

Silicon diode: $\mu_d = 500{,}000$.
Silicon transistor: $\mu_t = 100{,}000$.

Composition resistor: $\mu_r = 1{,}000{,}000$.
Ceramic capacitor: $\mu_c = 500{,}000$.

Then (6.6) yields the conservative predicted system reliability:

$$\bar{F}(t) \geq e^{-.0001t} \qquad \text{for} \quad t < 100{,}000,$$

where, as in Bazovsky's example, .0001 represents the failure rate computed on a parts-count basis, and 100,000 represents the minimum mean life among components.

Next we obtain upper bounds on reliability for IFRA, and alternately DFR units, with the mean known.

6.9. Theorem. Let F be IFRA with mean μ_1. Then for fixed $t > 0$,

$$\bar{F}(t) \leq \begin{cases} 1 & \text{for} \quad t \leq \mu_1, \\ e^{-wt} & \text{for} \quad t > \mu_1, \end{cases} \tag{6.7}$$

where $w > 0$, a function of t, satisfies

$$1 - w\mu_1 = e^{-wt}. \tag{6.8}$$

The bound is sharp.

PROOF. Let t be a fixed value $> \mu_1$. Define

$$\bar{G}(x) = \begin{cases} e^{-wx} & \text{for} \quad x < t, \\ 0 & \text{for} \quad x \geq t, \end{cases}$$

where $w > 0$ satisfies

$$\int_0^t e^{-wx}\, dx = \mu_1, \tag{6.9}$$

equivalent to (6.8). (See Figure 4.6.1.) Since $t > \mu_1$, such a value exists uniquely.

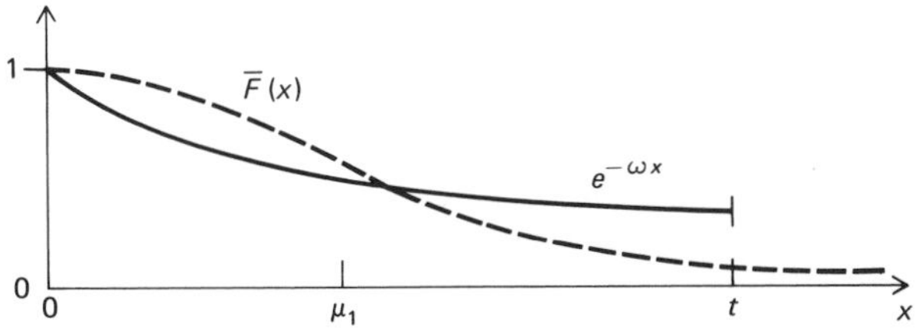

Figure 4.6.1.
Construction of upper bound.

By Theorem 2.18, $\bar{F}$ crosses $\bar{G}$ at most once in $(0, t)$. Since F and G have the same mean, by choice of w, and $\bar{F}$ dominates $\bar{G}$ on $[t, \infty)$, then $\bar{F}$ cannot strictly dominate $\bar{G}$ on $[0, t]$. Thus on $[0, t]$, $\bar{F}$ either crosses $\bar{G}$ exactly once and from above, or $\bar{F}$ lies entirely below $\bar{G}$. In either case $\bar{F}(t) \leq \bar{G}(t)$.

Demonstrating the sharpness of the bound is left to Exercise 8. ||

Since the bound is attained by an IFR distribution, this upper bound is actually a sharp upper bound for IFR distributions.

Next we present an upper bound on a DFR distribution with known mean. We refer the reader to B–P (1965), pp. 31–32, for the proof.

6.10. Theorem. Let F be DFR with mean μ_1. Then

$$\bar{F}(t) \leq \begin{cases} e^{-t/\mu_1} & \text{for } t \leq \mu_1, \\ \dfrac{\mu_1 e^{-1}}{t} & \text{for } t \geq \mu_1. \end{cases} \tag{6.10}$$

The bound is sharp.

Similarly, we may obtain lower bounds on reliability for IFRA distributions. The proof is given in Barlow-Marshall (1967).

6.11. Theorem. If F is IFRA and $\int_0^\infty x^r \, dF(x) = \mu_r$ $(r > 0)$, then

$$\bar{F}(x) \geq \begin{cases} \min[e^{-b_s s}, e^{-cs}] & \text{for } x < \mu_r^{1/r}, \\ 0 & \text{for } x \geq \mu_r^{1/r}, \end{cases}$$

where b_s is determined by

$$s^r(1 - e^{-b_s s}) + \int_s^\infty x^r b_s e^{-b_s x} \, dx = \mu_r,$$

and $c = [\mu_r/\Gamma(r + 1)]^{1/r}$.

Numerical values of the IFRA lower bound for $\mu_1 = 1$ are given in Table 6.1. A graphical comparison of the IFR and IFRA lower bounds is provided in Figure 4.6.2.

Bounds on Moments

Assuming a known mean, we may obtain bounds on the moments of IFRA (DFRA) distributions by comparison with the exponential.

6.12. Theorem. Let F be IFRA (DFRA) with known mean μ_1. Bounds for the rth moment $\mu_r = \int_0^\infty x^r \, dF(x)$ are given by:

$$\begin{aligned} \mu_r &\geq (\leq)\Gamma(r + 1)\mu_1{}^r \quad \text{for } 0 < r \leq 1, \\ \mu_r &\leq (\geq)\Gamma(r + 1)\mu_1{}^r \quad \text{for } 1 \leq r < \infty. \end{aligned} \tag{6.11}$$

The bounds are sharp.

Table 6.1. IFRA Lower Bound for $\bar{F}(t)$ with Mean One

t	Lower Bound	t	Lower Bound
.00	1.0000	.50	.5671
.01	.9900	.51	.5589
.02	.9801	.52	.5506
.03	.9704	.53	.5424
.04	.9607	.54	.5341
.05	.9511	.55	.5258
.06	.9416	.56	.5175
.07	.9322	.57	.5091
.08	.9228	.58	.5007
.09	.9136	.59	.4923
.10	.9043	.60	.4839
.11	.8952	.61	.4754
.12	.8861	.62	.4668
.13	.8771	.63	.4582
.14	.8681	.64	.4496
.15	.8592	.65	.4409
.16	.8504	.66	.4321
.17	.8416	.67	.4233
.18	.8329	.68	.4144
.19	.8242	.69	.4055
.20	.8155	.70	.3964
.21	.8069	.71	.3873
.22	.7983	.72	.3781
.23	.7898	.73	.3688
.24	.7813	.74	.3594
.25	.7728	.75	.3499
.26	.7644	.76	.3403
.27	.7560	.77	.3306
.28	.7476	.78	.3207
.29	.7393	.79	.3107
.30	.7310	.80	.3005
.31	.7227	.81	.2902
.32	.7144	.82	.2796
.33	.7062	.83	.2689
.34	.6979	.84	.2580
.35	.6897	.85	.2468
.36	.6815	.86	.2354
.37	.6733	.87	.2237
.38	.6651	.88	.2117
.39	.6570	.89	.1993
.40	.6488	.90	.1865
.41	.6406	.91	.1733
.42	.6325	.92	.1595
.43	.6243	.93	.1452
.44	.6162	.94	.1301
.45	.6080	.95	.1142
.46	.5998	.96	.0971
.47	.5917	.97	.0786
.48	.5835	.98	.0580
.49	.5753	.99	.0341

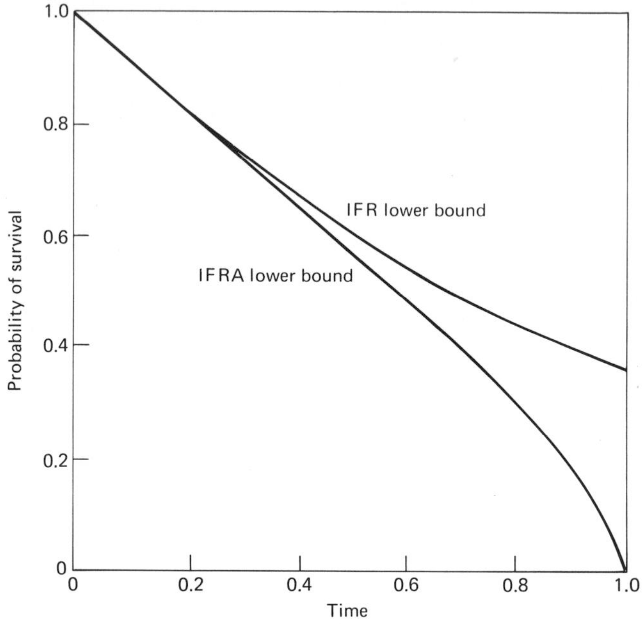

Figure 4.6.2.
Comparison of IFR and IFRA lower bounds on survival probability ($\mu = 1$).

PROOF. By Corollary 6.5, if F is IFRA (DFRA), $\mu_r = \Gamma(r+1)\lambda_r \geq (\leq)\Gamma(r+1)\lambda_1{}^r = \Gamma(r+1)\mu_1{}^r$ for $0 < r \leq 1$. The inequalities are reversed for $1 \leq r < \infty$.

To verify that the bounds are sharp, choose F exponential. ||

6.13. Coefficient of Variation. If we choose $r = 2$, Theorem 6.12 tells us that for F IFRA (DFRA), $\mu_2 \leq (\geq) 2\mu_1{}^2$, or, equivalently, $\sigma^2 \leq (\geq)\mu_1{}^2$, where $\sigma^2 = \mu_2 - \mu_1{}^2$, denotes the variance of F. Thus for F IFRA (DFRA), the coefficient of variation $\sigma/\mu \leq (\geq) 1$, that is, the distribution is more (less) peaked than the exponential distribution, for which $\sigma/\mu = 1$.

EXERCISES

1. A component having increasing failure rate has a .99 probability of surviving for 1000 hours.
   (a) Give a lower bound on its probability of surviving for 900 hours.
   (b) Give an upper bound on its probability of surviving for 1500 hours.

2. A system is composed of two IFRA components in parallel, each component having .95 probability of surviving for 500 hours. Obtain a lower bound on system survival probability for 400 hours using each of two methods:

   (a) First compute a lower bound for each component. Then compute the parallel system survival probability based on this component lower bound.
   (b) First compute the probability of survival of the parallel system for 500 hours. Then use the lower bound of Theorem 6.1.

   Which method gives a better lower bound?

   Does this comparison generalize from parallel systems to all coherent systems? How do methods (a) and (b) compare when the system is a series system?

3. Let F be a distribution formed by mixing DFRA distributions, $F_1, \ldots, F_n$, each having a .98 probability of surviving 1000 hours. Obtain an upper bound on $\bar{F}(750)$ using each of two methods:

   (a) First compute upper bounds on $\bar{F}_1(750), \ldots, \bar{F}_n(750)$, and then mix upper bounds.
   (b) Mix $\bar{F}_1(1000), \ldots, \bar{F}_n(1000)$ to obtain $\bar{F}(1000)$, and then obtain an upper bound on $\bar{F}(750)$.

   Do the results differ?

4. Let X (Y) have continuous distribution F (G), with G strictly increasing. Then $Y \overset{\text{st}}{=} G^{-1}F(X)$.
5. Show that the lower bound given in (6.2) is sharp for $t > \mu_r^{1/r}$.
6. Let $F(0^-) = 0$ and $\mu_r = \int_0^\infty t^r \, dF(t)$. Then $\mu_r^{1/r}$ is increasing in $r \geq 0$.
7. Assume the system shown in the accompanying diagram. The three components are mutually independent, IFR, with mean lives of 1000, 1200, and 1600 hours, respectively. Obtain a lower bound on system reliability for an 800-hour mission.

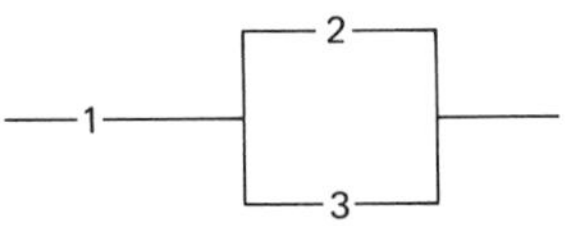

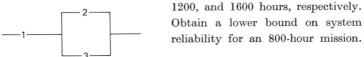

8. Establish that the bound of Theorem 6.9 is sharp.
9. Suppose the mean of the distribution F is known. Compare the upper bounds on $\bar{F}(t)$ in the following two cases:

   (a) F IFR,
   (b) F DFR.

10. Show that the rth moment $\mu_r = \int_0^\infty t^r(1/\mu_1)e^{-t/\mu}\,dt$ of the exponential is given by $\mu_r = \Gamma(r+1)\mu_1{}^r$ for $0 < r < \infty$.

11. Let X be a random variable with distribution F, where $F(0^-) = 0$. Then if EX^r exists and is finite, $EX^r = r\int_0^\infty x^{r-1}\bar{F}(x)\,dx$.

12. Show that if F and G have the same first moment, then they must cross at least once. Show that in general if F and G have the same first s moments, then they must cross at least s times.

*13. Let $F \underset{*}{<} G$ show that

$$\left\{\frac{\int_0^\infty {}^r x\,dG(x)}{\int_0^\infty x^r\,dF(x)}\right\}^{1/r}\uparrow \qquad \text{in} \quad r \geq 0.$$

In particular, if F has mean μ_F and variance $\sigma_F{}^2$, and G has mean μ_G and variance $\sigma_G{}^2$, then

$$\frac{\sigma_F}{\mu_F} \leq \frac{\sigma_G}{\mu_G}.$$

7. MEAN LIFE OF SERIES AND PARALLEL SYSTEMS

In this section we obtain useful bounds and comparisons for the expected life of series and parallel systems under various assumptions on component distributions.

7.1. Lemma. (a) Let $W(x)$ be a Lebesgue-Stieltjes measure, not necessarily positive, for which $\int_t^\infty dW(x) \geq 0$ for all t, and let $h(x) \geq 0$ be increasing. Then

$$\int_{-\infty}^{\infty} h(x)\,dW(x) \geq 0.$$

(b) Let $W(x)$ be a Lebesgue-Stieltjes measure, not necessarily positive, for which $\int_{-\infty}^t dW(x) \geq 0$ for all t, and let $h(x) \geq 0$ be decreasing. Then

$$\int_{-\infty}^{\infty} h(x)\,dW(x) \geq 0.$$

PROOF. (a) The hypothesis says that $\int_{-\infty}^\infty I_{[t,\infty)}(x)\,dW(x) \geq 0$, where $I_{[t,\infty)}$ is the indicator function of the set $[t, \infty)$. Any nonnegative increasing function $h(x)$ can be approximated from below by functions of the form $\sum_{i=1}^n a_i I_{[t_i,\infty)}(x)$, where each $a_i > 0$. The result then follows from the Lebesgue monotone convergence theorem.

(b) The proof is similar. ||

Note that the conclusion of Lemma 7.1(a) can be rewritten as $\int_t^\infty h(x)\,dW(x) \geq 0$ for all t, because the function $h_t(x)$ which is 0 for $x < t$

and $h(x)$ for $x \geq t$ satisfies the conditions of the lemma. Similarly, the conclusion of Lemma 7.1(b) can be rewritten as $\int_{-\infty}^{t} h(x)\, dW(x) \geq 0$ for all t.

The two results can be combined to yield the following corollary.

7.2. Corollary. Let $\int_t^\infty dW(x) \geq 0$ for all t, $\int_{-\infty}^{\infty} dW(x) = 0$, and $h(x)$ be increasing. Then $\int_{-\infty}^{\infty} h(x)\, dW(x) \geq 0$.

We may now study series systems using these tools. We assume in the remainder of this section that F and G (with or without subscripts) are probability distributions such that $F(0) = 0 = G(0)$.

7.3. Theorem. Let $\int_0^t \bar{F}_i(x)\, dx \geq \int_0^t \bar{G}_i(x)\, dx$ for all $t > 0$, $i = 1, 2, \ldots, n$. Then

$$\int_0^t \prod_{i=1}^n \bar{F}_i(x)\, dx \geq \int_0^t \prod_{i=1}^n \bar{G}_i(x)\, dx \qquad \text{for all} \quad t > 0. \tag{7.1}$$

PROOF. It clearly suffices to prove (7.1) for $n = 2$. Note that

$$\int_0^t \bar{F}_1(x)\bar{F}_2(x)\, dx - \int_0^t \bar{G}_1(x)\bar{G}_2(x)\, dx$$
$$= \int_0^t [\bar{F}_1(x) - \bar{G}_1(x)]\bar{F}_2(x)\, dx + \int_0^t [\bar{F}_2(x) - \bar{G}_2(x)]\bar{G}_1(x)\, dx.$$

Each of the last two integrals is nonnegative by Lemma 7.1(b). ||

Similar results hold for parallel systems.

7.4. Theorem. Let $\int_t^\infty \bar{F}_i(x)\, dx \leq \int_t^\infty \bar{G}_i(x)\, dx$ for all $t \geq 0$, $i = 1, \ldots, n$. Then

$$\int_t^\infty \left\{1 - \prod_{i=1}^n F_i(x)\right\} dx \leq \int_t^\infty \left\{1 - \prod_{i=1}^n G_i(x)\right\} dx \qquad \text{for all} \quad t \geq 0. \tag{7.2}$$

PROOF. The proof is similar to that of Theorem 7.3 except that Lemma 7.1(a) is used. ||

7.5. Lemma. Let F and G have the same mean. Suppose that $\bar{F}$ crosses $\bar{G}$ at most once, and that if such a crossing occurs, $\bar{F}$ crosses $\bar{G}$ from above. Then

$$\int_0^t \bar{F}(x)\, dx \geq \int_0^t \bar{G}(x)\, dx \quad \text{and} \quad \int_t^\infty \bar{F}(x)\, dx$$
$$\leq \int_t^\infty \bar{G}(x)\, dx \qquad \text{for all} \quad t \geq 0. \tag{7.3}$$

PROOF. Since F and G have the same mean, $\bar{F}$ must cross $\bar{G}$ at least once; by hypothesis, there is at most one such crossing. Hence $\bar{F}$ crosses $\bar{G}$ exactly once, from above—say, at x_0. Thus $\bar{F}(x) \geq \bar{G}(x)$ for $0 \leq x < x_0$ and $\bar{F}(x) \leq \bar{G}(x)$ for $x_0 \leq x < \infty$. Since $\int_0^\infty \bar{F}(x)\,dx = \int_0^\infty \bar{G}(x)\,dx$, the result follows. ||

We may now apply these results to obtain bounds for the mean life of series and parallel systems.

7.6. Theorem. Let X_i (Y_i), the life length of component i, have distribution F_i (G_i) with mean μ_i, and $F_i \underset{*}{<} G_i$, $i = 1, \ldots, n$. Let $X_1, \ldots, X_n$ ($Y_1, \ldots, Y_n$) be independent. Then the mean life of a series system using components with lives $X_1, \ldots, X_n$ is greater than the corresponding system mean life using components with lives $Y_1, \ldots, Y_n$, that is,

$$E \min(X_1, \ldots, X_n) \geq E \min(Y_1, \ldots, Y_n), \tag{7.4a}$$

while the reverse inequality holds for a parallel system, that is,

$$E \max(X_1, \ldots, X_n) \leq E \max(Y_1, \ldots, Y_n). \tag{7.4b}$$

PROOF OF (7.4a). Clearly, $P[\min(X_1, \ldots, X_n) > t] = \prod_{i=1}^n \bar{F}_i(t)$. By 5.5, $\bar{F}_i$ crosses $\bar{G}_i$ exactly once and from above. By Lemma 7.5, (7.3) holds for each component. By Theorem 7.3, (7.1) follows. Finally, choosing $t = \infty$ in (7.1), we obtain

$$\begin{aligned} E \min(X_1, \ldots, X_n) &= \int_0^\infty P[\min(X_1, \ldots, X_n) > t]\,dt = \int_0^\infty \prod_{i=1}^n \bar{F}_i(t)\,dt \\ &\geq \int_0^\infty \prod_{i=1}^n \bar{G}_i(t)\,dt = \int_0^\infty P[\min(Y_1, \ldots, Y_n) > t]\,dt \\ &= E \min(Y_1, \ldots, Y_n). \end{aligned}$$

PROOF OF (7.4b). The proof is similar:

$$\begin{aligned} E \max(X_1, \ldots, X_n) &= \int_0^\infty P[\max(X_1, \ldots, X_n) > t]\,dt \\ &= \int_0^\infty \left\{1 - \prod_{i=1}^n F_i(t)\right\} dt \leq \int_0^\infty \left\{1 - \prod_{i=1}^n G_i(t)\right\} dt \\ &= \int_0^\infty P[\max(Y_1, \ldots, Y_n) > t]\,dt \\ &= E \max(Y_1, \ldots, Y_n). \quad || \end{aligned}$$

Remark. Actually, the hypothesis of Theorem 7.6 can be weakened. To obtain (7.4a) and (7.4b) it suffices that $X_1, \ldots, X_n$ be merely associated, rather than that $X_1, \ldots, X_n$ be independent. This follows from

$$P[\min(X_1, \ldots, X_n) > t] \geq \prod_{i=1}^{n} P[X_i > t]$$

and

$$P[\max(X_1, \ldots, X_n) > t] \leq 1 - \prod_{i=1}^{n} P[X_i \leq t],$$

a consequence of Theorem 3.1 of Chapter 2.

By choosing G_i exponential with mean μ_i, $i = 1, \ldots, n$, we obtain the following corollary.

7.7. Corollary. Let μ_s (μ_p) be the mean life of a series (parallel) system of n associated components. Let the ith component have an IFRA marginal distribution with mean μ_i. Then

$$\mu_s \geq \int_0^\infty \prod_{i=1}^{n} \bar{G}_i(t)\, dt \equiv \left(\sum_{i=1}^{n} \mu_i^{-1}\right)^{-1}, \tag{7.5a}$$

and

$$\mu_p \leq \int_0^\infty \coprod_{i=1}^{n} \bar{G}_i(t)\, dt, \tag{7.5b}$$

where $\bar{G}_i(t) \equiv e^{-t/\mu_i}$ for $t \geq 0$, $i = 1, \ldots, n$.

Example. In Bazovsky's Example following Theorem 2.11, Chapter 3, we considered a series system of 10 diodes, 4 transistors, 20 resistors, and 10 capacitors, mutually independent and exponentially distributed with respective failure rates: $\lambda_d = .000{,}002$, $\lambda_t = .000{,}010$, $\lambda_r = .000{,}001$, and $\lambda_c = .000{,}002$.

Suppose instead we know only that the components are associated rather than independent, and that the components have IFRA marginal distributions, with respective means: $\mu_d = 500{,}000$, $\mu_t = 100{,}000$, $\mu_r = 1{,}000{,}000$, and $\mu_c = 500{,}000$. Then by Corollary 7.7, a conservative prediction or lower bound for system mean life μ_s is given by

$$\mu_s \geq \left(\sum_{i=1}^{n} \mu_i^{-1}\right)^{-1} = [10(.000{,}002) + 4(.000{,}010) + 20(.000{,}001) + 10(.000{,}002)]^{-1} = 10{,}000.$$

7.8. Corollary. Let μ_s (μ_p) be the mean life of a series (parallel) system of n independent components. Let the ith component have a DFRA marginal distribution with mean μ_i. Then

$$\mu_s \le \int_0^\infty \prod_{i=1}^n \bar{G}_i(t)\,dt = \left(\sum_{i=1}^n \mu_i^{-1}\right)^{-1}, \tag{7.6a}$$

$$\mu_p \ge \int_0^\infty \coprod_{i=1}^n \bar{G}_i(t)\,dt. \tag{7.6b}$$

7.9. Remark. In the special case $\mu_1 = \cdots = \mu_n = \mu$, the bound for μ_p given in (7.5b) and (7.6b) may be expressed in convenient closed form as follows:

$$\int_0^\infty \{1 - (1 - e^{-t/\mu})^n\}\,dt = \mu\left(1 + \frac{1}{2} + \cdots + \frac{1}{n}\right), \tag{7.7}$$

since

$$\int_0^\infty [1 - (1 - e^{-t/\mu})^n]\,dt = \int_0^\infty \sum_{i=0}^{n-1} e^{-t/\mu}(1 - e^{-t/\mu})^i\,dt$$

$$= \mu\left[1 + \sum_{i=1}^{n-1} \frac{1}{i+1}(1 - e^{-t/\mu})^{i+1}\Big|_0^\infty\right] = \mu \sum_{i=1}^n \frac{1}{i}.$$

Using Corollary 7.7, we may obtain bounds for the mean life of a coherent system in terms of the mean lives of components, as stated in Theorem 7.10.

7.10. Theorem. Suppose ϕ is a coherent system with min cut sets $K_1, \ldots, K_k$ and min path sets $P_1, \ldots, P_p$. Assume component lifetimes are associated and marginal distributions are IFRA with means $\mu_1, \ldots, \mu_n$. Let T be the system lifetime. Then

$$\max_{1 \le r \le p}\left[\sum_{i \in P_r} \mu_i^{-1}\right]^{-1} \le ET \le \min_{1 \le s \le k} \int_0^\infty \coprod_{i \in K_s} \bar{G}_i(t)\,dt,$$

where $\bar{G}_i(t) = e^{-t/\mu_i}$ for $i = 1, \ldots, n$.

The proof is left for Exercise 7.

EXERCISES

1. Give a detailed proof of part (b) of Lemma 7.1.
2. Let F_i be the gamma (Weibull) distribution with shape parameter α_i, G_i the gamma (Weibull) distribution with shape parameter $\beta_i \le \alpha_i$, with both distributions having the same mean μ_i, $i = 1, \ldots, n$. Then

$$\int_0^\infty \prod_{i=1}^n \bar{F}_i(x)\,dx \ge \int_0^\infty \prod_{i=1}^n \bar{G}_i(x)\,dx.$$

3. Let G have decreasing density with mean μ. Obtain an upper bound for $\int_0^\infty \bar{G}^n(t)\,dt$ and a lower bound for $\int_0^\infty [1 - G^n(t)]\,dt$.

*4. Theorem 14 of Bessler and Veinott (1966) states that

$$\int_t^\infty \bar{F}_i(x)\,dx \le \int_t^\infty \bar{G}_i(x)\,dx \qquad \text{for all } t, \quad i = 1, \ldots, n$$

$$\Leftrightarrow \int \cdots \int \phi(x_1, \ldots, x_n)\, dF_1(x_1) \cdots dF_n(x_n)$$

$$\le \int \cdots \int \phi(x_1, \ldots, x_n)\, dG_1(x_1) \cdots dG_n(x_n)$$

for every increasing convex function ϕ. Using this result, prove the following:

(a) Theorem 7.4;
(b) $\int_t^\infty \bar{F}_i(x)\,dx \le \int_t^\infty \bar{G}_i(x)\,dx$ for $t \ge 0$, $i = 1, \ldots, n \Rightarrow \int_t^\infty \bar{F}(x)\,dx \le \int_t^\infty \bar{G}(x)\,dx$, where $F = F_1 * \cdots * F_n$ and $G = G_1 * \cdots * G_n$.

5. Prove Corollary 7.2.
6. Provide a counterexample to the following statement: Let life distribution F_i have greater mean than life distribution G_i. Then a series system of independent components with respective life distributions $F_1, \ldots, F_n$ has greater mean life than a series system of independent components with respective life distributions $G_1, \ldots, G_n$.
7. Prove Theorem 7.10.

8. NOTES AND REFERENCES

Section 2. This section is based largely on the results of Birnbaum-Esary-Marshall (1966). However, the simpler proof of the IFRA Closure Theorem 2.6 is due to Ross (1972).

Example 2.1 of a coherent system of IFR components whose life distribution is not IFR was first presented in Esary-Proschan (1963b). Properties of star-shaped functions and their relationships with other classes of functions are studied in Bruckner and Ostrow (1962). Asymptotically minimax tests for exponentiality versus IFRA (DFRA) alternatives are discussed in Chapter 5 of Barlow, Bartholomew, Bremner, and Brunk (1972).

Section 3. The results of this section are based on a part of Esary-Marshall-Proschan (1973). The shock models discussed in this paper furnish further corroboration of the fundamental role that the IFR, DFR, IFRA, and DFRA classes play in reliability theory.

The theory of total positivity has been applied in a variety of fields such as inventory theory, econometrics, mechanics, statistics, various other branches of mathematics, as well as reliability theory. An authoritative comprehensive treatment of the subject is given by Karlin (1968), who is largely responsible for the development of the field. A generalization of the result stated in Exercise 5 and its ramifications are presented in Karlin-Proschan (1960). Further generalizations to Markov processes and other stochastic processes appear in Karlin (1964).

The shock models discussed in this section and again in Section 2 of Chapter 6 may also be viewed as models for use in risk and actuarial analysis. See Buhlmann (1970) and Seal (1969). To interpret (3.1) as an equation in risk analysis, identify $e^{-\lambda t}[(\lambda t)^k/k!]$ as the Poisson probability of k claims in $[0, t]$, F as the distribution of the amount of a single claim, and $\bar{H}(t)$ as the probability that total claims during $[0, t]$ do not exceed the insurance company's capital available x.

Section 4. The results concerning preservation of IFR and DFR under appropriate reliability operations first appeared in Barlow-Marshall-Proschan (1963). The proof of preservation of DFR and DFRA under mixtures presented here is based on Esary-Marshall-Proschan (1970). A somewhat different proof for the DFR case is given in Barlow-Marshall-Proschan (1963). Estimation procedures for IFR and DFR distributions are discussed in Chapter 5 of Barlow, Bartholomew, Bremner, and Brunk (1972).

Section 5. Convex ordering was introduced by Van Zwet (1964); moment inequalities were developed and applications to statistics were shown. Star-shaped ordering is discussed in Barlow-Proschan (1966a, b) and Marshall-Olkin-Proschan (1967). The superadditive ordering of Exercise 5 is new. Some of its properties are used in Hollander-Proschan (1972) in testing to determine whether a new device has stochastically greater life length than a used device. The results of Exercises 6 through 9 are based on Esary-Proschan (1963b).

Section 6. The lower bound on reliability given in 6.7 is proved directly using Jensen's inequality in B-P (1965), Chapter 2. The moment inequalities of Theorem 6.12 were originally obtained under the stronger hypothesis that the underlying density is PF_2 (Karlin-Proschan-Barlow, 1961). The result is given in Exercise 13 and its ramifications are developed in Marshall-Olkin-Proschan (1967).

Section 7. Lemma 7.1 has been known for a long time. A discrete version is given in Hardy-Littlewood-Pólya (1952), p. 89. A more general result is given in Karlin-Novikoff (1963), p. 1264. Results on the expected life of series or parallel systems of the type presented in this section are given in B-P (1965), Chapter 2. Solovyev-Ushakov (1967) obtain further results of this type. The material presented in this section is based on Marshall-Proschan (1970).

5

MULTIVARIATE DISTRIBUTIONS FOR DEPENDENT COMPONENTS

In most reliability analyses, components are assumed to have independent life distributions. However, as pointed out in the discussion of association (Section 2 of Chapter 2), in many reliability situations it is more realistic to assume some form of positive dependence among components. This positive dependence among component life lengths arises from common environmental stresses and shocks, from components depending on common sources of power, and so on.

In this chapter we consider several basic multivariate parametric families of distributions such as the multivariate exponential, Weibull, gamma, and normal distributions, and shock models that give rise to them. In addition, we study various notions of positive dependence and the relationships among them. Finally, we give a multivariate version of increasing failure rate distribution retaining certain essential properties of the univariate case.

Until recently, the study of bivariate distributions was mostly confined to the normal case. Partly for this reason, general properties of bivariate distributions are not often discussed in textbooks. Since we will be using these properties in discussing the bivariate exponential and other models, we summarize these properties now.

(T_1, T_2) will be jointly distributed with bivariate distribution $F(t_1, t_2)$ and marginals $F_1(t_1)$, $F_2(t_2)$ if

$$F(t_1, \infty) = F_1(t_1); F(\infty, t_2) = F_2(t_2), \tag{0.1}$$

$$\begin{aligned} F(-\infty, t_2) = F(t_1, -\infty) = F(-\infty, -\infty) = 0, \\ F(\infty, \infty) = 1, \end{aligned} \tag{0.2}$$

and the content of rectangles $[x_1, x_2] \times [y_1, y_2]$ are nonnegative; that is,

$$P[x_1 \le T_1 \le x_2, y_1 \le T_2 \le y_2] \\ = F(x_2, y_2) - F(x_2, y_1) - F(x_1, y_2) + F(x_1, y_1) \ge 0. \quad (0.3)$$

If the partial derivative $\partial^2 F(t_1, t_2)/\partial t_1 \, \partial t_2$ exists everywhere, the bivariate distribution is said to have a density $f(t_1, t_2)$ defined by

$$f(t_1, t_2) = \frac{\partial^2 F(t_1, t_2)}{\partial t_1 \, \partial t_2}. \quad (0.4)$$

In this case, (0.3) holds if and only if $f(t_1, t_2) \ge 0$. The conditions (0.1), (0.2), and (0.3) are necessary and sufficient conditions that $F(t_1, t_2)$ be a bivariate probability distribution function with marginals $F_1(t_1)$ and $F_2(t_2)$.

The random variables (T_1, T_2) are independent if and only if for all t_1, t_2,

$$F(t_1, t_2) = F_1(t_1)F_2(t_2). \quad (0.5)$$

The joint probability

$$P[T_1 > t_1, T_2 > t_2] = \bar{F}(t_1, t_2) = 1 - F_1(t_1) - F_2(t_2) + F(t_1, t_2). \quad (0.6)$$

Even if the conditions (0.1), (0.2), and (0.3) are met, a bivariate probability function is *not* uniquely determined by its marginals. On the contrary, Fréchet (1951) has shown that there is an infinite number of solutions to the problem of determining a distribution from its marginals. Fréchet obtained the condition

$$\max[F_1(t_1) + F_2(t_2) - 1, 0] \le F(t_1, t_2) \le \min[F_1(t_1), F_2(t_2)]. \quad (0.7)$$

These upper and lower bounds are themselves bivariate distributions with the given marginals, and so constitute solutions to the problem.

1. THE BIVARIATE EXPONENTIAL DISTRIBUTION

First we consider a bivariate exponential distribution for the life lengths of two *nonindependent* components. Suppose three independent sources of shocks are present in the environment. A shock from source 1 destroys component 1; it occurs at a random time U_1, where $P[U_1 > t] = e^{-\lambda_1 t}$. A shock from source 2 destroys component 2; it occurs at a random time U_2, where $P[U_2 > t] = e^{-\lambda_2 t}$. Finally, a shock from source 3 destroys *both* components; it occurs at a random time U_{12}, where $P[U_{12} > t] = e^{-\lambda_{12} t}$. Thus the random life length T_1 of component 1 satisfies

$$T_1 = \min(U_1, U_{12}),$$

while the random life length T_2 of component 2 satisfies

$$T_2 = \min(U_2, U_{12}).$$

Hence the joint survival probability

$$\bar{F}(t_1, t_2) = P[T_1 > t_1, T_2 > t_2] = e^{-\lambda_1 t_1 - \lambda_2 t_2 - \lambda_{12} \max(t_1, t_2)} \tag{1.1}$$

for $t_1 \geq 0$, $t_2 \geq 0$.

The joint distribution $F(t_1, t_2)$ with survival probability given by (1.1) is called the bivariate exponential distribution (BVE). Note that, conversely, if T_1, T_2 have BVE survival probability given by (1.1), then there exist independent exponential random variables U_1, U_2, U_{12} such that $T_1 = \min(U_1, U_{12})$ and $T_2 = \min(U_2, U_{12})$.

The BVE has exponential marginal distributions with survival probabilities given by:

$$\begin{aligned} \bar{F}_1(t_1) &= P[T_1 > t_1] = e^{-(\lambda_1 + \lambda_{12})t_1} \quad \text{for} \quad t_1 \geq 0, \\ \bar{F}_2(t_2) &= P[T_2 > t_2] = e^{-(\lambda_2 + \lambda_{12})t_2} \quad \text{for} \quad t_2 \geq 0. \end{aligned} \tag{1.2}$$

Note that T_1 and T_2 are associated random variables since they are increasing functions of U_1, U_2, U_{12}. (See Theorem 2.2 and property P_3 of Section 2, Chapter 2.) It follows by Theorem 3.1 of Chapter 2 that

$$\bar{F}(t_1, t_2) \geq \bar{F}_1(t_1)\bar{F}_2(t_2), \tag{1.3a}$$

$$F(t_1, t_2) \geq F_1(t_1)F_2(t_2). \tag{1.3b}$$

Actually, (1.3a) may be verified directly from (1.1) and (1.2). The reader is asked to verify that (1.3a) and (1.3b) are equivalent in Exercise 3 at the end of this section.

The univariate exponential distribution is characterized by a property which makes it of great interest in reliability theory. If T is the life length of a component having exponential life distribution, then

$$P[T > s + t \mid T > s] = P[T > t] \qquad \text{for all} \quad s \geq 0, \quad t \geq 0;$$

this means that *the probability of survival for an additional t units of time for a component of age s is the same as that of a new component*. See Section 1 of Chapter 3. We may verify immediately that the bivariate exponential distribution (1.1) enjoys the corresponding bivariate property:

$$P[T_1 > s_1 + t, T_2 > s_2 + t \mid T_1 > t, T_2 > t] = P[T_1 > s_1, T_2 > s_2] \tag{1.4}$$

for all $s_1 \geq 0$, $s_2 \geq 0$, $t \geq 0$. Equation (1.4) asserts that *the joint survival probability of a pair of components each of age t is the same as that of a pair of new components*.

In terms of the joint survival probability $\bar{F}$ given in (1.1), we may write equivalently:

$$\bar{F}(s_1 + t, s_2 + t) = \bar{F}(s_1, s_2)\bar{F}(t, t) \tag{1.5}$$

for all $s_1 \geq 0$, $s_2 \geq 0$, $t \geq 0$.

Next we show that if a bivariate survival probability $\overline{F}$ with exponential marginals satisfies (1.5), then it *must* be of the form (1.1). We first show the following lemma.

1.1. Lemma. If (1.5) holds, then

$$\overline{F}(x, y) = \begin{cases} e^{-\theta y}\overline{F}_1(x - y) & \text{for } x \geq y, \\ e^{-\theta x}\overline{F}_2(y - x) & \text{for } x \leq y, \end{cases} \tag{1.6}$$

for some $\theta > 0$, where $F_1(t) = F(t, \infty)$ and $F_2(t) = F(\infty, t)$ are the marginal distributions.

PROOF. Setting $s_1 = s = s_2$ in (1.5) yields $\overline{F}(s + t, s + t) = \overline{F}(s, s)\overline{F}(t, t)$, which implies that $\overline{F}(s, s) = e^{-\theta s}$ for some $\theta > 0$. (See Theorem 2.2, Chapter 3.) Next set $s_2 = 0$ in (1.5) to get

$$\overline{F}(s_1 + t, t) = \overline{F}(s_1, 0)\overline{F}(t, t) = \overline{F}_1(s_1)e^{-\theta t}.$$

Finally, let $x = s_1 + t$, $y = t$, so that

$$\overline{F}(x, y) = e^{-\theta y}\overline{F}_1(x - y) \quad \text{for } x \geq y.$$

By a similar argument, we see that

$$\overline{F}(x, y) = e^{-\theta x}\overline{F}_2(y - x) \quad \text{for } x \leq y. \ \|$$

1.2. Theorem. The BVE is the only bivariate distribution $F(x, y)$ with exponential marginals satisfying (1.5).

PROOF. Let $\overline{F}_1(x) = e^{-\delta_1 x}$ and $\overline{F}_2(x) = e^{-\delta_2 x}$ for some $\delta_1 > 0$, $\delta_2 > 0$. By Lemma 1.1,

$$\overline{F}(x, y) = \begin{cases} e^{-\theta y - \delta_1(x-y)} & \text{for } x \geq y, \\ e^{-\theta x - \delta_2(y-x)} & \text{for } x \leq y, \end{cases} \tag{1.7}$$

and some $\theta > 0$. Since $\overline{F}(x, y)$ is decreasing in y, then $\theta \geq \delta_1$. Thus we may let $\lambda_2 = \theta - \delta_1$ and be assured $\lambda_2 \geq 0$. Since $\overline{F}(x, y)$ is decreasing in x, then $\theta \geq \delta_2$. Thus we may let $\lambda_1 = \theta - \delta_2$ and be assured $\lambda_1 \geq 0$.

Next, let $\lambda_{12} = \delta_1 + \delta_2 - \theta$. To insure $\lambda_{12} > 0$, we must show that $\delta_1 + \delta_2 \geq \theta$. Consider the univariate distribution

$$G(x) = F(x, x) = 1 - e^{-\delta_1 x} - e^{-\delta_2 x} + e^{-\theta x}$$

[see Equation (0.6)], with density

$$g(x) = \delta_1 e^{-\delta_1 x} + \delta_2 e^{-\delta_2 x} - \theta e^{-\theta x} \geq 0.$$

Letting x tend to zero, we see that $\lambda_{12} = \delta_1 + \delta_2 - \theta \geq 0$, as required.

From the choice

$$\lambda_1 = \theta - \delta_2, \lambda_2 = \theta - \delta_1, \quad \text{and} \quad \lambda_{12} = \delta_1 + \delta_2 - \theta,$$

we obtain

$$\theta = \lambda_1 + \lambda_2 + \lambda_{12}, \delta_1 = \lambda_1 + \lambda_{12}, \quad \text{and} \quad \delta_2 = \lambda_2 + \lambda_{12}.$$

Substituting in (1.7), we obtain the BVE given in (1.1). ‖

1.3. Remark. An alternate interpretation of the functional equation (1.5) may be obtained by rewriting it as

$$\frac{\bar{F}(s_1 + t, s_2 + t)}{\bar{F}(s_1, s_2)} = \bar{F}(t, t). \tag{1.8}$$

Equation (1.8) states that the survival probability of a series system of two components of ages s_1 and s_2 respectively is the same as that of a new system. Thus we conclude that *the life distribution of a series system of two used components each having marginal exponential life distribution is independent of component ages if and only if the joint distribution of the two components is BVE.*

The following theorem presents some interesting properties of the BVE. We will use them to estimate parameters in the planned volume on reliability inference.

1.4. Theorem. Let T_1, T_2 have the BVE survival probability given in (1.1). Then

(a) $P[\min(T_1, T_2) \le t] = 1 - e^{-\lambda t}$ for $t \ge 0$, where $\lambda = \lambda_1 + \lambda_2 + \lambda_{12}$.
(b) $\min(T_1, T_2)$ is independent of each of the events $[T_1 < T_2]$, $[T_1 > T_2]$, and $[T_1 = T_2]$.
(c) $\min(T_1, T_2)$ is independent of $|T_1 - T_2| \equiv \max(T_1, T_2) - \min(T_1, T_2)$.

PROOF. (a) Note that $\min(T_1, T_2) = \min(U_1, U_2, U_{12})$, where the U's are the (independent) times to shock from the three sources. Thus (a) follows.

(b) Let $h(t) = \lambda e^{-\lambda t}$, the density of $\min(T_1, T_2)$. Using the representation $T_1 = \min(U_1, U_{12})$ and $T_2 = \min(U_2, U_{12})$, we see that, neglecting terms of higher order,

$$\begin{aligned} h(t \mid T_1 < T_2)\, dt &= \frac{P[t < T_1 < t + dt < T_2]}{P[T_1 < T_2]} \\ &= \frac{P[t < U_1 < t + dt < \min(U_2, U_{12})]}{P[U_1 < \min(U_2, U_{12})]} \\ &= \frac{\lambda_1 e^{-\lambda_1 t} e^{-(\lambda_2 + \lambda_{12})t}}{\lambda_1/\lambda}\, dt, \end{aligned}$$

where the value of the denominator is established in Exercise 1. Thus

$$h(t \mid T_1 < T_2) = \lambda e^{-\lambda t} = h(t).$$

In a similar fashion, we may show that

$$h(t \mid T_2 < T_1) = h(t).$$

Finally, neglecting terms of higher order,

$$h(t \mid T_1 = T_2)\, dt = \frac{P[t < U_{12} < t + dt < \min(U_1, U_2)]}{P[U_{12} < \min(U_1, U_2)]}$$

$$= \frac{\lambda_{12} e^{-\lambda_{12} t} e^{-(\lambda_1 + \lambda_2) t}}{\lambda_{12}/\lambda}\, dt,$$

so that

$$h(t \mid T_1 = T_2) = h(t).$$

(c) Compute

$$\begin{aligned}
P[\min(T_1, T_2) \le t_1 \quad &\text{and} \quad |T_1 - T_2| > t_2] \\
&= P[T_1 \le t_1 \quad \text{and} \quad T_2 > t_2 + T_1] + P[T_2 \le t_1 \quad \text{and} \quad T_1 > t_2 + T_2] \\
&= P[U_1 \le t_1 \quad \text{and} \quad \min(U_2, U_{12}) > t_2 + U_1] \\
&\quad + P[U_2 \le t_1 \quad \text{and} \quad \min(U_1, U_{12}) > t_2 + U_2] \\
&= \int_0^{t_1} \lambda_1 e^{-\lambda_1 \tau_1} e^{-(\lambda_2 + \lambda_{12})(t_2 + \tau_1)}\, d\tau_1 + \int_0^{t_1} \lambda_2 e^{-\lambda_2 \tau_2} e^{-(\lambda_1 + \lambda_{12})(t_2 + \tau_2)}\, d\tau_2 \\
&= (1 - e^{-\lambda t_1}) \left[\frac{\lambda_1}{\lambda} e^{-(\lambda_2 + \lambda_{12}) t_2} + \frac{\lambda_2}{\lambda} e^{-(\lambda_1 + \lambda_{12}) t_2} \right] \\
&= P[\min(T_1, T_2) \le t_1] \{ P[T_1 < T_2] P[T_2 - T_1 > t_2 \mid T_1 < T_2] \\
&\qquad + P[T_2 < T_1] P[T_1 - T_2 > t_2 \mid T_2 < T_1] \} \\
&= P[\min(T_1, T_2) \le t_1] P[|T_1 - T_2| > t_2]. \quad \|
\end{aligned}$$

Dependence of One Variable on the Other

Given that T_1, T_2 have the BVE distribution (1.1), then the conditional survival probability $P[T_2 > t_2 \mid T_1 = t_1]$ may be computed as

$$P[T_2 > t_2 \mid T_1 = t_1] = \begin{cases} e^{-\lambda_2 t_2} & \text{for } t_1 > t_2, \\ \dfrac{\lambda_1}{\lambda_1 + \lambda_{12}} e^{-\lambda_{12}(t_2 - t_1) - \lambda_2 t_2} & \text{for } t_1 \le t_2. \end{cases} \tag{1.9}$$

(See Exercise 9.) Note that this conditional survival probability is increasing in t_1. We say that T_2 is *stochastically increasing* in T_1. In Section 4 we discuss this and other concepts of bivariate dependence more fully.

From (1.9) we may also compute the regression $E[T_2 \mid T_1 = t_1]$ of T_2 on T_1 as

$$E[T_2 \mid T_1 = t_1] = \frac{1}{\lambda_2} - \frac{\lambda\lambda_{12}e^{-\lambda_2 t_1}}{\lambda_2(\lambda_1 + \lambda_{12})(\lambda_2 + \lambda_{12})}, \tag{1.10}$$

a concave increasing function of t_1. (See Exercise 9.)

Decomposition of the BVE Distribution into Absolutely Continuous and Singular Parts

The BVE F has an absolutely continuous part F_a and a singular part F_s, but no discrete part. Thus we may write

$$\bar{F}(t_1, t_2) = \alpha\bar{F}_a(t_1, t_2) + (1 - \alpha)\bar{F}_s(t_1, t_2) \qquad \text{for all} \quad t_1, \quad t_2 \geq 0,$$

where $0 \leq \alpha \leq 1$, F_a is absolutely continuous, and $\partial^2 F_s/\partial t_1\, \partial t_2 = 0$ for almost all (t_1, t_2) with respect to two-dimensional Lebesgue measure. In the next theorem we determine α, $\bar{F}_a$, and $\bar{F}_s$ in terms of the parameters λ_1, λ_2, λ_{12} of the BVE F.

1.5. Theorem. If $\bar{F}$ is the BVE of (1.1) and $\lambda = \lambda_1 + \lambda_2 + \lambda_{12}$, then

$$\bar{F}(t_1, t_2) = \frac{\lambda_1 + \lambda_2}{\lambda}\bar{F}_a(t_1, t_2) + \frac{\lambda_{12}}{\lambda}\bar{F}_s(t_1, t_2), \tag{1.11}$$

where

$$\bar{F}_s(t_1, t_2) = e^{-\lambda \max(t_1, t_2)}$$

is singular, and

$$\bar{F}_a(t_1, t_2) = \frac{\lambda}{\lambda_1 + \lambda_2} e^{-\lambda_1 t_1 - \lambda_2 t_2 - \lambda_{12}\max(t_1, t_2)} - \frac{\lambda_{12}}{\lambda_1 + \lambda_2} e^{-\lambda \max(t_1, t_2)}$$

is absolutely continuous.

PROOF. We may write

$$\begin{aligned} \bar{F}(t_1, t_2) = {} & P[T_1 > t_1, T_2 > t_2 \mid U_{12} > \min(U_1, U_2)] \\ & \times P[U_{12} > \min(U_1, U_2)] \\ & + P[T_1 > t_1, T_2 > t_2 \mid U_{12} \leq \min(U_1, U_2)] \\ & \times P[U_{12} \leq \min(U_1, U_2)]. \end{aligned} \tag{1.12}$$

By Exercise 1,

$$P[U_{12} \leq \min(U_1, U_2)] = \frac{\lambda_{12}}{\lambda},$$

so that

$$P[U_{12} > \min(U_1, U_2)] = \frac{\lambda_1 + \lambda_2}{\lambda}.$$

Also

$$P[T_1 > t_1, T_2 > t_2 \mid U_{12} \le \min(U_1, U_2)]$$
$$= P[\min(U_1, U_{12}) > t_1, \min(U_2, U_{12}) > t_2 \mid U_{12} \le \min(U_1, U_2)]$$
$$= P[U_{12} > \max(t_1, t_2) \mid U_{12} \le \min(U_1, U_2)] = e^{-\lambda \max(t_1, t_2)},$$

since $\min(U_1, U_2, U_{12})$ is an exponential random variable with failure rate λ.
Thus (1.11) holds, where we identify $\bar{F}_s(t_1, t_2)$ as

$$P[T_1 > t_1, T_2 > t_2 \mid U_{12} \le \min(U_1, U_2)] \quad \text{and} \quad \bar{F}_a(t_1, t_2)$$
$$\text{as} \quad P[T_1 > t_1, T_2 > t_2 \mid U_{12} > \min(U_1, U_2)].$$

Since we know all the terms in (1.12) except $\bar{F}_a(t_1, t_2)$, we obtain $\bar{F}_a(t_1, t_2)$ by subtraction, thus yielding the conclusion. ||

EXERCISES

1. Let $Y_1, \ldots, Y_n$ be independent exponential random variables with respective failure rates $\lambda_1, \ldots, \lambda_n$. Then

$$P[Y_1 \le \min(Y_2, \ldots, Y_n)] = \frac{\lambda_1}{\lambda_1 + \cdots + \lambda_n}.$$

2. Verify (1.3b) from (1.1) and (1.2).
3. Let F be a bivariate distribution with marginals F_1 and F_2. Then

$$\bar{F}(x, y) - \bar{F}_1(x)\bar{F}_2(y) = F(x, y) - F_1(x)F_2(y).$$

   Using this result, show that (1.3a) implies (1.3b).
4. Show that the only bivariate distribution satisfying

$$\bar{F}(s_1 + t_1, s_2 + t_2) = \bar{F}(s_1, s_2)\bar{F}(t_1, t_2)$$

   for all $s_1, s_2, t_1, t_2 \ge 0$ is the joint distribution of independent exponential random variables.
5. Using the marginal distributions of the BVE, show that

$$ET_1 = \frac{1}{\lambda_1 + \lambda_{12}}, \qquad \text{var } T_1 = \frac{1}{(\lambda_1 + \lambda_{12})^2},$$
$$ET_2 = \frac{1}{\lambda_2 + \lambda_{12}}, \qquad \text{var } T_2 = \frac{1}{(\lambda_2 + \lambda_{12})^2}.$$

6. The following result on integration by parts can be used to obtain the moments of the BVE. Let F and G be bivariate distributions such that

$G(0, y) \equiv 0 \equiv G(x, 0)$ and G is of bounded variation on finite intervals. Then

$$\int_0^\infty \int_0^\infty G(x, y)\, dF(x, y) = \int_0^\infty \int_0^\infty \bar{F}(x, y)\, dG(x, y). \tag{1.13}$$

7. Using the method of Exercise 6, show that the moment generating function for the BVE of (1.1) is given by

$$\int_0^\infty \int_0^\infty e^{-z_1 t_1 - z_2 t_2}\, dF(t_1, t_2)$$
$$= \frac{(\lambda + z_1 + z_2)(\lambda_1 + \lambda_{12})(\lambda_2 + \lambda_{12}) + z_1 z_2 \lambda_{12}}{(\lambda + z_1 + z_2)(\lambda_1 + \lambda_{12} + z_1)(\lambda_2 + \lambda_{12} + z_2)}.$$

8. (a) From (1.13) show that $\iint x^i y^j\, dF(x, y) = \iint ij x^{i-1} y^{j-1} \bar{F}(x, y)\, dx\, dy$ for $i, j > 0$.

   (b) Using (1.13), show that for i and j positive integers, the moments of the BVE are given by

$$EX^i Y^j = j\Gamma(i + 1) \sum_{k=0}^{i-1} \frac{\Gamma(j + k)}{\Gamma(k + 1)(\lambda_1 + \lambda_{12})^{i-k} \lambda^{j+k}}$$
$$+ i\Gamma(j + 1) \sum_{k=0}^{j-1} \frac{\Gamma(i + k)}{\Gamma(k + 1)(\lambda_2 + \lambda_{12})^{j-k} \lambda^{i+k}}.$$

   (c) Letting $i = 1, j = 1$ in (b), show that for the BVE

$$\operatorname{cov}(X, Y) = \frac{\lambda_{12}}{\lambda(\lambda_1 + \lambda_{12})(\lambda_2 + \lambda_{12})},$$

   and the correlation $\rho(T_1, T_2) = \lambda_{12}/\lambda$.

9. Verify (1.9) and (1.10).

10. Compute the density f_a corresponding to the absolutely continuous distribution F_a given in (1.11).

11. Find a distribution $F(x, y)$ with nonexponential marginals satisfying (1.5).

12. Compute a lower bound on the correlation between two correlated exponential random variables (not necessarily BVE). Use Equation (0.7) and Exercise 4, Section 2 of Chapter 2.

2. SHOCK MODELS YIELDING BIVARIATE DISTRIBUTIONS

Recall that the BVE of (1.1) describes the joint survival probability for the life lengths of two components subject to fatal shocks from three independent sources. The time until the first shock from source i follows an exponential distribution with failure rate λ_i. From a slightly different, though equivalent, point of view, we can consider the shocks from source i as governed by a Poisson process $\{Z_i(t), t \geq 0\}$ with parameter λ_i, where $Z_i(t)$ indicates the number of shocks from source i experienced during $[0, t]$. When $i = 3$, we write λ_{12} and $Z_{12}(t)$.

2.1. The BVE Derived from a Nonfatal Shock Model. Suppose now that the shocks from the three sources are not necessarily fatal. Rather a shock from source 1 causes the failure of component 1 with probability q_1, a shock from source 2 causes the failure of component 2 with probability q_2. Finally, a shock from source 3 causes the failure (a) of both components, with probability q_{11}, (b) of component 1 only, with probability q_{10}, (c) of component 2 only, with probability q_{01}, (d) of neither component, with probability q_{00}, where, of course, $q_{11} + q_{10} + q_{01} + q_{00} = 1$. Assume that each shock represents an independent opportunity for failure.

Then we may write the joint survival probability for T_1, the life length of component 1, and for T_2, the life length of component 2, as

$$P[T_1 > t_1, T_2 > t_2]$$

$$= \left\{\sum_{k=0}^{\infty} e^{-\lambda_1 t_1} \frac{(\lambda_1 t_1)^k}{k!} (1 - q_1)^k\right\} \left\{\sum_{l=0}^{\infty} e^{-\lambda_2 t_2} \frac{(\lambda_2 t_2)^l}{l!} (1 - q_2)^l\right\}$$

$$\times \left\{\sum_{n=0}^{\infty} \sum_{m=0}^{\infty} \left[e^{-\lambda_{12} t_1} \frac{(\lambda_{12} t_1)^m}{m!} q_{00}{}^m\right]\right.$$

$$\left.\times \left[e^{-\lambda_{12}(t_2 - t_1)} \frac{(\lambda_{12}(t_2 - t_1))^n}{n!} (q_{00} + q_{10})^n\right]\right\}$$

when $0 \leq t_1 \leq t_2$. Summing series and simplifying, we obtain

$$P[T_1 > t_1, T_2 > t_2] = e^{-t_1[\lambda_1 q_1 + \lambda_{12} q_{10}] - t_2[\lambda_2 q_2 + \lambda_{12}(1 - q_{00} - q_{10})]}.$$

For $0 \leq t_2 \leq t_1$, by symmetry,

$$P[T_1 > t_1, T_2 > t_2] = e^{-t_1[\lambda_1 q_1 + \lambda_{12}(1 - q_{00} - q_{01}] - t_2[\lambda_2 q_2 + \lambda_{12} q_{01})]}.$$

Combining the two survival probabilities, we obtain the BVE

$$P[T_1 > t_1, T_2 > t_2] = e^{-\lambda_1{}^* t_1 - \lambda_2{}^* t_2 - \lambda_{12}{}^* \max(t_1, t_2)}, \tag{2.1}$$

where

$$\lambda_1{}^* = \lambda_1 q_1 + \lambda_{12} q_{10}, \qquad \lambda_2{}^* = \lambda_2 q_2 + \lambda_{12} q_{01}, \qquad \lambda_{12}{}^* = \lambda_{12} q_{11}.$$

Note that the BVE of (2.1) can be obtained directly by using Exercises 3 and 4 below. Note further that for $q_1 = q_2 = q_{11} = 1$, (2.1) reduces to the fatal shock model with BVE given in (1.1).

2.2 The Bivariate Poisson Process Derived from a Shock Model. Assume the same shock model as just above. But assume now that failed components are immediately replaced. We wish to determine the joint probability $P[N_1(t) = n_1, N_2(t) = n_2]$ that the number $N_i(t)$ of replacements of component i during $[0, t]$ is n_i, $i = 1, 2$.

Note that $Z(t)$, $t \geq 0$, where $Z(t) = Z_1(t) + Z_2(t) + Z_{12}(t)$, is a Poisson process with parameter $\lambda = \lambda_1 + \lambda_2 + \lambda_{12}$; it governs the sequence of shocks of the three types combined. Given that a shock has occurred according to process $\{Z(t)\}$, the probability is

$p_{11} = (\lambda_{12}/\lambda)q_{11}$ that both components are replaced;
$p_{10} = (\lambda_1/\lambda)q_1 + (\lambda_{12}/\lambda)q_{10}$ that component 1 only is replaced;
$p_{01} = (\lambda_2/\lambda)q_2 + (\lambda_{12}/\lambda)q_{01}$ that component 2 only is replaced;
$p_{00} = (\lambda_1/\lambda)(1 - q_1) + (\lambda_2/\lambda)(1 - q_2) + (\lambda_{12}/\lambda)q_{00}$ that neither component is replaced.

Note that $p_{11} + p_{10} + p_{01} + p_{00} = (\lambda_1/\lambda) + (\lambda_2/\lambda) + (\lambda_{12}/\lambda) = 1$.

We may verify immediately that

$$P[N_1(t) = n_1, N_2(t) = n_2]$$

$$= \sum_{m=0}^{\min(n_1, n_2)} \sum_{k=n_1+n_2-m}^{\infty} e^{-\lambda t} \times \frac{(\lambda t)^k}{k!} \frac{k!\, p_{11}{}^m p_{10}^{n_1-m} p_{01}^{n_2-m} p_{00}^{k-n_1-n_2+m}}{m!\,(n_1 - m)!\,(n_2 - m)!\,(k - n_1 - n_2 + m)!}.$$

Performing the inner summation, we obtain the *bivariate Poisson* frequency function

$$P[N_1(t) = n_1, N_2(t) = n_2]$$

$$= e^{-\lambda t(p_{11}+p_{10}+p_{01})} \sum_{m=0}^{\min(n_1, n_2)} \frac{(\lambda t p_{11})^m (\lambda t p_{10})^{n_1-m} (\lambda t p_{01})^{n_2-m}}{m!\,(n_1 - m)!\,(n_2 - m)!}. \tag{2.2}$$

The corresponding bivariate process $\{N_1(t), N_2(t), t \geq 0\}$ is called the *bivariate Poisson process*.

It is easy to verify that the bivariate Poisson distribution specified in (2.2) has Poisson marginal distributions (Exercise 1). Dwass and Teicher (1957) have shown that this is the only bivariate distribution with Poisson marginals that is infinitely divisible. [A distribution F is infinitely divisible if for each positive integer n, we may write $F = F_n{}^{(n)}$, an n-fold convolution of a distribution function F_n.]

In the special case that each shock is fatal [with each shock from process $\{Z_{12}(t), t \geq 0\}$ being fatal to both components], the parameters in (2.2) reduce to $p_{00} = 0$, $p_{01} = \lambda_2/\lambda$, $p_{10} = \lambda_1/\lambda$, $p_{11} = \lambda_{12}/\lambda$.

2.3. A Bivariate Distribution with Gamma Marginals. Using the fatal shock model of Section 1, we may write for $t_1 < t_2$,

$$\begin{aligned}\bar{F}^{(k)}(t_1, t_2) &= P[Z_1(t_1) + Z_{12}(t_1) \le k - 1;\, Z_2(t_2) + Z_{12}(t_2) \le k - 1] \\ &= \sum_{l+m \le k-1} P[Z_{12}(t_1) = l,\, Z_{12}(t_2) = l + m] \\ &\quad \times P[Z_1(t_1) \le k - 1 - l]P[Z_2(t_2) \le k - 1 - l - m] \\ &= \sum_{l+m=0}^{k-1} e^{-\lambda_{12}t_1} \frac{(\lambda_{12}t_1)^l}{l!} e^{-\lambda_{12}(t_2-t_1)} \frac{[\lambda_{12}(t_2 - t_1)]^m}{m!} \sum_{i=0}^{k-1-l} e^{-\lambda_1 t_1} \frac{(\lambda_1 t_1)^i}{i!} \\ &\quad \times \sum_{j=0}^{k-1-l-m} e^{-\lambda_2 t_2} \frac{(\lambda_2 t_2)^j}{j!}.\end{aligned}$$

Thus

$$\begin{aligned}\bar{F}^{(k)}(t_1, t_2) &= e^{-\lambda_1 t_1 - \lambda_2 t_2 - \lambda_{12} t_2} \sum_{l+m=0}^{k-1} \lambda_{12}^{l+m} \frac{t_1{}^l}{l!} \frac{(t_2 - t_1)^m}{m!} \sum_{i=0}^{k-1-l} \frac{(\lambda_1 t_1)^i}{i!} \\ &\quad \times \sum_{j=0}^{k-1-l-m} \frac{(\lambda_2 t_2)^j}{j!}.\end{aligned} \tag{2.3}$$

An alternate form may be obtained by making use of Exercise 5, Section 2, Chapter 3.

$$\begin{aligned}\bar{F}^{(k)}(t_1, t_2) &= e^{-\lambda_{12}t_2} \sum_{l+m=0}^{k-1} \lambda_{12}^{l+m} \frac{t_1{}^l}{l!} \frac{(t_2 - t_1)^m}{m!} \\ &\quad \times \int_{\lambda_1 t_1}^{\infty} \frac{x^{k-1-l}e^{-x}}{(k - 1 - l)!}\, dx \int_{\lambda_2 t_2}^{\infty} \frac{y^{k-1-l-m}e^{-y}}{(k - 1 - l - m)!}\, dy,\end{aligned}$$

where $t_1 < t_2$. The $F^{(k)}$ specified in (2.3) is a *bivariate distribution* with gamma marginals. In the special case in which $k = 2$, (2.3) reduces to:

$$\begin{aligned}\bar{F}^{(2)}(t_1, t_2) = e^{-\lambda_1 t_1 - \lambda_2 t_2 - \lambda_{12} t_2}[(1 + \lambda_1 t_1)(1 + \lambda_2 t_2) \\ + \lambda_{12} t_1 + \lambda_{12}(t_2 - t_1)(1 + \lambda_1 t_1)].\end{aligned}$$

EXERCISES

1. Show that the bivariate Poisson distribution has Poisson marginal distributions.

2. Show that for the bivariate Poisson process

$$P[N_1(t) \le k_1, N_2(t) \le k_2] \ge P[N_1(t) \le k_1]P[N_2(t) \le k_2]$$

and

$$P[N_1(t) > k_1, N_2(t) > k_2] \ge P[N_1(t) > k_1]P[N_2(t) > k_2].$$

(Hint: Use association as defined in Section 2 of Chapter 2.)

3. Formulate a multivariate generalization of the bivariate Poisson process.

3. MULTIVARIATE DISTRIBUTIONS

In this section we consider two interesting families of multivariate distributions, the exponential and the normal.

3.1. The Multivariate Exponential Distribution. To fix ideas, consider first an extension of the fatal shock model to a three-component system. Assume the Poisson process $Z_i(t)$ with rate λ_i governs the occurrence of shocks fatal to component i for $i = 1, 2, 3$, the Poisson process $Z_{ij}(t)$ with rate λ_{ij} governs the occurrence of shocks fatal to components i and j simultaneously for $1 \leq i \leq j \leq 3$, and the Poisson process $Z_{123}(t)$ with rate λ_{123} governs the occurrence of shocks fatal to all three components simultaneously. Assume all the Poisson processes are independent. Let T_i denote the life length of component i for $i = 1, 2, 3$. Then the joint survival probability

$$\begin{aligned}\bar{F}(t_1, t_2, t_3) &= P[T_1 > t_1, T_2 > t_2, T_3 > t_3] \\ &= P[Z_i(t_i) = 0, \qquad i = 1, 2, 3; \\ &\qquad Z_{ij}(\max(t_i, t_j)) = 0, \qquad 1 \leq i \leq j \leq 3; \\ &\qquad Z_{123}(\max(t_1, t_2, t_3)) = 0].\end{aligned}$$

Thus

$$\begin{aligned}&\bar{F}(t_1, t_2, t_3) \\ &\quad = \exp[-\lambda_1 t_1 - \lambda_2 t_2 - \lambda_3 t_3 - \lambda_{12} \max(t_1, t_2) - \lambda_{13} \max(t_1, t_3) \\ &\qquad - \lambda_{23} \max(t_2, t_3) - \lambda_{123} \max(t_1, t_2, t_3)]. \qquad (3.1)\end{aligned}$$

By similar arguments we obtain the n-dimensional *multivariate exponential distribution* (MVE) with joint survival probability:

$$\begin{aligned}\bar{F}(t_1, \ldots, t_n) = \exp\Bigg[&-\sum_{i=1}^{n} \lambda_i t_i - \sum_{i<j} \lambda_{ij} \max(t_i, t_j) \\ &- \sum_{i<j<k} \lambda_{ijk} \max(t_i, t_j, t_k) - \cdots - \lambda_{12\cdots n} \max(t_1, \ldots, t_n)\Bigg]. \qquad (3.2)\end{aligned}$$

By setting $t_i = 0$, we obtain an $(n-1)$-dimensional MVE in the remaining variables. By iterating this process, we see that the marginal distributions of all orders are MVE; in particular, the two-dimensional marginals are BVE, and the one-dimensional marginals are exponential.

In the bivariate case, the BVE results when shocks are nonfatal, as well as fatal. (See 2.1.) In a similar fashion, the MVE may be obtained when shocks are nonfatal. The calculations are tedious but direct.

Recall that in the bivariate case, the BVE is the only joint life distribution with exponential marginals for which the joint survival probability of a pair of components each of age t is the same as that of a pair of new components. Similarly, we now show *that the n-dimensional MVE is the only joint life distribution with (n − 1)-dimensional MVE marginals for which the joint survival probability of a set of n components each of age t is the same as that of a set of new components.*

Assume then that

$$P[T_1 > s_1 + t, \ldots, T_n > s_n + t \mid T_1 > t, \ldots, T_n > t] = P[T_1 > s_1, \ldots, T_n > s_n],$$

or, equivalently, that

$$\bar{F}(s_1 + t, \ldots, s_n + t) = \bar{F}(s_1, \ldots, s_n)\bar{F}(t, \ldots, t). \tag{3.3}$$

Setting $s_1 = \cdots = s_n = s$, we have $\bar{F}(s, \ldots, s) = e^{-\theta s}$ for some $\theta > 0$. Next set $s_n = 0$, yielding

$$\begin{aligned}\bar{F}(s_1 + t, \ldots, s_{n-1} + t, t) &= \bar{F}(s_1, \ldots, s_{n-1}, 0)\bar{F}(t, \ldots, t) \\ &= e^{-\theta t}\bar{F}_{n-1}(s_1, \ldots, s_{n-1}),\end{aligned}$$

where F_{n-1} is an $(n - 1)$-dimensional marginal, and so by hypothesis is MVE. Thus (3.2) holds on the domain $t_i \geq t_n$ for $i = 1, \ldots, n - 1$. By symmetry (3.2) holds for all $t_i \geq 0$, $i = 1, \ldots, n$. Hence F is an n-dimensional MVE. ||

3.2. Remark. Just as in the bivariate case (Remark 1.3), an alternate interpretation of (3.3) is available. *Assume the joint life distribution F of n components has (n − 1)-dimensional marginals which are MVE. Then the life distribution of a series system of these n components is independent of component ages if and only if F is MVE.*

A very important property of the MVE random vector is that it can be represented in terms of independent exponential random variables. As in the bivariate case, we obtain the result from the fatal shock model. If $T_1, \ldots, T_n$ are MVE, then there exist independent exponential random variables $U_{\mathbf{s}}$, $\mathbf{s} \in S$, where S is the set of vectors $\mathbf{s} = (s_1, \ldots, s_n)$ with each $s_j = 0$ or 1, but $(s_1, \ldots, s_n) \neq (0, \ldots, 0)$ such that $T_i = \min_{s_i = 1} U_{\mathbf{s}}$.

3.3. The Associated Multivariate Normal Distribution. All multivariate distributions considered thus far have a singular part, reflecting the fact that two or more components can fail simultaneously with positive probability. However, for some applications, this may not be an appropriate model. Thus we are led to consider the multivariate normal distribution with a positive definite covariance matrix, which does not have a singular part. This generalizes the univariate normal (or truncated normal) distribution which is often used to describe the life of items, such as light bulbs, under close production control. Its multivariate extension (or a truncated version) may likewise be a useful model for component lifetimes in some systems.

Let $\mathbf{T} = (T_1, \ldots, T_n)$ have a multivariate normal distribution with mean vector $\boldsymbol{\mu} = (\mu_1, \ldots, \mu_n)$ and positive definite covariance matrix $\Sigma = (\sigma_{ij})_{i,j=1}^n$. Since Σ is positive definite, $|\Sigma| > 0$ and $\Sigma^{-1} \equiv R = (r_{ij})_{i,j=1}^n$ exists. Thus we may write the density function of $\mathbf{T}$ as

$$f(t_1, \ldots, t_n) = \frac{1}{(2\pi)^{n/2}|\Sigma|^{1/2}} \exp\left[-\frac{1}{2}\sum_{i,j=1}^{n} r_{ij}(t_i - \mu_i)(t_j - \mu_j)\right] \tag{3.4}$$

for $-\infty < t_i < \infty$, $i = 1, \ldots, n$.

As proved in 4.16 below, *multivariate normal* $\mathbf{T}$ *are associated if* $r_{ij} \leq 0$ *for all* $i \neq j$.

3.4. Example: The Bivariate Normal Distribution. The simplest and most popular case corresponds to $n = 2$. Explicitly, we have

$$\boldsymbol{\mu} = (\mu_1, \mu_2), \Sigma = \begin{pmatrix} \sigma_1^2 & \rho\sigma_1\sigma_2 \\ \rho\sigma_1\sigma_2 & \sigma_2^2 \end{pmatrix},$$

and

$$R = \frac{1}{1-\rho^2}\begin{pmatrix} \dfrac{1}{\sigma_1^2} & -\dfrac{\rho}{\sigma_1\sigma_2} \\ -\dfrac{\rho}{\sigma_1\sigma_2} & \dfrac{1}{\sigma_2^2} \end{pmatrix},$$

where σ_i^2 is the variance of T_i, $i = 1, 2$, and ρ is the correlation coefficient. The probability density (3.4) becomes

$$f(t_1, t_2) = \frac{1}{2\pi\sigma_1\sigma_2\sqrt{1-\rho^2}} \exp\left\{-\frac{1}{2(1-\rho^2)}\left[\frac{(t_1-\mu_1)^2}{\sigma_1^2} - 2\rho\frac{(t_1-\mu_1)(t_2-\mu_2)}{\sigma_1\sigma_2} + \frac{(t_2-\mu_2)^2}{\sigma_2^2}\right]\right\} \tag{3.5}$$

for $-\infty < t_1 < \infty$, $-\infty < t_2 < \infty$.

From the definition of association, 2.1 of Chapter 2, we see immediately that if T_1 and T_2 are associated, then $\text{cov}[T_1, T_2] \geq 0$, so that $\rho \geq 0$. Conversely, if $\rho \geq 0$, then T_1 and T_2 are associated, as follows from the fact that $\rho \geq 0 \Leftrightarrow r_{12} = -\rho/[(1-\rho^2)\sigma_1\sigma_2] \leq 0$.

EXERCISES

1. Show that if $T_1, \ldots, T_n$ have the MVE distribution, then $T_1, \ldots, T_n$ are associated.

2. Show that if an n-component coherent system has component lifetimes jointly distributed according to the MVE, then the system life distribution is IFRA. (Recall that the IFRA Closure Theorem in Chapter 4 requires *independent* components.)

4. RELATIONSHIPS AMONG SOME NOTIONS OF MULTIVARIATE DEPENDENCE

We have seen repeatedly that the multivariate notion of association among random variables, introduced in Chapter 2, is a useful concept in reliability theory. Many basic results ordinarily obtained in the case of independent components have been shown to hold in the more general case of associated components. In addition, comparisons have been obtained in which the independent component case represents a bound for the associated components case.

However, the notion of association among random variables is just one among many notions of multivariate dependence. In this section we consider a number of alternative notions of positive dependence and study the relationships among them. In applications it is often easier to verify one of these alternative notions which imply association, rather than to verify association directly.

The Bivariate Case

As might be expected, the notions of dependence are simpler and their relationships are more readily exposed in the bivariate case.

4.1. Definitions. Given random variables S and T, we say the following:

(a) S and T are *positively quadrant dependent* if

$$P[S \leq s, T \leq t] \geq P[S \leq s]P[T \leq t] \quad \text{for all} \quad s, t. \tag{4.1}$$

We write PQD(S, T).

(b) T is *left tail decreasing* in S if

$$P[T \leq t \mid S \leq s] \tag{4.2}$$

is decreasing in s for all t. We write LTD($T \mid S$). Symmetrically, we have:

(c) T is *right tail increasing* in S if

$$P[T > t \mid S > s] \tag{4.3}$$

is increasing in s for all t. We write RTI($T \mid S$).

(d) T is *stochastically increasing* in S if

$$P[T > t \mid S = s] \tag{4.4}$$

is increasing in s for all t. We write SI($T \mid S$).

(e) Let S, T have joint probability density (or, in the discrete case, joint frequency function) $f(s, t)$. Then recall from Definition 3.2 of Chapter 4 that $f(s, t)$ is *totally positive of order* 2 if

$$\begin{vmatrix} f(s_1, t_1) & f(s_1, t_2) \\ f(s_2, t_1) & f(s_2, t_2) \end{vmatrix} \geq 0 \tag{4.5}$$

for all $s_1 < s_2$, $t_1 < t_2$ in the domain of S and T. We write $f(s, t)$ is TP_2 or, alternately, $\mathrm{TP}_2(S, T)$.

We are able to arrange these bivariate notions of positive dependence into a hierarchy as follows. [The notation $\mathrm{A}(S, T)$ signifies that S and T are associated random variables.]

4.2. Theorem. $\mathrm{TP}_2(S, T) \Rightarrow \mathrm{SI}(T \mid S) \Rightarrow \mathrm{RTI}(T \mid S) \Rightarrow \mathrm{A}(S, T) \Rightarrow \mathrm{PQD}(S, T)$. The sequence of implications remains true when $\mathrm{RTI}(T \mid S)$ is replaced by $\mathrm{LTD}(T \mid S)$.

PROOF OF $\mathrm{TP}_2(S, T) \Rightarrow \mathrm{SI}(T \mid S)$. From (4.5), we obtain

$$\begin{vmatrix} \int_t^\infty f(s_1, \tau)\, d\tau & \int_t^\infty f(s_2, \tau)\, d\tau \\ \int_0^t f(s_1, \tau)\, d\tau & \int_0^t f(s_2, \tau)\, d\tau \end{vmatrix} \leq 0,$$

for $s_1 < s_2$. Adding the top row to the bottom row and converting to ratios, we obtain the inequality for conditional probabilities:

$$P[T > t \mid S = s_1] \leq P[T > t \mid S = s_2].$$

Thus $\mathrm{SI}(T \mid S)$ holds.

PROOF OF $\mathrm{SI}(T \mid S) \Rightarrow \mathrm{RTI}(T \mid S)$. From (4.4) we obtain for $s_1' \leq s_2'$:

$$\begin{vmatrix} \int_t^\infty f(s_1', \tau)\, d\tau & \int_t^\infty f(s_2', \tau)\, d\tau \\ f_1(s_1') & f_1(s_2') \end{vmatrix} \leq 0,$$

where $f_1(s)$ is the marginal density of S. Integrating the first column between s_1 and s_2, and the second column between s_2 and ∞, we obtain

$$\begin{vmatrix} \int_{s_1}^{s_2} \int_t^\infty f(s, \tau)\, d\tau\, ds & \int_{s_2}^\infty \int_t^\infty f(s, \tau)\, d\tau\, ds \\ \int_{s_1}^{s_2} f_1(s)\, ds & \int_{s_2}^\infty f_1(s)\, ds \end{vmatrix} \leq 0.$$

Adding the second column to the first yields:

$$\begin{vmatrix} \int_{s_1}^{\infty}\int_{t}^{\infty} f(s, \tau)\, d\tau\, ds & \int_{s_2}^{\infty}\int_{t}^{\infty} f(s, \tau)\, d\tau\, ds \\ \int_{s_1}^{\infty} f_1(s)\, ds & \int_{s_2}^{\infty} f_1(s)\, ds \end{vmatrix} \le 0,$$

which is equivalent to (4.3).

The proof that SI($T \mid S$) $\Rightarrow$ LTD($T \mid S$) is left to Exercise 1.

PROOF OF RTI($T \mid S$) $\Rightarrow$ A(S, T). The proof is long, and so we refer the reader to Esary-Proschan (1972), Lemmas 1 and 2, and the theorem. The proof of LTD($T \mid S$) $\Rightarrow$ A(S, T) is, of course, similar.

Although the implication SI($T \mid S$) $\Rightarrow$ A($S \mid T$) is a consequence of SI($T \mid S$) $\Rightarrow$ RTI($T \mid S$) $\Rightarrow$ A($S \mid T$), we give a direct proof, since it is an implication which finds practical applications in establishing useful inequalities.

PROOF OF SI($T \mid S$) $\Rightarrow$ A($S \mid T$). Let $F_t(s) = P[S \le s \mid T = t]$. Define $h(u, t) = \inf\{s\colon u \le F_t(s)\}$. Since S is stochastically increasing in T, it follows that $h(u, t)$ is increasing in both arguments. Let U be a uniform random variable independent of S. Then $(S, T) \stackrel{\text{st}}{=} (S, h(U, S))$. Since S and U are associated, S and $h(U, S)$ are also associated. Thus A(S, T) follows.

PROOF OF A(S, T) $\Rightarrow$ PQD(S, T). Let $X_1 = 1$ if $S > s$, 0 otherwise; let $X_2 = 1$ if $T > t$, 0 otherwise. Then X_1 and X_2 are increasing functions of S and T, and so are associated by P_3 of Section 2 of Chapter 2. Hence

$$\begin{aligned} 0 \le \operatorname{cov}[X_1, X_2] &= \operatorname{cov}[1 - X_1, 1 - X_2] \\ &= E[1 - X_1)(1 - X_2) - E(1 - X_1)E(1 - X_2) \\ &= P[S \le s, T \le t] - P[S \le s]P[T \le t]. \; \| \end{aligned}$$

Remark. If S and T are binary variables, then all of the above conditions for bivariate dependence are equivalent. To verify this, it suffices to show that PQD(S, T) $\Rightarrow$ TP_2(S, T), or, equivalently, that

$$D \equiv \begin{vmatrix} P[S = 0, T = 0] & P[S = 0, T = 1] \\ P[S = 1, T = 0] & P[S = 1, T = 1] \end{vmatrix} \ge 0.$$

But, adding the bottom row to the top row, and then the second column to the first column, yields,

$$D = \begin{vmatrix} 1 & P[T = 1] \\ P[S = 1] & P[S = 1, T = 1] \end{vmatrix},$$

which is nonnegative when PQD(S, T) holds.

More generally, the implications of Theorem 4.2 are strict, that is, no two of the conditions are equivalent. The verification of this statement is left to Exercise 2.

Next we define an additional notion of bivariate dependence which does not fit directly into the hierarchy described in Theorem 4.2, but whose multivariate version has been used in formulating the concept of a multivariate increasing failure rate distribution. (See Harris, 1970.)

4.3. Definition. Random variables S and T are said to be *right corner set increasing* if $P[S > s,\ T > t \mid S > s', T > t']$ is increasing in s' and t' for each fixed s, t. We write $\mathrm{RCSI}(S, T)$.

4.4. Theorem. $\mathrm{TP}_2(S, T) \Rightarrow \mathrm{RCSI}(S, T) \Rightarrow \mathrm{RTI}(T \mid S)$ and $\mathrm{RTI}(S \mid T)$.

PROOF OF $\mathrm{TP}_2(S, T) \Rightarrow \mathrm{RCSI}(S, T)$. Let

$$P(s', t') = P[S > s, T > t \mid S > s', T > t'].$$

Note that if $s' < s$, $t' < t$, then $P(s', t')$ is increasing in s', t'.

Assume then that $s' \geq s$, $t' < t$. Then $P(s', t') = P[T > t \mid S > s', T > t']$, which is increasing in t' for all (s, t). To show that $P(s', t')\uparrow$ in s', write for $s' < s''$,

$$D = \begin{vmatrix} P[S > s', T > t] & P[S > s'', T > t] \\ P[S > s', T > t'] & P[S > s'', T > t'] \end{vmatrix}$$

$$= \begin{vmatrix} P[s' < S \leq s'', T > t] & P[S > s'', T > t] \\ P[s' < S \leq s, t' < T \leq t] & P[S > s'', t' < T \leq t] \end{vmatrix} \leq 0$$

since $\mathrm{TP}_2(S, T)$.

A similar argument applies for $s' < s$, $t' \geq t$.

Finally, note that if $s' \geq s$, $t' \geq t$, then $P(s', t') = 1$, and so is increasing in s', t'.

PROOF OF $\mathrm{RCSI}(S, T) \Rightarrow \mathrm{RTI}(T \mid S)$. Choose $s = -\infty, t' = -\infty$. Then $P(s', t') = P[T > t \mid S > s']$ is increasing in s'.

PROOF OF $\mathrm{RCSI}(S, T) \Rightarrow \mathrm{RTI}(S \mid T)$. The proof is by symmetry.

Although both RCSI and SI lie between TP_2 and RTI, neither RCSI nor SI implies the other. See Exercise 17.

In Figure 5.4.1, we summarize the implications among the notions of bivariate dependence presented in this section. $\mathrm{I}(S, T)$ means that random variables S and T are independent.

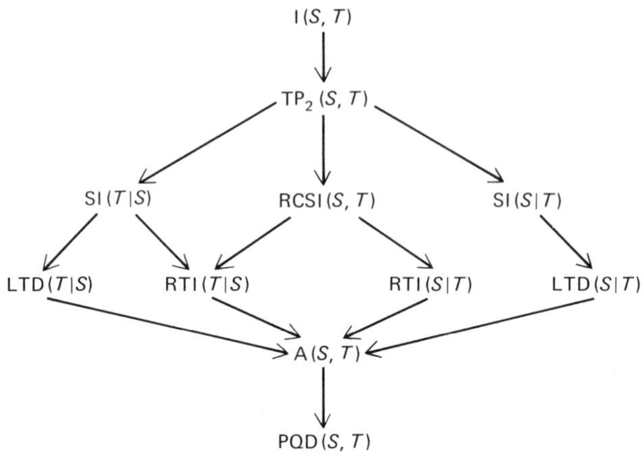

Figure 5.4.1.
Implications among notions of bivariate dependence.

The Multivariate Case

The notions of positive dependence in the multivariate case are more numerous, more complex, and their interrelationships are less well understood. In this subsection we do not attempt to give a comprehensive coverage; rather we select a few of the more applicable notions and show how they may be used in reliability theory.

4.5. Definition. A random variable T is *stochastically increasing in random variables* $S_1, \ldots, S_k$ if $P[T > t \mid S_1 = s_1, \ldots, S_k = s_k]$ is increasing in $s_1, \ldots, s_k$. We write $T \uparrow st$ in $s_1, \ldots, s_k$.

This is a direct generalization of the corresponding bivariate notion of 4.1(d).

Notation. We write $\mathbf{U}_{\mathbf{V}=\mathbf{v}}$ to denote the conditional random vector $\mathbf{U}$ given that the random vector $\mathbf{V}$ takes on the value $\mathbf{v}$.

4.6. Definition. Random variables $T_1, \ldots, T_n$ are *conditionally increasing in sequence* if T_i is stochastically increasing in $T_1, \ldots, T_{i-1}$ for $i = 2, \ldots, n$.

As our main result we will prove the following theorem.

4.7. Theorem. Let $T_1, \ldots, T_n$ be conditionally increasing in sequence. Then $T_1, \ldots, T_n$ are associated.

The following lemma leads to a proof of Theorem 4.7.

4.8. Lemma. Let $S \uparrow st$ in $\mathbf{T} = (T_1, \ldots, T_n)$. Then there exists an increasing function $h(u, \mathbf{t})$ and a random variable U independent of $\mathbf{T}$ such that $(S, \mathbf{T}) \stackrel{\text{st}}{=} (h(U, \mathbf{T}), \mathbf{T})$. (We say that $\mathbf{X} \stackrel{\text{st}}{=} \mathbf{Y}$ if $\mathbf{X}$ and $\mathbf{Y}$ have the same probability distribution.)

PROOF. Let $F_{\mathbf{t}}$ be the distribution function of $S_{\mathbf{T}=\mathbf{t}}$; that is, $F_{\mathbf{t}}(s) = P[S \le s \mid \mathbf{T} = \mathbf{t}]$. Define $h(u, \mathbf{t}) = \inf\{s\colon u \le F_{\mathbf{t}}(s)\}$. Then h is increasing in u by its definition. Since $S \uparrow st$ in $\mathbf{T}$, then $h(u, \mathbf{t}^{(1)}) \le h(u, \mathbf{t}^{(2)})$ for $\mathbf{t}^{(1)} \le \mathbf{t}^{(2)}$. Thus h is increasing in $\mathbf{t}$. Since $F_{\mathbf{t}}(s)$ is continuous from the right in s, $h(u, \mathbf{t}) \le s \Leftrightarrow u \le F_{\mathbf{t}}(s)$.

Next select U as the random variable uniformly distributed on $[0, 1]$. Then $P[h(U, \mathbf{t}) \le s] = P[U \le F_{\mathbf{t}}(s)] = F_{\mathbf{t}}(s)$; that is, $h(U, \mathbf{t}) \stackrel{\text{st}}{=} S_{\mathbf{T}=\mathbf{t}}$. Let U be independent of $\mathbf{T}$. Then $h(U, \mathbf{t})_{\mathbf{T}=\mathbf{t}} \stackrel{\text{st}}{=} h(U, \mathbf{t})$. Thus

$$S_{\mathbf{T}=\mathbf{t}} \stackrel{\text{st}}{=} h(U, \mathbf{t}) \stackrel{\text{st}}{=} h(U, \mathbf{t})_{\mathbf{T}=\mathbf{t}} \stackrel{\text{st}}{=} h(U, \mathbf{T})_{\mathbf{T}=\mathbf{t}}.$$

It follows that $(S, \mathbf{T}) \stackrel{\text{st}}{=} (h(U, \mathbf{T}), \mathbf{T})$. ||

PROOF OF THEOREM 4.7. By hypothesis, T_2 is stochastically increasing in T_1. Hence by Theorem 4.2, T_1, T_2 are associated.

Next assume $T_1, \ldots, T_k$ are associated and T_{k+1} is stochastically increasing in $T_1, \ldots, T_k$. Then by Lemma 4.8,

$$(T_1, \ldots, T_k, T_{k+1}) \stackrel{\text{st}}{=} (T_1, \ldots, T_k, h(U, T_1, \ldots, T_k)),$$

where h is increasing and U is a uniform random variable independent of $T_1, \ldots, T_k$. Since $U, T_1, \ldots, T_k$ are associated and h is increasing, then $T_1, \ldots, T_k, h(U, T_1, \ldots, T_k)$ are associated. Hence $T_1, \ldots, T_k, T_{k+1}$ are associated.

The conclusion now follows by induction. ||

We will find the following concept helpful.

4.9. Definition. $\mathbf{S}$ is stochastically less than $\mathbf{S}'$ if $f(\mathbf{S}) \stackrel{\text{st}}{\le} f(\mathbf{S}')$ for all increasing functions $f(\mathbf{s})$. We write $\mathbf{S} \stackrel{\text{st}}{\le} \mathbf{S}'$.

A generalization of the concept introduced in Definition 4.5 is presented in Definition 4.10.

4.10. Definition. $\mathbf{S}$ is *stochastically increasing* in $\mathbf{T}$ if $f(\mathbf{S}) \uparrow st$ in $\mathbf{T}$ for all increasing functions $f(\mathbf{s})$. We write $\mathbf{S} \uparrow st$ in $\mathbf{T}$.

Next we give sufficient conditions for one vector to be stochastically less than a second vector.

4.11. Definition. A set S is said to be increasing if the indicator function I_S for S is increasing.

4.12. Theorem. $\mathbf{X} \overset{\text{st}}{\leq} \mathbf{Y} \Leftrightarrow P_{\mathbf{X}}(S) \leq P_{\mathbf{Y}}(S)$ for each increasing set S.

PROOF. $\Leftarrow$ Let $h(\mathbf{x})$ be an increasing function. Then $P[h(\mathbf{X}) > u] = P_{\mathbf{X}}(S_u)$ and $P[h(\mathbf{Y}) > u] = P_{\mathbf{Y}}(S_u)$, where $S_u = [\mathbf{x} \mid h(\mathbf{x}) > u]$. Note that S_u is an increasing set since $\mathbf{x} \in S_u$, $\mathbf{x}' \geq \mathbf{x} \Rightarrow h(\mathbf{x}') \geq h(\mathbf{x}) > u \Rightarrow \mathbf{x}' \in S_u$. Thus $P_{\mathbf{X}}(S_u) \leq P_{\mathbf{Y}}(S_u)$, and so $P[h(\mathbf{X}) > u] \leq P[h(\mathbf{Y}) > u]$ for each u. Hence $\mathbf{X} \overset{\text{st}}{\leq} \mathbf{Y}$.

$\Rightarrow$ Let S be an increasing set. Define $I_S(\mathbf{x}) = 1$ if $\mathbf{x} \in S$, 0 otherwise. Then $I_S(\mathbf{x})$ is increasing in $\mathbf{x}$. Hence $P_{\mathbf{X}}(S) = P[I_S(\mathbf{X}) = 1] \leq P[I_S(\mathbf{Y}) = 1] = P_{\mathbf{Y}}(S)$. ‖

A second sufficient condition may be stated in terms of conditional distributions.

4.13. Theorem. Let (a) $\mathbf{X} = (X_1, \ldots, X_n)$ be conditionally increasing in sequence, and (b) $X_1 \overset{\text{st}}{\leq} Y_1$ and

$$X_{j,X_1=x_1,\ldots,X_{j-1}=x_{j-1}} \overset{\text{st}}{\leq} Y_{j,Y_1=x_1,\ldots,Y_{j-1}=x_{j-1}}$$

for each $x_1, \ldots, x_{j-1}$, $j = 2, \ldots, n-1$. Then $\mathbf{X} \overset{\text{st}}{\leq} \mathbf{Y}$.

PROOF. By hypothesis (b), $X_1 \overset{\text{st}}{\leq} Y_1$. Thus the conclusion holds for $n = 1$.

Assume the theorem holds for $n - 1$. Let h be an increasing function of n arguments. By Theorem 4.12 it suffices to show $Eh(X_1, \ldots, X_n) \leq Eh(Y_1, \ldots, Y_n)$.

$$Eh(X_1, \ldots, X_n)$$

$$= \int \cdots \int h(x_1, \ldots, x_{n-1}, x_n)$$

$$dF_{X_n}(x_n \mid X_1 = x_1, \ldots, X_{n-1} = x_{n-1})\, dF_{X_1,\ldots,X_{n-1}}(x_1, \ldots, x_{n-1})$$

$$\leq \int \cdots \int h(x_1, \ldots, x_{n-1}, x_n)$$

$$dF_{Y_n}(x_n \mid Y_1 = x_1, \ldots, Y_{n-1} = x_{n-1})\, dF_{X_1,\ldots,X_{n-1}}(x_1, \ldots, x_{n-1})$$

since $h \uparrow$ in x_n and (b) holds. However, the latter integral

$$\leq \int \cdots \int h(x_1, \ldots, x_{n-1}, x_n)$$

$$dF_{Y_n}(x_n \mid Y_1 = x_1, \ldots, Y_{n-1} = x_{n-1})\, dF_{Y_1,\ldots,Y_{n-1}}(x_1, \ldots, x_{n-1})$$

since by (a),

$$\int h(x_1, \ldots, x_{n-1}, x_n)$$

$$dF_{Y_n}(x_n \mid Y_1 = x_1, \ldots, Y_{n-1} = x_{n-1}) \uparrow \text{ in } x_1, \ldots, x_{n-1}$$

and since the inductive hypothesis holds for $n - 1$. Since the last integral equals $Eh(\mathbf{Y})$, the theorem holds for n. ||

We showed in the bivariate case that if random variables S, T are TP_2, then T is stochastically increasing in S (Theorem 4.2). Next we obtain a multivariate generalization.

4.14. Theorem. Let $X_1, \ldots, X_n$ have joint density $f_n(x_1, \ldots, x_n)$ which is TP_2 in each pair of arguments for fixed values of the remaining arguments. Then $X_1, \ldots, X_n$ are conditionally increasing in sequence.

PROOF. Define $f_i(x_1, \ldots, x_i)$ as the marginal density of $X_1, \ldots, X_i$. Then we may write $f_{n-1}(x_1, \ldots, x_{n-1}) = \int_{-\infty}^{\infty} f_n(x_1, \ldots, x_n) g(x_1, \ldots, x_n)\, dx_n$, where $g(x_1, \ldots, x_n) \equiv 1$. Since f_n and g are each TP_2 in each pair of arguments, then f_{n-1} is TP_2 in each pair of arguments for fixed values of the remaining arguments (Theorem 5.1 and Supplement 5.1 of Karlin, 1968, pp. 123–124). Repetition of this argument shows that f_i is TP_2 in each pair of arguments, $i = 2, 3, \ldots, n$.

Since $f_2(x_1, x_2)$ is TP_2, then $P[X_2 > x_2 \mid X_1 = x_1] \uparrow$ in x_1 by Theorem 4.2. Similarly, for fixed x_2, $f(x_1, x_2, x_3)$ is TP_2 in x_1, x_3, so that

$$P[X_3 > x_3 \mid X_1 = x_1, X_2 = x_2] \uparrow \quad \text{in} \quad x_1.$$

By symmetry, for fixed x_1,

$$P[X_3 > x_3 \mid X_1 = x_1, X_2 = x_2] \uparrow \quad \text{in} \quad x_2.$$

It follows that

$$P[X_3 > x_3 \mid X_1 = x_1, X_2 = x_2] \uparrow \quad \text{in} \quad x_1, x_2.$$

Repetition of this argument proves that $X_i \uparrow st$ in $X_1, \ldots, X_{i-1}$ for $i = 2, \ldots, n$. ||

4.15. Corollary. Let $X_1, \ldots, X_n$ have joint density $f(x_1, \ldots, x_n)$ which is TP_2 in each pair of arguments for fixed values of the remaining arguments. Then $X_1, \ldots, X_n$ are associated.

PROOF. By Theorem 4.14, $X_1, \ldots, X_n$ are conditionally increasing in sequence. It follows immediately from Theorem 4.7 that $X_1, \ldots, X_n$ are associated. ||

4.16. Example: TP_2 Normal Variates. *The multivariate normal density of (3.4) is TP_2 in each pair of arguments for fixed values of the remaining arguments if and only if $r_{ij} \leq 0$ for all $i \neq j$.*

For any pair (i, j), (3.4) reduces to $C_1(\mathbf{t})C_2(\mathbf{t})e^{-r_{ij}t_it_j}$ where C_1 does *not* depend on t_i, and C_2 does *not* depend on t_j. By Exercise 18, f is TP_2 in t_i, t_j for fixed values of the remaining arguments if and only if $r_{ij} \leq 0$.

By Corollary 4.15, we conclude that *if $T_1, \ldots, T_n$ have a multivariate normal density with $r_{ij} \leq 0$ for all $i \neq j$, then $T_1, \ldots, T_n$ are associated.*

Bounds on System Reliability

If it can be verified, by use of any of the previous results, that component lifetimes $T_1, T_2, \ldots, T_n$ are associated, then we can obtain useful bounds on the system life distribution. Recall that if coherent structure ϕ with lifetime T has min path sets $P_1, P_2, \ldots, P_p$ and min cut sets $K_1, K_2, \ldots, K_k$, then

$$T = \min_{1\leq s\leq k} \max_{i\in K_s} T_i \tag{4.6}$$

$$= \max_{1\leq r\leq p} \min_{i\in P_r} T_i \tag{4.7}$$

by (3.7) of Chapter 1. We will prove the following theorem using (4.6) and (4.7). It can also be viewed as basically a restatement of Theorem 3.9 in Chapter 2.

4.17. Theorem. If $T_1, T_2, \ldots, T_n$ are associated and coherent structure ϕ has min paths $P_1, P_2, \ldots, P_p$ and min cuts $K_1, K_2, \ldots, K_k$, then

$$\max_{1\leq r\leq p} \prod_{i\in P_r} P[T_i > t] \leq P[T > t] \leq \min_{1\leq s\leq k} \coprod_{i\in K_s} P[T_i > t] \tag{4.8}$$

where T is the system lifetime.

PROOF. Since $T_1, T_2, \ldots, T_n$ are associated,

$$P[\min_{i\in P_r} T_i > t] \geq \prod_{i\in P_r} P[T_i > t] \tag{4.9}$$

and

$$P[\max_{i\in K_s} T_i > t] \leq \coprod_{i\in K_s} P[T_i > t] \tag{4.10}$$

by (3.5) and (3.6) of Corollary 3.3 of Chapter 2. Hence

$$\prod_{i\in P_r} P[T_i > t] \leq P[T > t] \leq \coprod_{i\in K_s} P[T_i > t]$$

for all r $(1 \leq r \leq p)$ and s $(1 \leq s \leq k)$ by (4.6), (4.7), (4.9), and (4.10). Equation (4.8) follows easily. ||

EXERCISES

1. Prove that $\mathrm{SI}(T \mid S) \Rightarrow \mathrm{LTD}(T \mid S)$.
2. Give examples to show that no two of the conditions (4.1) through (4.5) are equivalent.
3. Prove that S and T positive quadrant dependent $\Leftrightarrow P[S > s, T > t] \geq P[S > s]P[T > t]$ for all s, t.
4. Let $(X_1, Y_1), \ldots, (X_n, Y_n)$ be independent pairs of random variables, with X_i, Y_i positive quadrant dependent for $i = 1, \ldots, n$. Let $X = r(X_1, \ldots, X_n)$, $Y = s(Y_1, \ldots, Y_n)$, with r and s increasing. Then X, Y are positive quadrant dependent.
5. If X, Y are positive quadrant dependent, then $\mathrm{cov}(X, Y) \geq 0$ if it exists.
6. Let $Y = \alpha + \beta X + U$, where X and U are independent, and $\beta \geq 0$. Then Y is stochastically increasing in X.
7. Let $U_1, \ldots, U_k$ have a multinomial distribution corresponding to n trials and success probabilities $p_1, \ldots, p_k$. Then U_i is stochastically increasing in $-U_j$ for $i \neq j$.
8. $X_{1:n} \leq \cdots \leq X_{n:n}$ be order statistics from a DFR distribution. Then $X_{s:n} - X_{r:n}$ is stochastically increasing in $X_{r:n}$ for $r < s$.
9. Show that if a multivariate normal density is TP_2 in each pair, then each correlation coefficient $\rho_{ij} \geq 0$, but not conversely.
10. Let $X_{i:n}$ and $X_{j:n}$ be order statistics from an arbitrary distribution. Then the following properties hold: (a) $\mathrm{TP}_2(X_{i:n}, X_{j:n})$, (b) $X_{1:n}, X_{2:n}, \ldots, X_{n:n}$ are conditionally increasing in sequence.
11. Let $f(x, y, z) = e^{xyz}$. (a) Is f TP_2 in each pair for x, y, z real? (b) On what domain is f TP_2 in each pair?
12. Let $X_{1:n} \leq \cdots \leq X_{n:n}$ be the order statistics in a sample of size n from a DFR distribution. Let $D_1 = X_{1:n}$ and $D_i = X_{i:n} - X_{(i-1):n}$, $i = 2, \ldots, n$. Then (a) $D_1, \ldots, D_n$ are conditionally increasing in sequence, and (b) $D_1, \ldots, D_n$ are associated.
13. Let $\mathbf{X} = (X_1, \ldots, X_n)$ and $\mathbf{Y} = (Y_1, \ldots, Y_n)$. Let $\mathbf{X}$ have joint distribution F and $\mathbf{Y}$ have joint distribution G. Show that $\bar{F}(\mathbf{x}) \geq \bar{G}(\mathbf{x})$ for each $\mathbf{x}$ does not imply $\mathbf{X} \overset{\mathrm{st}}{\geq} \mathbf{Y}$.
14. Find a counterexample to the following statement: Let $X_1, \ldots, X_n$ be associated; let $Y_1, \ldots, Y_n$ be mutually independent with $X_i \overset{\mathrm{st}}{=} Y_i$ for $i = 1, \ldots, n$. Then $\mathbf{X} \overset{\mathrm{st}}{\geq} \mathbf{Y}$.
15. Show that $\mathbf{S} \overset{\mathrm{st}}{\leq} \mathbf{S}' \Leftrightarrow Eh(\mathbf{S}) \leq Eh(\mathbf{S}')$ for all increasing functions h.
16. Give examples to show that neither of $\mathrm{RCSI}(S, T)$ and $\mathrm{SI}(T \mid S)$ implies the other.

17. Show that TP_2 is preserved under translation and change of scale. Is it preserved under rotation of axes?

18. Show that (a) $\bar{F}(t, t, \ldots, t) \geq \prod_{i=1}^{n} \bar{F}_i(t)$ if and only if (b) $F(t, t, \ldots, t) \geq \prod_{i=1}^{n} F_i(t)$. (Hint: Use Exercise 3, Section 1.) Hence if either (a) or (b) holds for all marginal distributions corresponding to subsets of $T_1, T_2, \ldots, T_n$, then (4.8) holds.

*19. A stochastic process $\{X(t); t \geq 0\}$ is *associated in time* if $X(t_1), \ldots, X(t_n)$ are associated for all choices of $t_1, \ldots, t_n$ and all $n \geq 1$. Show that if a process is Markov and $P[X(t) = y \mid X(s) = x]$ is TP_2 in x and y for all $s \leq t$, then the process is associated in time.

*20. Use Exercise 19 to show that a two-state birth and death process is associated. (Chapter 7, Section 3 discusses birth and death processes.)

5. MULTIVARIATE MONOTONE FAILURE RATE DISTRIBUTIONS

In this section we define a multivariate version of increasing (decreasing) failure rate distributions due to Thompson (1971), and develop some of its desirable properties. We mention also the stronger version of multivariate IFR due to Harris (1970), and make a brief comparison between the two concepts.

5.1. Definition. A distribution $F(x_1, \ldots, x_n)$ is said to be a *multivariate increasing failure rate* (MIFR) *distribution* if its marginal distribution $F_{i_1,\ldots,i_k}(x_{i_1}, \ldots, x_{i_k})$ satisfies

$$\frac{\bar{F}_{i_1,\ldots,i_k}(x_{i_1} + t, \ldots, x_{i_k} + t)}{\bar{F}_{i_1,\ldots,i_k}(x_{i_1}, \ldots, x_{i_k})} \downarrow \tag{5.1}$$

in $x_{i_1} > -\infty, \ldots, x_{i_k} > -\infty$ for $t > 0$, for each subset $\{i_1, \ldots, i_k\} \subset \{1, 2, \ldots, n\}$. In addition we assume that F puts positive probability only on the nonnegative orthant although it is defined for all n-tuples.

Equation (5.1) tells us that the life length of a series system of any subset of components decreases stochastically as the ages of the components increase. Note that for $n = 1$, (5.1) defines the usual univariate IFR distribution, as given in (1.7) of Chapter 3. In this sense, the MIFR distribution represents the multivariate generalization of the univariate IFR distribution. In a similar fashion we form the following definition.

5.2. Definition. A distribution $F(x_1, \ldots, x_n)$ on the nonnegative orthant is said to be a *multivariate decreasing failure rate distribution* (MDFR) if its marginal distribution $F_{i_1,\ldots,i_k}(x_{i_1}, \ldots, x_{i_k})$ satisfies

$$\frac{\bar{F}_{i_1,\ldots,i_k}(x_{i_1} + t, \ldots, x_{i_k} + t)}{\bar{F}_{i_1,\ldots,i_k}(x_{i_1}, \ldots, x_{i_k})} \uparrow \text{ in } x_{i_1} \geq 0, \ldots, x_{i_k} \geq 0 \quad \text{for } t > 0, \tag{5.2}$$

for each subset $\{i_1, \ldots, i_k\} \subset \{1, \ldots, n\}$.

Remarks similar to those following Definition 5.1 apply to the MDFR distribution, with the direction of monotonicity reversed.

MIFR and MDFR distributions enjoy the following desirable multivariate properties.

5.3. Theorem. (a) A univariate MIFR (MDFR) random variable is IFR(DFR) in the usual sense.

(b) The union of two mutually independent sets of MIFR (MDFR) random variables is MIFR (MDFR).

(c) Any subset of MIFR (MDFR) random variables is MIFR (MDFR).

(d) Minima of subsets of MIFR (MDFR) random variables are MIFR (MDFR).

(e) Random variables $X_1, \ldots, X_n$ are both MIFR and MDFR if and only if they are MVE.

PROOF. We prove (a)–(d) in the case of MIFR. The proof in the MDFR case is similar.

(a) By (1.7) of Chapter 3, (a) follows immediately.

(b) The proof is left for Exercise 1.

(c) To show a given subset is MIFR, set the x_i in the complementary set all equal to $-\infty$ in (5.1).

(d) Let $Y_1, \ldots, Y_m$ be MIFR. Let $X_i = \min_{j \in J_i} Y_j$, where J_i is a subset of $\{1, \ldots, m\}$, $i = 1, \ldots, n$. Given $t > 0$ and real $x_1, \ldots, x_n$, we write

$$\begin{aligned} P[X_1 > x_1 + t, \ldots, X_n > x_n + t \mid X_1 > x_1, \ldots, X_n > x_n] \\ = P[Y_1 > y_1 + t, \ldots, Y_m > y_m + t \mid Y_1 > y_1, \ldots, Y_m > y_m], \end{aligned}$$

where $y_j = \max_{j \in J_i} x_i$, $j = 1, \ldots, m$. But since the last conditional probability is decreasing in each y_j, and each y_j is increasing in every x_i, it follows that the first conditional probability is decreasing in each x_i. Thus $X_1, \ldots, X_n$ is MIFR. ||

(e) Let F be MVE as given in (3.2). Then

$$\frac{\bar{F}(x_1 + t, \ldots, x_n + t)}{\bar{F}(x_1, \ldots, x_n)} = \exp\left\{-t\left[\sum_1^n \lambda_i + \sum_{i<j} \lambda_{ij} + \cdots + \lambda_{12\cdots n}\right]\right\},$$

constant in $x_1 \geq 0, \ldots, x_n \geq 0$ for $t > 0$. A similar result holds for each marginal of F. Thus F is both MIFR and MDFR.

Conversely, let $X_1, \ldots, X_n$ be both MIFR and MDFR. Then by (5.1) and (5.2),

$$\frac{\bar{F}(x_1 + t, \ldots, x_n + t)}{\bar{F}(x_1, \ldots, x_n)} = c(t) \qquad \text{for} \quad x_1 \geq 0, \ldots, x_n \geq 0, \quad t > 0, \tag{5.3}$$

where $c(t)$ depends on t but not on $x_1, \ldots, x_n$. In (5.3), set $x_1 = \cdots = x_n = 0$ to obtain $c(t) = \bar{F}(t, \ldots, t)$. Thus

$$\bar{F}(x_1 + t, \ldots, x_n + t) = \bar{F}(x_1, \ldots, x_n)\bar{F}(t, \ldots, t) \tag{5.4}$$

for all $\mathbf{x} \geq \mathbf{0}$, $t \geq 0$. In a similar fashion, if $F_{i_1,\ldots,i_k}(x_{i_1},\ldots,x_{i_k})$ is a k-dimensional marginal of F, then we may show that $\bar{F}_{i_1,\ldots,i_k}$ satisfies an equation of the form (5.4). It follows by the arguments of 3.1 that F is MVE. ||

From (a) and (b) of Theorem 5.3, we immediately deduce Corollary 5.4.

5.4. Corollary. Let $X_1, \ldots, X_n$ be independent IFR (DFR) random variables. Then $X_1, \ldots, X_n$ have a MIFR (MDFR) distribution.

From (e) of Theorem 5.3, we see that the MVE lies on the boundary of both the MIFR class and the MDFR class of distributions. This is analogous to the situation in the univariate case, in which the univariate exponential is the only distribution which is both IFR and DFR.

Next we show that under certain scale and translation transformations, the MIFR (MDFR) property is preserved.

5.5. Theorem. (a) If $X_1, \ldots, X_n$ are MIFR (MDFR) and $a > 0$, then $aX_1, \ldots, aX_n$ are MIFR (MDFR).

(b) If $X_1, \ldots, X_n$ are MIFR, and $b_1 \geq 0, \ldots, b_n \geq 0$, then $X_1 + b_1, \ldots, X_n + b_n$ are MIFR.

PROOF. (a) Suppose $X_1, \ldots, X_n$ are MIFR. Then

$$P[aX_1 > x_1 + t, \ldots, aX_n > x_n + t \mid aX_1 > x_1, \ldots, aX_n > x_n]$$
$$= P\left[X_1 > \frac{x_1}{a} + \frac{t}{a}, \ldots, X_n > \frac{x_n}{a} + \frac{t}{a} \mid X_1 > \frac{x_1}{a}, \ldots, X_n > \frac{x_n}{a}\right] \downarrow$$

in $x_1 > -\infty, \ldots, x_n > -\infty$ for $t > 0$, since $X_1, \ldots, X_n$ are MIFR. The proof is similar in the MDFR case.

(b)

$$P[X_1 + b_1 > x_1 + t, \ldots, X_n + b_n > x_n + t \mid X_1 + b_1 > x_1, \ldots, X_n + b_n > x_n]$$
$$= P[X_1 > x_1 - b_1 + t, \ldots, X_n > x_n - b_n + t \mid X_1 > x_1 - b_1, \ldots, X_n > x_n - b_n] \downarrow$$

in $x_1 > -\infty, \ldots, x_n > -\infty$ for $t > 0$, since $X_1, \ldots, X_n$ are MIFR. ||

In (a), note that the scale factor a must be the same for all X_i. In (b), note that the $b_1, \ldots, b_n$ may differ; however, they must all be nonnegative since the random variables are constrained to be nonnegative. Note also that no dual result for MDFR exists in (b). This follows from that the fact that for $x_i - b_i + t < 0$, $i = 1, \ldots, n$,

$$P[X_1 > x_1 - b_1 + t, \ldots, X_n > x_n - b_n + t \mid X_1 > x_1 - b_1, \ldots, X_n > x_n - b_n] = 1,$$

and thus this conditional probability cannot increase in $x_1, \ldots, x_n$.

An alternate formulation of the notion of multivariate increasing failure rate distributions is presented by Harris (1970). We first need the following generalization of the bivariate notion of right corner set increasing given in 4.3.

5.6. Definition. A set of random variables $X_1, \ldots, X_n$ is said to be *right corner set increasing* (RCSI) if

$$P[X_1 > x_1, \ldots, X_n > x_n \mid X_1 > x_1', \ldots, X_n > x_n']$$

is increasing in $x_1', \ldots, x_n'$ for every choice of $x_1, \ldots, x_n$. We write RCSI $(X_1, \ldots, X_n)$. We also refer to the joint distribution, F, of $X_1, \ldots, X_n$ as RCSI.

Note that the set consisting of a single random variable (that is, $n = 1$) is always RCSI. For $n \geq 2$, RCSI represents a form of positive dependence.

Harris' formulation of multivariate increasing failure rate requires that the set of random variables $X_1, \ldots, X_n$ be MIFR as defined in 5.1 and *also* be RCSI. Thus the Harris class of multivariate increasing failure rate distributions is a subclass of the Thompson MIFR. One disadvantage of the Harris formulation is that there seems to be no dual notion of decreasing failure rate distributions appropriate for reliability applications.

We do *not* claim that Definition 5.1 is the only multivariate version of univariate IFR (DFR) distributions of practical interest. However, the properties presented in Theorem 5.3 are intuitively desirable for any such class.

5.7. Example. Let U, V, and W be independent IFR random variables with distributions F_1, F_2, and F_3, respectively. Let $X_1 = \min(U, W)$ and $X_2 = \min(V, W)$. Then (X_1, X_2) is bivariate IFR according to Definition 5.1, since

$$\overline{F}(x_1, x_2) = \overline{F}_1(x_1)\overline{F}(x_2)\overline{F}_3(\max(x_1, x_2)).$$

EXERCISES

1. Prove that the union of two mutually independent sets of MIFR random variables is MIFR.
2. Using Harris' definition of multivariate IFR, prove that properties (a), (b), (c), and (d) of Theorem 5.3 hold. Show that the MVE satisfies Harris' definition.
3. Let $X_1, \ldots, X_n$ be RCSI. Then

$$P[X_1 > x_1, \ldots, X_n > x_n] \geq \prod_{i=1}^{n} P[X_i > x_i].$$

4. Does a MIFR set of random variables satisfy the inequality of Exercise 3?
5. Gumbel (1960) presents a bivariate exponential with joint survival function $\bar{G}(x, y) = e^{-x-y-\delta xy}$ for $x \geq 0$, $y \geq 0$, where $0 \leq \delta \leq 1$. Show that (a) the marginal distributions are exponential, and (b) the joint distribution is MIFR, but not multivariate IFR according to Harris.
6. Let $\bar{F}(x_1 + t_1, \ldots, x_n + t_n)/\bar{F}(x_1, \ldots, x_n)$ be decreasing in $x_1 \geq 0, \ldots, x_n \geq 0$ for $t_1 \geq 0, \ldots, t_n \geq 0$. Then $\bar{F}(x_1, \ldots. x_n) \leq \prod_{i=1}^n \bar{F}_i(x_i)$, where $F_1, \ldots, F_n$ are the marginal distributions of F.
7. Let $\mathbf{X} \sim F(x_1, \ldots, x_n)$ be MIFR. Let $F_{\mathbf{y}}(x_1, \ldots, x_n)$ be the conditional distribution of $\mathbf{X}$ given that $X_i > y_i$, $i = 1, \ldots, n$. Then, for fixed $\mathbf{y}$, $F_{\mathbf{y}}(x_1, \ldots, x_n)$ is MIFR.
8. In analogy to the shock model derivation of the MVE (Section 3), consider shocks to an n-component system. Let H_i, $i = 1, 2, \ldots, n$ be the IFR distributions of occurrence times of independent shocks to components $i = 1, 2, \ldots, n$. To each subset of components we assign an IFR distribution for shock occurrence time; H_{ijk}, for example, will denote the distribution assigned to components i, j, and k. All shock occurrence times are assumed independent so that the multivariate distribution of component survival is

$$\bar{F}(t_1, t_2, \ldots, t_n)$$
$$= \prod_{i=1}^{n} \bar{H}_i(t_i) \prod_{i<j} H_{ij}[\max(t_i, t_j)] \cdots H_{1,2,3,\ldots,n}[\max(t_1, \ldots, t_n)].$$

Show that F is MIFR according to Definition 5.1.

6. NOTES AND REFERENCES

Section 1. The bivariate exponential distribution is discussed in Marshall and Olkin (1967a). A more general bivariate exponential distribution is treated in Marshall and Olkin (1967b).

Section 2. The derivation of the BVE and of the bivariate Poisson process from a nonfatal shock model may be found in Marshall and Olkin (1967a). The multivariate Poisson process is treated by Teicher (1954).

Section 3. The multivariate exponential distribution of this section is derived from the shock model point of view in Marshall and Olkin (1967a). A generalization of the MVE is proposed by Arnold (1967); he shows how to generate multivariate distributions with marginals in a prescribed family, such as the geometric, uniform, or normal family.

The sufficient condition that a set of normal random variables be associated stated in this section is derived by Sarkar (1969). His proof differs from the one given in Section 4.

Section 4. The discussion of bivariate dependence is based largely on Esary and Proschan (1972). Lehmann (1966) shows the relationship of TP_2 dependence to the other forms of bivariate dependence. Harris (1970) introduces right corner set increasing dependence in order to define multivariate increasing failure rate distributions.

The notion of random variables conditionally increasing in sequence and its implication of association (Theorem 4.7) are presented in Esary and Proschan (1968). The stochastic comparison of two random vectors and a sufficient condition for its existence (Theorem 4.11) are presented in Veinott (1965). The sufficient condition of Theorem 4.14 that a set of random variables be conditionally increasing in sequence is new. The necessary and sufficient condition that a multivariate normal density is TP_2 in each pair of arguments given in 4.16 is first derived in Sarkar (1969); our proof is simpler.

Section 5. This multivariate version of increasing failure rate distributions is presented in Thompson (1971); our discussion, although based on his paper, differs slightly in technical detail. The alternate version of multivariate IFR appeared earlier in Harris (1970); Harris' definition is not quite correct as given since it permits mass at the origin or on the coordinate axes. In 5.1 we correct this weakness by requiring the domain of monotonicity to be the entire n-dimensional Euclidean space.

6

CONCEPTS HELPFUL IN THE STUDY OF MAINTENANCE POLICIES

1. INTRODUCTION

Maintenance policies are followed to reduce the incidence of system failure or to return a failed system to the operating state. In this chapter we develop theory helpful in the study of maintenance policies. We consider classes of probability distributions especially applicable in maintenance theory and obtain useful and interesting properties of them. We also review renewal theory and show how it can be applied to solve certain maintenance problems. Finally, we compare various replacement policies stochastically.

In Chapter 7 we apply the tools developed to solve certain problems in maintenance, and to arrive at optimum policies.

We will be studying the following types of replacement policy.

1.1. Definition. Under an *age replacement* policy, a unit is replaced upon failure or at age T, whichever comes first.

Throughout this chapter T is nonrandom.

1.2. Definition. Under a *block replacement* policy the unit in operation is replaced upon failure and at times $T, 2T, 3T, \ldots$.

By its nature, age replacement is administratively more difficult to implement, since the age of the unit must be recorded. On the other hand, block replacement, although simpler to administer since the age of the unit need not be recorded, leads to more frequent replacement of relatively new items. The word "block" is used to imply that in practice a *block* or set of components will be replaced at times $T, 2T, 3T, \ldots$, regardless of which have failed.

2. CLASSES OF DISTRIBUTIONS APPLICABLE IN REPLACEMENT

Certain classes of distribution arise quite naturally in considering replacement policies.

New Better (Worse) Than Used

A distribution F is NBU (NWU) if

$$\overline{F}(x + y) \leq (\geq)\overline{F}(x)\overline{F}(y) \qquad \text{for} \quad x \geq 0, \quad y \geq 0. \tag{2.1}$$

This is equivalent to stating that the conditional survival probability $\overline{F}(x + y)/\overline{F}(x)$ of a unit of age x is less (greater) than the corresponding survival probability $\overline{F}(y)$ of a new unit. Equality in (2.1) holds if and only if F is the exponential distribution.

New Better (Worse) Than Used in Expectation

A life distribution F is NBUE (NWUE) if

(a) F has finite (finite or infinite) mean;

(b) $$\int_t^\infty \overline{F}(x)\, dx \leq (\geq)\mu\overline{F}(t) \qquad \text{for} \quad t \geq 0. \tag{2.2}$$

Note that $\int_t^\infty [\overline{F}(x)/\overline{F}(t)]\, dx$ represents the conditional mean remaining life of a unit of age t. Hence (2.2) implies that a used unit of age t has smaller mean remaining life than a new unit if F is NBUE.

The chains of implication

$$\text{IFR} \Rightarrow \text{IFRA} \Rightarrow \text{NBU} \Rightarrow \text{NBUE} \tag{2.3}$$

$$\text{DFR} \Rightarrow \text{DFRA} \Rightarrow \text{NWU} \Rightarrow \text{NWUE} \tag{2.4}$$

are readily established. The proof of the second implication in (2.3) and (2.4) is left to Exercise 1. To obtain the last implication in (2.3) and (2.4), integrate both sides of (2.1) with respect to y from 0 to ∞ and use Exercise 11 of Chapter 3, Section 6.

Next we show that the NBU, NBUE, NWU, and NWUE classes of life distributions may arise from a consideration of shock models similar to those of Section 3 of Chapter 4.

2.1. Shock Model Leading to NBU. Assume a device is subject to shocks occurring in time according to a Poisson process with rate λ. Suppose the probability of surviving k shocks is $\bar{P}_k$, where $1 = \bar{P}_0 \geq \bar{P}_1 \geq \cdots$. Then, as in (3.6) of Chapter 4, the survival probability $\bar{H}(t)$ of the device corresponding to the time interval $[0, t]$ is given by

$$\bar{H}(t) = \sum_{k=0}^{\infty} \bar{P}_k \frac{(\lambda t)^k}{k!} e^{-\lambda t} \quad \text{for} \quad 0 \leq t < \infty. \tag{2.5}$$

If we now assume that the probability of surviving k additional shocks is smaller when the device has already absorbed one or more shocks than before it has absorbed any, then we may conclude that the life distribution H is NBU, as proved in Theorem 2.2.

2.2. Theorem. Let $\bar{P}_k$ satisfy

$$\bar{P}_{k+l} \leq (\geq) \bar{P}_k \bar{P}_l \qquad \text{for} \quad k = 0, 1, 2, \ldots; \quad l = 0, 1, 2, \ldots. \tag{2.6}$$

Then H is NBU (NWU).

PROOF.

$$\bar{H}(x)\bar{H}(t) = \sum_{k=0}^{\infty} \sum_{l=0}^{\infty} \bar{P}_k \bar{P}_l e^{-\lambda(x+t)} \frac{(\lambda x)^k}{k!} \frac{(\lambda t)^l}{l!}$$

$$= \sum_{j=0}^{\infty} \sum_{k=0}^{j} \bar{P}_k \bar{P}_{j-k} e^{-\lambda(x+t)} \frac{(\lambda x)^k}{k!} \frac{(\lambda t)^{j-k}}{(j-k)!}$$

$$\geq (\leq) \sum_{j=0}^{\infty} \frac{\bar{P}_j}{j!} e^{-\lambda(x+t)} \sum_{k=0}^{j} \binom{j}{k} (\lambda x)^k (\lambda t)^{j-k} = \bar{H}(x + t). \quad \|$$

Note that the discrete NBU (NWU) property of (2.6) is transformed into the continuous NBU (NWU) property of (2.1) under the transformation (2.5). The same transformation carries the discrete NBUE (NWUE) property into the continuous NBUE (NWUE) property, as proved in Theorem 2.3.

2.3. Theorem. Let $\bar{P}_k$ satisfy

$$\bar{P}_k \sum_{j=0}^{\infty} \bar{P}_j \geq (\leq) \sum_{j=k}^{\infty} \bar{P}_j, \qquad k = 0, 1, 2, \ldots. \tag{2.7}$$

Then H is NBUE (NWUE).

PROOF.

$$\bar{H}(t)\mu - \int_t^\infty \bar{H}(x)\,dx = \bar{H}(t)\int_0^\infty \bar{H}(x)\,dx - \int_t^\infty \bar{H}(x)\,dx$$

$$= \sum_{k=0}^\infty \bar{P}_k e^{-\lambda t}\frac{(\lambda t)^k}{k!}\sum_{j=0}^\infty \bar{P}_j \int_0^\infty e^{-\lambda x}\frac{(\lambda x)^j}{j!}\,dx - \sum_{k=0}^\infty \bar{P}_k \int_t^\infty e^{-\lambda x}\frac{(\lambda x)^k}{k!}\,dx$$

$$= \sum_{k=0}^\infty \bar{P}_k e^{-\lambda t}\frac{(\lambda t)^k}{k!}\sum_{j=0}^\infty \bar{P}_j \frac{1}{\lambda} - \sum_{k=0}^\infty \bar{P}_k \frac{1}{\lambda}\sum_{j=0}^k e^{-\lambda t}\frac{(\lambda t)^j}{j!}$$

[by (5.4) of Chapter 3]

$$= \frac{1}{\lambda}\sum_{k=0}^\infty \bar{P}_k e^{-\lambda t}\frac{(\lambda t)^k}{k!}\sum_{j=0}^\infty \bar{P}_j - \frac{1}{\lambda}\sum_{k=0}^\infty \sum_{j=k}^\infty \bar{P}_j e^{-\lambda t}\frac{(\lambda t)^j}{j!}$$

(summation by parts)

$$= \frac{1}{\lambda}\sum_{k=0}^\infty e^{-\lambda t}\frac{(\lambda t)^k}{k!}\left[\bar{P}_k \sum_{j=0}^\infty \bar{P}_j - \sum_{j=k}^\infty \bar{P}_j\right] \geq (\leq) 0 \qquad \text{by (2.7).} \;\|$$

EXERCISES

1. Prove that F IFRA (DFRA) $\Rightarrow F$ NBU (NWU).
2. Show that a distribution is NBU (NWU) if and only if its hazard function $-\log \bar{F}(t)$ is superadditive (subadditive) on $(0, \infty)$.
3. Prove that F NBUE $\Leftrightarrow \int_0^t \bar{F}(x)\,dx \geq \mu F(t)$ for $t \geq 0$.
4. Construct a distribution which is (a) NBU but not IFRA, (b) NBUE but not NBU.

3. RENEWAL THEORY USEFUL IN REPLACEMENT MODELS

A *renewal process* is a sequence of independent, identically distributed, nonnegative random variables $X_1, X_2, \ldots$, which, with probability 1, are not all zero. Let F be the distribution of X_1; F is called the *underlying distribution* of the renewal process. $F^{(k)}$, the k-fold convolution of F with itself, is the distribution of $S_k \equiv X_1 + X_2 + \cdots + X_k$. For convenience, we define

$$F^{(0)}(t) = \begin{cases} 1 & \text{for } t \geq 0, \\ 0 & \text{for } t < 0. \end{cases}$$

As a simple example of a renewal process, consider a system operating over an indefinite period of time. Upon failure, repair (or replacement) is performed, requiring negligible time. The successive intervals between failures are the independent, identically distributed random variables $X_1, X_2, \ldots$ of a renewal process.

Renewal theory is primarily concerned with the number of renewals $N(t)$ in $[0, t]$. $N(t)$, the *renewal random variable,* is the maximum value of k for which $S_k \le t$, with the understanding that $N(t) = 0$ if $X_1 > t$. The process $\{N(t); t \ge 0\}$ is known as a *renewal counting process.*

Note that $P[N(t) \ge n] = P[S_n \le t] = F^{(n)}(t)$. It follows that $P[N(t) = n] = P[N(t) \ge n] - P[N(t) \ge n + 1]$, so that

$$P[N(t) = n] = F^{(n)}(t) - F^{(n+1)}(t). \tag{3.1}$$

It is fairly easy to show that $N(t)$ has finite moments of all orders.

3.1. Example. Let F be IFR with mean μ. Then $\bar{F}(t) \ge e^{-t/\mu}$ for $0 \le t < \mu$, as shown in 6.7 of Chapter 4. It follows that

$$F^{(n)}(t) \le 1 - \sum_{j=0}^{n-1} \frac{(t/\mu)^j}{j!} e^{-t/\mu} \qquad \text{for} \quad 0 \le t < \mu.$$

Therefore,

$$P[N(t) \ge n] \le \sum_{j=n}^{\infty} \frac{(t/\mu)^j}{j!} e^{-t/\mu} \qquad \text{for} \quad 0 \le t < \mu. \tag{3.2}$$

Thus we have the elementary but important result that under the IFR assumption for component life, the Poisson distribution provides a conservative estimate of the probability of n or more failures in $[0, t]$ for t less than the mean life of a component.

NBU Bounds on $P[N(t) < n]$

The Poisson bound, (3.2), is very convenient since we need only specify the mean of the failure distribution to calculate $P[N(t) \ge n]$. However, this bound is not valid if we assume only that F is NBU or if $t \ge \mu$. A bound in this case is given below.

3.2. Theorem. (a) Let F be continuous, $F(0) = 0$, and $R(t) = -\log \bar{F}(t)$. If F is NBU (NWU), then

$$P[N(t) < n] \ge (\le) \sum_{j=0}^{n-1} \frac{[R(t)]^j}{j!} e^{-R(t)} \qquad \text{for} \quad t \ge 0, \quad n = 1, 2, \ldots. \tag{3.3a}$$

(b) If F is IFR (DFR), then

$$P[N(t) < n] \le (\ge) \sum_{j=0}^{n-1} \frac{[nR(t/n)]^j}{j!} e^{-nR(t/n)} \qquad \text{for} \quad t \ge 0, \quad n = 1, 2, \ldots. \tag{3.3b}$$

PROOF. (a) Let F be NBU. Then $R(t)$ is superadditive by Exercise 2 of Section 2. It follows that $R^{-1}(t)$ is subadditive, that is,

$$R^{-1}(x + y) \le R^{-1}(x) + R^{-1}(y) \qquad \text{for} \quad x, \quad y \ge 0.$$

Let Y_j $(j = 1, 2, \ldots, n + 1)$ be i.i.d. random variables with distribution $G(x) = 1 - e^{-x}$ for $x \ge 0$. Then $X_j = R^{-1}(Y_j)$ $(j = 1, 2, \ldots, n)$ are i.i.d. with distribution F. Since R^{-1} is subadditive,

$$\sum_{j=1}^{n} R^{-1}(Y_j) \ge R^{-1}\left(\sum_{j=1}^{n} Y_j\right).$$

Then

$$P[N(t) < n] = P\left[\sum_{j=1}^{n} X_j > t\right] \ge P\left[R^{-1}\left(\sum_{j=1}^{n} Y_j\right) > t\right]$$

$$= P\left[\sum_{j=1}^{n} Y_j > R(t)\right]$$

$$= \sum_{j=0}^{n-1} \frac{[R(t)]^j}{j!} e^{-R(t)}.$$

A similar argument applies if F is NWU.

(b) Let F be IFR. Clearly, it is enough to show that

$$F^{(n)}(t) \ge G^{(n)}\left[nR\left(\frac{t}{n}\right)\right] \qquad \text{for} \quad t \ge 0, \quad n = 1, 2, \ldots.$$

The proof proceeds by induction. For $n = 1$, equality actually holds by definition of $R(t)$. Assume the above inequality holds for $n - 1$. Then

$$F^{(n)}(t) = \int_0^{\infty} F^{(n-1)}(t - x)\, dF(x)$$

$$\ge \int_0^{\infty} G^{(n-1)}\left[(n - 1)R\left(\frac{t - x}{n - 1}\right)\right] dG[R(x)].$$

Using the convexity of $R(t)$, we may write

$$R\left(\frac{t}{n}\right) \le \frac{n - 1}{n} R\left(\frac{t - x}{n - 1}\right) + \frac{1}{n} R(x)$$

since

$$\frac{t}{n} = \frac{(n - 1)}{n} \frac{(t - x)}{(n - 1)} + \frac{x}{n}.$$

Hence

$$(n - 1)R\left(\frac{t - x}{n - 1}\right) \ge nR\left(\frac{t}{n}\right) - R(x),$$

which implies

$$F^{(n)}(t) \geq \int_0^\infty G^{(n-1)}\left[nR\left(\frac{t}{n}\right) - R(x)\right] d\,[R(x)]$$

$$= \int_0^\infty G^{(n-1)}\left[nR\left(\frac{t}{n}\right) - u\right] dG(u) = G^{(n)}\left[nR\left(\frac{t}{n}\right)\right].$$

To show the reverse inequality when F is DFR, we use the fact that F DFR implies R concave. ‖

If F is IFR and $t < \mu$, then $R(t) \leq t/\mu$ by (6.4) of Chapter 3. Thus

$$\sum_{j=0}^{n-1} \frac{[R(t)]^j}{j!} e^{-R(t)} \geq \sum_{j=0}^{n-1} \frac{(t/\mu)^j}{j!} e^{-t/\mu},$$

so that (3.3a) is an improvement on (3.2) when F is IFR.

Equations (3.3a) and (3.3b) are very convenient since we need have only a Poisson table to compute bounds on $F^{(n)}(t)$ from known $R(t)$.

Remark. Theorem 3.2 can easily be generalized to the case in which (a) F is superadditive with respect to G, and (b) F is convex with respect to G. See Exercises 11 and 12.

3.3. Example: Bounds on the Weibull Renewal Distribution. If $F(t) = 1 - e^{-(\lambda t)^\alpha}$ is the Weibull distribution (see Chapter 3, Section 5, and Chapter 8, Section 1), then $R(t) = (\lambda t)^\alpha$. By (3.3a), for $\alpha \geq 1$,

$$P[N(t) \leq n] \geq \sum_{j=0}^{n} \frac{[\lambda t]^{\alpha j}}{j!} e^{-(\lambda t)^\alpha}$$

for $t \geq 0$ and $n = 1, 2, \ldots$. Since $F^{(n)}$ can be computed only numerically for this distribution, the bounds are very convenient.

3.4. Example. The tires in the landing gear of an airplane are much more likely to fail on takeoff and landing than at other times. The survival probability as a function of operating time looks like a step function, where the steps correspond to the scheduled takeoffs and landings.

Intuitively, it seems reasonable to assume that a new tire is better than a used tire. One model for the survival probability is

$$\bar{F}(t) = e^{-\alpha[t/h]} \qquad \text{for} \quad t \geq 0,$$

where $[t]$ denotes the largest integer less than or equal to t, and h represents the interval between flights, assumed constant for simplicity. It is easy to

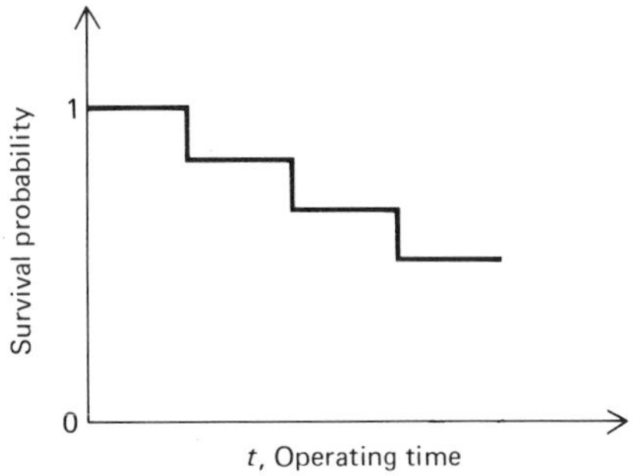

Figure 6.3.1.
Survival probability for airplane tire.

verify that this distribution is NBU but neither IFR nor IFRA. Using (3.3a), it follows that

$$P[N(t) \leq n] \geq \sum_{j=0}^{n} \frac{[t/h]^j}{j!} e^{-[t/h]} \qquad \text{for} \quad t \geq 0.$$

Numerical Example. Assume each of the eight tires of a certain airplane have survival probability

$$\bar{F}(t) = e^{-.002[t/2]} \qquad \text{for} \quad t \geq 0,$$

corresponding to the fact that individual flights are about 2 hours long and the additional fact that the mean time to failure of a tire is known to be

$$\cong \int_0^\infty e^{-.002[t/2]}\, dt = 2[1 + e^{-.002} + e^{-.004} + \cdots] = \frac{2}{1 - e^{-.002}}$$

$$\cong 1000 \text{ hours.}$$

Determine the number of spare tires to be stocked so that with assurance $\alpha \geq .95$, no shortage of tires is experienced during a period of 200 operating hours.

Let N_i be the number of failures experienced by tire i during [0, 200], $i = 1, \ldots, 8$. Then from (3.3), for $i = 1, \ldots, 8$,

$$N_i \overset{\text{st}}{\leq} M_i,$$

where $M_1, \ldots, M_8$ are independent Poisson random variables with mean $.002[\frac{1}{2}(200)] = .2$. Assuming the $N_1, \ldots, N_8$ are also mutually independent, it follows that $\sum_1^8 N_i \overset{\text{st}}{\leq} \sum_1^8 M_i$, a Poisson random variable with mean $8 \times .2 = 1.6$. Hence we have

$$P\left[\sum_1^8 N_i \leq n\right] \geq P\left[\sum_1^8 M_i \leq n\right] = \sum_{j=0}^{n} e^{-1.6} \frac{1.6^j}{j!}.$$

The smallest value of n for which this cumulative Poisson with mean 1.6 first exceeds .95 is $n = 4$; specifically, $P[\sum_1^8 N_i \leq 4] \geq .976$.

The Renewal Function

The *renewal function* $M(t)$ is defined as the expected number of renewals in $[0, t]$; that is,

$$M(t) = EN(t).$$

This function plays a central role in renewal theory and has many applications in reliability, especially in maintenance models. By Exercise 11 of Section 6, Chapter 4,

$$EN(t) = \sum_{k=1}^{\infty} P[N(t) \geq k],$$

so that

$$M(t) = \sum_{k=1}^{\infty} F^{(k)}(t). \tag{3.4}$$

We obtain an integral equation for the renewal function as follows. Replacing $F^{(k+1)}(t)$ in (3.4) by the equivalent $\int_0^t F^{(k)}(t - x)\, dF(x)$, we obtain

$$M(t) = F(t) + \sum_{k=1}^{\infty} \int_0^t F^{(k)}(t - x)\, dF(x),$$

or

$$M(t) = F(t) + \int_0^t M(t - x)\, dF(x), \tag{3.5}$$

known as the *fundamental renewal equation*.

If F has a density f, then differentiating (3.4) yields

$$m(t) = \sum_{k=1}^{\infty} f^{(k)}(t), \tag{3.6}$$

where $m(t) \equiv dM(t)/dt$, known as the *renewal density*. From (3.6), we may interpret $m(t)\, dt$ as the probability of a renewal occurring in $[t, t + dt]$. Alternately, differentiating (3.5) we obtain

$$m(t) = f(\;) + \int_0^t m(t - x) f(x)\, dx. \tag{3.7}$$

The Laplace-Stieltjes transform

$$M^*(s) = \int_0^{\infty} e^{-sx}\, dM(x)$$

plays a useful role in renewal theory. Taking transforms in (3.5), we obtain

$$M^*(s) = F^*(s) + M^*(s)F^*(s),$$

so that

$$M^*(s) = \frac{F^*(s)}{1 - F^*(s)},$$

or, equivalently,

$$F^*(s) = \frac{M^*(s)}{1 + M^*(s)}.$$

Thus we see that $M(t)$ is determined by $F(t)$, and, conversely, $F(t)$ is determined by $M(t)$.

3.5. Definition. A statement is true *almost surely* (a.s.), if it is true with probability 1.

3.6. Theorem. Let F have mean μ. Then

$$\frac{N(t)}{t} \overset{\text{a.s.}}{\to} \frac{1}{\mu} \quad \text{as} \quad t \to \infty.$$

(We interpret $1/\mu$ as 0 when $\mu = \infty$.)

PROOF. First suppose $\mu < \infty$. Since $S_{N(t)} \leq t \leq S_{N(t)+1}$, we may write

$$\frac{S_{N(t)}}{N(t)} \leq \frac{t}{N(t)} \leq \frac{S_{N(t)+1}}{N(t)+1} \frac{N(t)+1}{N(t)}.$$

As $t \to \infty$, $N(t) \to \infty$ a.s., and so $S_{N(t)}/N(t) \to \mu$ a.s. by the strong law of large numbers. We conclude that $t/N(t) \overset{\text{a.s.}}{\to} \mu$.

Next, supposing $\mu = \infty$, we can show $S_{N(t)}/N(t) \overset{\text{a.s.}}{\to} \infty$ using a truncation argument. ||

Remark. The related *elementary renewal theorem*, proved in Ross (1970), pp. 40–41, states that if F has mean μ, then $\lim_{t\to\infty} M(t)/t = 1/\mu$. Note the difference between the types of convergence in Theorem 3.8 and in the elementary renewal theorem.

Recall from Chapter 3 that a *Poisson process* is a renewal process with an underlying exponential distribution $F(t) = 1 - e^{-t/\mu}$. The corresponding renewal function $M(t)$ is linear; more specifically, $M(t) = t/\mu$ for $t \geq 0$. [See Exercise 3(a).] Thus $M(t + h) - M(t) = h/\mu$ for $t \geq 0$, $h \geq 0$. Hence, for the Poisson process, the expected number of renewals in any interval of length h is simply h divided by the mean life.

Intuitively, we might expect this to be the case for any renewal process after an indefinitely long time. This is in fact true under mild restrictions, as shown in Blackwell's Theorem below.

3.7. Definition. A random variable X is said to be *lattice* (or periodic) if there exists $h > 0$ such that

$$P[X = nh, n = 0, 1, 2, \text{ or } \cdots] = 1.$$

The distribution of X is also called lattice.

3.8. Theorem (Blackwell). If F is a nonlattice distribution with mean μ, then

$$\lim_{t \to \infty} [M(t + h) - M(t)] = \frac{h}{\mu}.$$

An elementary proof of this fundamental theorem may be found in Feller (1966), Chapter XI.

Remark. The *Key Renewal Theorem* of Smith (1958), equivalent to Blackwell's Theorem, states: If F is a nonlattice distribution with mean μ, and if $Q(t)$ is directly Riemann-integrable (Feller, 1966, p. 348), then

$$\lim_{t \to \infty} \int_0^t Q(t - x)\, dM(x) = \frac{1}{\mu} \int_0^\infty Q(t)\, dt.$$

We define the *age* random variable $\delta(t)$ in a renewal process by

$$\delta(t) = t - S_{N(t)}; \tag{3.8}$$

intuitively $\delta(t)$ is the age of the unit in use at time t. The *remaining life* random variable $\gamma(t)$ is defined by

$$\gamma(t) = S_{N(t)+1} - t; \tag{3.9}$$

$\gamma(t)$ is the remaining life of the unit in use at time t. Note that

$$P[\delta(t) \leq t] = 1,$$

and

$$P[\delta(t) = t] = \bar{F}(t).$$

Note that

$$P[\gamma(t) > u] = \bar{F}(t + u) + \sum_{n=1}^{\infty} \int_0^t \bar{F}(t - x + u)\, dF^{(n)}(x),$$

so that

$$P[\gamma(t) > u] = \bar{F}(t + u) + \int_0^t \bar{F}(t - x + u)\, dM(x). \tag{3.10}$$

From (3.10) we may obtain a lower bound for $P[\gamma(t) > u]$.

3.9. Example: Interval Reliability. If units are replaced at failure and replacement takes a negligible amount of time, then we may ask for the probability that the unit is in operation at time t and continues to operate without failure for an additional interval of length u. This probability is just (3.10).

3.10. Lemma. If F is NBU (NWU), then

$$P[\gamma(t) > u] \le (\ge) \bar{F}(u). \tag{3.11}$$

PROOF. Suppose F is NBU. By (3.10),

$$\begin{aligned} P[\gamma(t) > u] &= \bar{F}(t+u) + \int_0^t \bar{F}(t-x+u)\,dM(x) \\ &\le \bar{F}(u)\bar{F}(t) + \bar{F}(u)\int_0^t \bar{F}(t-x)\,dM(x) \\ &= \bar{F}(u)P[\gamma(t) \ge 0] = \bar{F}(u). \end{aligned}$$

If F is NWU, a similar argument applies with inequalities reversed. ||

Using Lemma 3.10 we may deduce the superadditivity of the renewal function when the underlying distribution is NBU.

3.11. Theorem. Let F be NBU (NWU). Then

$$M(h) \le (\ge) M(t+h) - M(t). \tag{3.12}$$

PROOF. Suppose F is NBU. Let $F_{\gamma(t)}(u) = P[\gamma(t) \le u]$. Then

$$M(t+h) - M(t) = \sum_{k=0}^{\infty} F_{\gamma(t)} * F^{(k)}(h) \ge \sum_{k=0}^{\infty} F * F^{(k)}(h)$$

by Lemma 3.10. Since $\sum_{k=0}^{\infty} F^{(k+1)}(h) = M(h)$, (3.12) follows.

If F is NWU, the argument is similar with the inequality reversed. ||

Note that if F is a nonlattice NBU (NWU) distribution with mean μ, then by Theorem 3.11,

$$M(h) \le (\ge) \lim_{t\to\infty} [M(t+h) - M(t)].$$

But by Theorem 3.8, this limit equals h/μ. Thus we conclude that

$$M(h) \le (\ge) \frac{h}{\mu}. \tag{3.13}$$

This inequality is actually true under a weaker assumption. First we need a useful result of Prokhorov and Kolmogorov.

3.12. Theorem (Wald's Equation). If $\{X_i\}_{i=1}^{\infty}$ is a sequence of independent and identically distributed random variables with $EX_i < \infty$, and if N is an integer-valued positive random variable such that the event $\{N = n\}$ is independent of $X_{n+1}, X_{n+2}, \ldots$ for all $n = 1, 2, \ldots$, then

$$E \sum_{i=1}^{N} X_i = ENEX_i.$$

PROOF. Letting

$$Z_i = \begin{cases} 1 & \text{if } N \geq i, \\ 0 & \text{if } N < i, \end{cases}$$

we see that

$$\sum_{i=1}^{N} X_i = \sum_{i=1}^{\infty} X_i Z_i.$$

Hence

$$E \sum_{i=1}^{N} X_i = E \sum_{i=1}^{\infty} X_i Z_i = \sum_{i=1}^{\infty} E(X_i Z_i).$$

However, Z_i is determined by $\{N < i\}$, and is thus independent of X_i. It follows that

$$E \sum_{i=1}^{N} X_i = \sum_{i=1}^{\infty} EX_i EZ_i = EX \sum_{i=1}^{\infty} EZ_i$$

$$= EX \sum_{i=1}^{\infty} P[N \geq i]$$

$$= EXEN. \; \|$$

Remark. From Theorem 3.12 and the fact that $N(t) + 1$ is independent of $X_{n+2}, X_{n+3}, \ldots$,

$$ES_{N(t)+1} = \mu[M(t) + 1]$$

follows from Theorem 3.12.

3.13. Definition. $\{\hat{X}_k\}_{k=1}^{\infty}$ is a *stationary renewal process* if all variables are independent, with $\hat{X}_2, \hat{X}_3, \ldots$ distributed according to F with mean μ (assumed finite), while $\hat{X}_1$ is distributed according to $\hat{F}(t) \equiv (1/\mu) \int_0^t \bar{F}(x)\, dx$. $\{\hat{N}(t); t \geq 0\}$ denotes the corresponding stationary renewal counting process.

As pointed out in Cox (1962), p. 46, the stationary renewal process has a simple physical interpretation. Suppose a renewal process is started at time $t = -\infty$, but observation of the process begins at $t = 0$. Then $\hat{X}_1$ has distribution $\hat{F}$. For a stationary renewal process $E\hat{N}(t) = t/\mu$.

By comparing a renewal process with its associated stationary process we can obtain (3.13) under the weaker assumption that F is NBUE (NWUE).

3.14. Theorem. (a) Let the underlying distribution F of a renewal process have mean $\mu < \infty$. Then

$$M(t) \geq \frac{t}{\mu} - 1 \qquad \text{for} \quad t \geq 0. \tag{3.14}$$

(b) If, in addition, F is NBUE (NWUE), then

$$M(t) \leq (\geq) \frac{t}{\mu} \qquad \text{for} \quad t \geq 0. \tag{3.13}$$

PROOF. (a)

$$0 \leq E\gamma(t) \equiv ES_{N(t)+1} - t = \mu[M(t) + 1] - t$$

by Theorem 3.12. Inequality (3.14) follows immediately.

(b) Suppose F is NBUE. Then by definition, $\hat{F}(t) = (1/\mu)\int_0^t \overline{F}(x)\,dx \geq F(t)$. It follows that

$$\frac{t}{\mu} = \hat{M}(t) = \sum_{n=1}^{\infty} \int_0^t F^{(n-1)}(t - x)\,d\hat{F}(x)$$

$$\geq \sum_{n=1}^{\infty} \int_0^t F^{(n-1)}(t - x)\,d\ \ (x) = M(t).$$

For F NWUE, the argument is similar with the inequalities reversed. ‖

Note that for F NBUE, from (3.13) and (3.14) we derive the close bounds

$$\frac{t}{\mu} - 1 \leq M(t) \leq \frac{t}{\mu} \qquad \text{for} \quad t \geq 0. \tag{3.15}$$

Thus we may estimate $M(t)$ by, say, $t/\mu - 1/2$ knowing only that the underlying distribution is NBUE with mean μ; the error of estimate is $\leq \frac{1}{2}$ uniformly for all $t \geq 0$. Since $M(t)$ is in general difficult to compute exactly, (3.15) constitutes a valuable pair of bounds.

By (3.13), if F is NBUE with mean μ and $\overline{G}(t) = \exp(-t/\mu)$, then

$$\sum_{n=1}^{\infty} F^{(n)}(t) \leq \frac{t}{\mu} \equiv \sum_{n=1}^{\infty} G^{(n)}(t) \qquad \text{for} \quad t \geq 0,$$

where $F^{(n)}(G^{(n)})$ is the n-fold convolution of $F(G)$.

This result can be generalized and the subsequent generalization can be applied to the spare parts problem.

3.15. Theorem. Let $F_i(i = 1, 2, \ldots)$ be NBUE with mean μ and $\overline{G}(t) = \exp(-t/\mu)$. Then

$$\sum_{i=j}^{\infty} F_1 * \cdots * F_i(t) \leq \sum_{n=j}^{\infty} G^{(n)}(t), \qquad j = 1, 2, \ldots; \quad t \geq 0.$$

PROOF. *Case 1:* $j = 1$. We first show that the conclusion holds for $j = 1$. By Theorem 3.14 the result holds when the F_i are identical and $j = 1$. Suppose that the result holds whenever $F_k \equiv F_{k+1} \equiv \cdots$ (that is, only the first k distributions can differ), and consider a renewal process where $F_{k+1} \equiv F_{k+2} \equiv \cdots$. Then by the induction hypothesis the result holds when inter-renewal time distributions are $F_2, F_3, \ldots, F_k, F_{k+1} = F_{k+2} = \cdots$ since only the first $k - 1$ distributions can differ. Thus

$$\sum_{i=1}^{\infty} F_1 * \cdots * F_i(t) = \int_0^t \left[1 + \sum_{j=1}^{\infty} F_2 * \cdots * F_{j+1}(t - x)\right] dF_1(x)$$

$$\leq \int_0^t \left[1 + \frac{t - x}{\mu}\right] dF_1(x)$$

$$= \frac{t}{\mu} + F_1(t) - \frac{1}{\mu}\int_0^t \bar{F}_1(x)\, dx \leq \frac{t}{\mu}$$

since F_1 is NBUE. The result follows by taking the limit as $k \to \infty$.

Case 2: $j > 1$. The proof is again by induction. The conclusion holds for $j = 1$ by Case 1. Assume now that it holds for $j = 1, 2, \ldots, j_0$. Then

$$\sum_{i=j_0+1}^{\infty} F_1 * \cdots * F_i(t)$$

$$= \sum_{i=j_0+1}^{\infty} \int_0^t F_2 * \cdots * F_i(t - x)\, dF_1(x)$$

$$= \int_0^t \left[\sum_{i=j_0}^{\infty} F_2 * \cdots * F_{i-1}(t - x)\right] dF_1(x) \leq \int_0^t \sum_{i=j_0}^{\infty} G^{(i)}(t - x)\, dF_1(x)$$

by inductive hypothesis. Commuting convolutions we have

$$\int_0^t \sum_{i=j_0}^{\infty} G^{(i)}(t - x)\, dF_1(x) = \int_0^t \sum_{i=j_0}^{\infty} (G^{(i-1)} * F_1)(t - x)\, dG(x)$$

$$\leq \int_0^t \sum_{i=j_0}^{\infty} G^{(i)}(t - x)\, dG(x) = \sum_{i=j_0+1}^{\infty} G^{(i)}(t).$$

The last inequality follows again by inductive hypothesis. ‖

From Theorem 3.15 it follows immediately that if $F_i \equiv F$ $(i = 1, 2, \ldots)$, then

$$\sum_{n=j}^{\infty} F^{(n)}(t) \leq \sum_{n=j}^{\infty} G^{(n)}(t), \qquad j = 1, 2, \ldots; \quad t \geq 0.$$

3.16. Corollary. Let F be NBUE with mean μ, $\bar{G}(t) = e^{-t/\mu}$, and $0 \le h(1) \le h(2) \le \cdots$. Then

$$\sum_{n=1}^{\infty} h(n)F^{(n)}(t) \le \sum_{n=1}^{\infty} h(n)G^{(n)}(t).$$

PROOF. Using summation by parts we see that

$$\begin{aligned}\sum_{n=1}^{\infty} h(n)F^{(n)}(t) &= h(1)\sum_{n=1}^{\infty} F^{(n)}(t) + [h(2) - h(1)]\sum_{n=2}^{\infty} F^{(n)}(t) \\ &\quad + [h(3) - h(2)]\sum_{n=3}^{\infty} F^{(n)}(t) + \cdots \\ &\le h(1)\sum_{n=1}^{\infty} G^{(n)}(t) + [h(2) - h(1)]\sum_{n=2}^{\infty} G^{(n)}(t) \\ &\quad + [h(3) - h(2)]\sum_{n=3}^{\infty} G^{(n)}(t) + \cdots \\ &= \sum_{n=1}^{\infty} h(n)\ {}^{(n)}(t).\end{aligned}$$

The last inequality follows from Theorem 3.15 and the hypothesis that $[h(i + 1) - h(i)] \ge 0$. ‖

3.17. Theorem. If F is NBUE with mean $\mu = 1/\lambda$, $c(k)$ is convex increasing, and $c(0) = 0$, then

$$\sum_{k=0}^{\infty} c(k)P[N(t) = k] \le \sum_{k=0}^{\infty} c(k)\frac{(\lambda t)^k}{k!}e^{-\lambda t}.$$

PROOF. Recall that $P[N(t) = k] = F^{(k)}(t) - F^{(k+1)}(t)$, so that

$$\sum_{k=0}^{\infty} c(k)P[N(t) = k] = \sum_{k=0}^{\infty} c(k)[F^{(k)}(t) - F^{(k+1)}(t)].$$

Note that

$$\sum_{k=0}^{\infty} a_k b_k = (a_1 - a_0)\sum_{k=1}^{\infty} b_k + (a_2 - a_1)\sum_{k=2}^{\infty} b_k + (a_3 - a_2)\sum_{k=3}^{\infty} b_k + \cdots,$$

where $a_0 = 0$, so that

$$\sum_{k=0}^{\infty} c(k)[F^{(k)}(t) - F^{(k+1)}(t)] = \sum_{k=1}^{\infty} [c(k) - c(k - 1)]F^{(k)}(t),$$

since $c(0) = 0$. Now $c(k)$ increasing implies

$$h(k) = c(k) - c(k - 1) \ge 0,$$

and $c(k)$ convex implies $h(k)$ increasing. Thus by Corollary 3.16,

$$\sum_{k=0}^{\infty} [c(k) - c(k-1)]F^{(k)}(t) \le \sum_{k=0}^{\infty} [c(k) - c(k-1)]G^{(k)}(t)$$

$$= \sum_{k=0}^{\infty} c(k) \frac{(\lambda t)^k e^{-\lambda t}}{k!},$$

where $G(t) = 1 - e^{-\lambda t}$. The last equality follows by resumming by parts. ||

3.18. Example: Minimizing the Expected Shortage. In the spare parts example (Chapter 4) we wished to determine the number of spares necessary to keep the *probability* of running out small. Suppose that instead we wish to keep the *expected* shortage small. The expected shortage if we stock N spares is

$$\sum_{k=N}^{\infty} [k - N]P[N(t) = k].$$

Note that the function

$$c(k) = \begin{cases} 0, & k \le N, \\ k - N, & k \ge N, \end{cases}$$

is convex on its domain of definition. By Theorem 3.17 if the life distribution of parts is NBUE with $\mu = 1/\lambda$, then for $N = 0, 1, 2, \ldots,$

$$\sum_{k=N}^{\infty} [k - N]P[N(t) = k] \le \sum_{k=N}^{\infty} [k - N] \frac{(\lambda t)^k}{k!} e^{-\lambda t}.$$

Note that the upper bound for expected shortage corresponds to an underlying exponential life distribution; it can be rewritten as

$$(\lambda t) \sum_{j=N-1}^{\infty} \frac{(\lambda t)^j}{j!} e^{-\lambda t} - N \sum_{j=N}^{\infty} \frac{(\lambda t)^j}{j!} e^{-\lambda t} \tag{3.16}$$

for $N \ge 1$. This can be evaluated for specified N using tables of the cumulative Poisson distribution. Figure 6.3.2 is a graph of the upper bound in (3.16) for $N = 1, 2,$ and 3.

3.19. Theorem. Let F be NBU (NWU). Then

$$\text{var } N(t) \le (\ge) M(t). \tag{3.17}$$

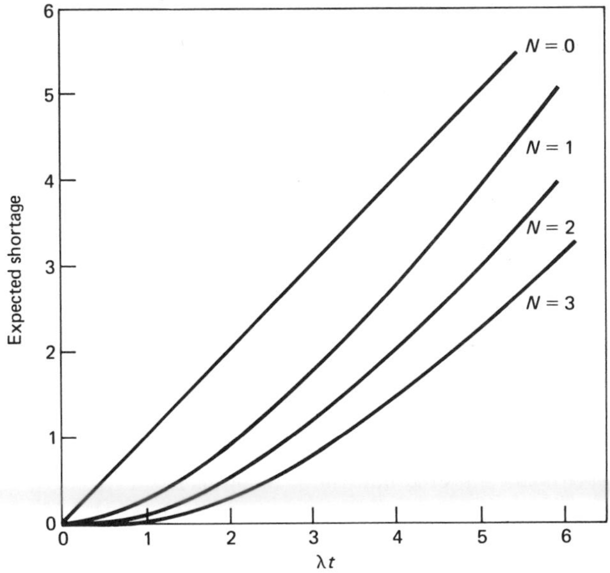

Figure 6.3.2.
Graph of upper bound (3.16) for expected shortage.

PROOF. Assume F is NBU. Then

$$\begin{aligned}\operatorname{var} N(t) - M(t) &= EN^2(t) - M^2(t) - M(t) = 2E\binom{N(t)}{2} - M^2(t)\\ &= 2M^{(2)}(t) - M^2(t) = \int_0^t [2M(t-x) - M(t)]\, dM(x),\end{aligned}$$

where $E\binom{N(t)}{2} = M^{(2)}(t)$ by Exercise 14.

By Theorem 3.11,

$$M(t) - M(t-x) \geq M(x),$$

so that

$$\operatorname{var} N(t) - M(t) \leq \int_0^t [M(t-x) - M(x)]\, dM(x).$$

Writing $\int_0^t [M(t-x) - M(x)]\, dM(x) = \int_0^{t/2} [M(t-x) - M(x)]\, dM(x) + \int_{t/2}^t [M(t-y) - M(y)]\, dM(y)$, and making the change of variable $x = t - y$ in the second integral, we obtain

$$\begin{aligned}\operatorname{var} N(t) - M(t) &\leq \int_0^{t/2} [M(t-x) - M(x)]\, dM(x)\\ &\quad + \int_0^{t/2} [M(t-x) - M(x)]\, d_x M(t-x)\\ &= \int_0^{t/2} [M(t-x) - M(x)]\, d_x[M(x) + M(t-x)].\end{aligned}$$

Note that $M(t-x) - M(x)$ is decreasing in x and is ≥ 0 for $0 \leq x \leq t/2$. Also $\int_0^x d_u[M(u) + M(t-u)] = M(x) + M(t-x) - M(t) \leq 0$. Thus by Lemma 7.1(b) of Chapter 4, $\int_0^{t/2} [M(t-x) - M(x)]\, d_x[M(x) + M(t-x)] \leq 0$, proving (3.17).

For F NWU, the proof is similar with inequalities reversed. ||

EXERCISES

1. Assume $X_i \overset{\text{st}}{\geq} Y_i$, $i = 1, 2$, and X_1, X_2 independent, Y_1, Y_2 independent. Then $X_1 + X_2 \overset{\text{st}}{\geq} Y_1 + Y_2$. Show that the assumption of independence cannot be dropped.

2. Consider a renewal process with underlying exponential distribution $F(t) = 1 - e^{-\lambda t}$, that is, a Poisson process. Compute

   (a) $M(t)$, the expected number of renewals in $[0, t]$;
   (b) $P[N(t) \leq n]$, the distribution of the number of renewals in $[0, t]$;
   (c) $M^*(s)$, the Laplace transform of $M(t)$;
   (d) The distribution of S_k.

3. Sockets 1, 2, and 3 of a system all use the same type of transistor. The life distribution for each component position is exponential, but with failure rate .001 for the first position, .002 for the second position, and .0015 for the third position. During a mission, the first component position is required to operate for 3000 hours, the second for 5000 hours, and the third for 1000 hours. How many spares should be stocked to provide an assurance of at least .95 of no shortage of spares during the mission?

4. (a) Show that age replacement generates two renewal processes, one corresponding to the sequence of failures, and the other corresponding to the sequence of removals of both failed and deliberately replaced items.
   (b) What are the underlying distributions of these renewal processes?
   (c) Does block replacement generate a renewal process?

5. Prove that F is NBUE (NWUE) if and only if $\hat{N}(t) \geq (\leq) N(t)$ for all $t \geq 0$. Use this result to provide an alternate proof of Theorem 3.14(b).

6. (a) Show that the lower bound for $M(t)$ given in (3.14) cannot be uniformly achieved for all $t \geq 0$ by a single distribution.
   (b) Show that given $\mu = 1$, $t > 1$, $\varepsilon > 0$, we may choose the underlying distribution of a renewal process to place mass $1/(t + \varepsilon)$ at $t + \varepsilon$ and mass $1 - (1/(t + \varepsilon))$ at 0, with the result that

$$M(t) = \frac{t}{\mu} - 1 + \varepsilon.$$

Thus at each point $t > \mu$, the lower bound $(t/\mu) - 1$ for $M(t)$ can be approached arbitrarily closely.

7. Estimate the expected number of renewals in $[0, 1000]$ in a renewal process with underlying density $f(x) = (.01)^2 x e^{-.01x}$. What is the maximum error of your estimate?

8. Consider the following system schematic diagram.

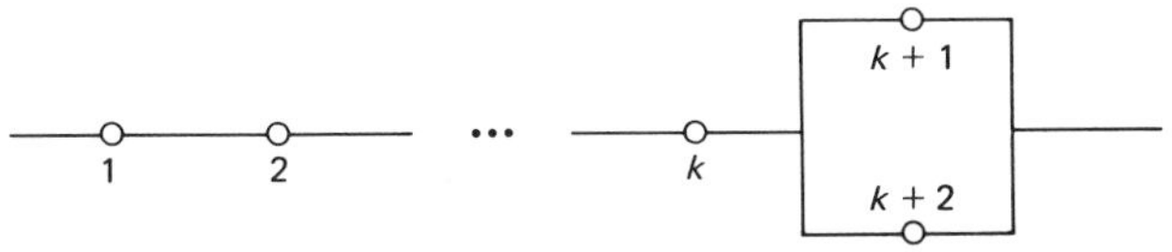

Suppose components are replaced at failure with negligible replacement time and that component i has distribution F_i. What is the probability that the system is operating at time t and continues to operate for an interval of length x? (This is the *interval reliability* of Example 3.9.)

*9. Let $m_\alpha(x)$ be the renewal density for a gamma density, f_α, with shape parameter α. Let r and p be integers. Show that

$$m_{r/p}(x) = \sum_{j=1}^{p-1} f_{(jr/p)} * m_r(x) + \sum_{j=1}^{p-1} f_{(jr/p)}(x) + m_r(x).$$

Hence, $m_{r/p}(x)$ can be computed using formula (3.6) for the renewal density when the underlying distribution is gamma with integer shape parameter.

*10. Let $N(t)$ denote the number of renewals during $[0, t]$ in a renewal process. Then for fixed $t > 0$, $N(t)$ satisfies the discrete IFRA property $P^{1/k}[N(t) \geq k] \downarrow$ in $k = 1, 2, \ldots$. (See Lemma 3.7 of Chapter 4.)

*11. Suppose $F \underset{\text{su}}{<} G$ where $F(0) = G(0) = 0$ and F and G are continuous on $[0, \infty)$. Let $R(t) = G^{-1}F(t)$. Show that

$$\bar{F}^{(n)}(t) \geq \bar{G}^{(n)}[R(t)] \quad \text{for} \quad t \geq 0, \quad n = 0, 1, 2, \ldots.$$

[Recall from Chapter 4, Section 5, that $F \underset{\text{su}}{<} G$ means $G^{-1}F(x)$ is superadditive for $x \geq 0$.]

*12. Suppose $F \underset{\text{c}}{<} G$ where $F(0) = G(0) = 0$ and F and G are continuous on $[0, \infty)$. Let $R(t) = G^{-1}F(t)$. Show that

$$\bar{F}^{(n)}(t) \leq \bar{G}^{(n)}\left[nR\left(\frac{t}{n}\right)\right]$$

for $t \geq 0$, $n = 0, 1, 2, \ldots$. (Recall from Chapter 4, Section 5, that $F \underset{\text{c}}{<} G$ means $G^{-1}F$ is convex.)

*13. Suppose $F \underset{c}{<} G$ where $G(x) = 1 - e^{-x}$ for $x \geq 0$, F is continuous on $[0, \infty)$ with mean $\mu = 1$, and $F(0) = 0$. Show that

$$\bar{F}^{(n)}(t) \leq \begin{cases} 1 & t < n, \\ \bar{G}^{(n)}[tW(t/n)] & t \geq n, \end{cases} \qquad (n \geq 1)$$

where $W(t/n)$ satisfies

$$\int_0^{t/n} e^{-xW(t/n)}\, dx = 1.$$

*14. Let $\{N(t); t \geq 0\}$ be a renewal counting process with renewal function $M(t)$. Let $B_k(t) = E\binom{N(t)}{k}$ be the kth binomial moment of $N(t)$, and $M^{(k)}(t)$ be the k-fold convolution of the renewal function. Show that

$$B_k(t) \equiv M^{(k)}(t)$$

for $k = 1, 2, \ldots,$ and $t \geq 0$.

15. Let F be NBUE (NWUE) with mean μ. Show that

$$EN^k(t) \leq (\geq) \sum_{j=0}^{\infty} \frac{j^k (t/\mu)^j}{j!} e^{-t/\mu}$$

for $k = 1, 2, \ldots,$ and $0 \leq t < \infty$.

16. Let F be IFR with mean μ. Then

$$\sum_{j=0}^{n-1} \frac{[R(t)]^j}{j!} e^{-R(t)} \leq \sum_{j=0}^{n-1} \frac{[nR(t/n)]^j}{j!} e^{-nR(t/n)}$$

for $t \geq 0$, $n = 1, 2, \ldots$. This shows that the lower bound in (3.3a) does not exceed the upper bound in (3.3b).

17. Let F be NBU and Y_t the remaining life random variable at time t; show that $EY_t \leq \mu = \int_0^\infty x\, dF(x)$.

4. REPLACEMENT POLICY COMPARISONS

In this section we compare several basic replacement policies assuming the underlying distribution F belongs to the appropriate class among those listed in Section 2.

Let

$N(t)$ = the number of renewals in $[0, t]$ for an ordinary renewal process,

$\hat{N}(t)$ = the number of renewals in $[0, t]$ for a stationary renewal process,

$N_A(t, T)$ = the number of failures in $[0, t]$ under an age replacement policy with replacement interval T,

$N_B(t, T)$ = the number of failures in $[0, t]$ under a block replacement policy with replacement interval T.

4.1. Theorem. $\hat{N}(t) \overset{\text{st}}{\geq} (\overset{\text{st}}{\leq})\ N(t)$ for all $t \geq 0 \Leftrightarrow F$ is NBUE (NWUE).

PROOF. $\Leftarrow$ Assume F is NBUE. Then $\hat{F}(t) = (1/\mu) \int_0^t \bar{F}(x)\, dx \geq F(t)$ by Exercise 3 of Section 2. Therefore,

$$P[\hat{N}(t) \geq n] = \int_0^t F^{(n-1)}(t - x)\, d\hat{F}(x)$$

$$\geq \int_0^t F^{(n-1)}(t - x)\, d\ \ (x) = P[N(t) \geq n],$$

so that $\hat{N}(t) \overset{\text{st}}{\geq} N(t)$.

For F NWUE, the proof is similar with the inequalities reversed.

$\Rightarrow$ Assume $\hat{N}(t) \overset{\text{st}}{\geq} N(t)$ for all $t \geq 0$. Then $P[\hat{N}(t) = 0] \leq P[N(t) = 0]$, so that $\int_t^\infty [\bar{F}(x)/\mu]\, dx \leq \bar{F}(t)$, implying F is NBUE.

Assuming $\hat{N}(t) \overset{\text{st}}{\leq} N(t)$ for all $t \geq 0$, we may prove F is NWUE in a similar fashion by reversing inequalities. ‖

4.2. Theorem. $N(t) \overset{\text{st}}{\geq} N_A(t, T)$ for all $t \geq 0$, $T \geq 0 \Leftrightarrow F$ is NBU.

PROOF. $\Leftarrow$ Let τ_i (τ_{iA}) be the interval between the $i - $1st and ith failure in an ordinary renewal process (under an age replacement policy). Then

$$P[\tau_i > x] = \bar{F}(x) \qquad \text{for} \quad x \geq 0,$$

$$P[\tau_{iA} > x] = [\bar{F}(T)]^j \bar{F}(x - jT) \qquad \text{for} \quad jT \leq x \leq (j+1)T, \quad j = 0, 1, 2, \ldots.$$

Since F is NBU, then $\bar{F}(x) \leq [\bar{F}(T)]^j \bar{F}(x - jT)$. Thus $\tau_i \overset{\text{st}}{\leq} \tau_{iA}$. Also $\tau_1, \tau_2, \ldots$ are independent, and $\tau_{1A}, \tau_{2A}, \ldots$ are independent. Thus $\tau_1 + \cdots + \tau_n \overset{\text{st}}{\leq} \tau_{1A} + \cdots + \tau_{nA}$ for $n = 1, 2, \ldots$. It follows that $P[N(t) \geq n] = P[\tau_1 + \cdots + \tau_n \leq t] \geq P[\tau_{1A} + \cdots + \tau_{nA} \leq t] = P[N_A(t, T) \geq n]$. Hence $N(t) \overset{\text{st}}{\geq} N_A(t, T)$.

$\Rightarrow$ Let $T < t < 2T$. Since $N(t) \overset{\text{st}}{\geq} N_A(t, T)$, then $\bar{F}(t) = P[N(t) = 0] \leq P[N_A(t, T) = 0] = \bar{F}(T)\bar{F}(t - T)$. Thus F is NBU. ‖

Theorem 4.2 states that the class of NBU distributions is the largest class for which age replacement diminishes stochastically the number of failures experienced in any particular time interval $[0, t]$, $0 < t < \infty$. In this sense, the NBU class of distributions is a natural class to consider in studying age replacement.

Next we show that a similar characterization applies in the case of block replacement. We first derive the following lemma.

4.3. Lemma. Let planned replacements occur at fixed time points $\{0 < t_1 < t_2 < \cdots\}$ under policy 1, and at time points $\{0 < t_1 < t_2 < \cdots\} \cup \{t_0\}$ under policy 2. Let $N_i(t)$ be the number of failures in $[0, t]$ under policy i, $i = 1, 2$. Then $N_1(t) \overset{\text{st}}{\geq} N_2(t)$ for each $t \geq 0 \Leftrightarrow$ the underlying life distribution F is NBU.

PROOF. $\Leftarrow$ For $t < t_0$, $N_1(t) \equiv N_2(t)$. Next assume that $t_0 < t \leq t_s$, where t_s is the smallest $t_i > t_0$. Let Z be the age of the unit in operation at time t_0^- under either policy. Let τ_i denote the interval between t_0 and the time of the first failure subsequent to time t_0 under policy i, $i = 1, 2$. Since F is NBU, then

$$P[\tau_1 > x \mid Z] \leq P[\tau_2 > x \mid Z] \qquad \text{for} \quad x \geq 0.$$

Let U_i denote the number of failures in (t_0, t) under policy i, $i = 1, 2$. Then

$$\begin{aligned} P[U_1 \geq n \mid Z] &= P[t_0 + \tau_1 + X_1 + \cdots + X_{n-1} \leq t \mid Z] \\ &\geq P[t_0 + \tau_2 + X_1 + \cdots + X_{n-1} \leq t \mid Z] \\ &= P[U_2 \geq n \mid Z] \end{aligned}$$

for $n = 0, 1, 2, \ldots$. By unconditioning on Z, we conclude that

$$P[U_1 \geq n] \geq P[U_2 \geq n] \qquad \text{for} \quad n = 0, 1, 2, \ldots.$$

Thus $N_1(t) \overset{\text{st}}{\geq} N_2(t)$.

Finally, assume that $t > t_s$. Let $N_i(t_s, t)$ denote the number of failures in (t_s, t) under policy i, $i = 1, 2$. Then $N_i(t_s) + N_i(t_s, t) = N_i(t)$, with $N_i(t_s)$ and $N_i(t_s, t)$ independent, $i = 1, 2$. Since $N_1(t_s) \overset{\text{st}}{\geq} N_2(t_s)$ and $N_1(t_s, t) \overset{\text{st}}{\equiv} N_2(t_s, t)$, we conclude that $N_1(t) \overset{\text{st}}{\geq} N_2(t)$.

$\Rightarrow$ Choose $0 < t_0 < t_1$. Then $\bar{F}(t_1) = P[N_1(t_1) = 0] \leq P[N_2(t_1) = 0] = \bar{F}(t_0)\bar{F}(t_1 - t_0)$ by hypothesis. Since $0 < t_0 < t_1$ are arbitrary, F is NBU. ||

4.4. Theorem. $N(t) \overset{\text{st}}{\geq} N_B(t, T)$ for all $t \geq 0$, $T \geq 0 \Leftrightarrow F$ is NBU.

PROOF. $\Leftarrow$ Apply Lemma 4.3 to compare successive pairs of a sequence of replacement policies, in which the ith policy calls for planned replacement at time points $\{0, T, \ldots, (i - 1)T\}$.

$\Rightarrow$ The argument is exactly the same as that used in the second half of the proof of Theorem 4.2. ||

Theorem 4.4 states that the class of NBU distributions is the largest class for which block replacement diminishes stochastically the number of failures experienced in any particular time interval $[0, t]$, $0 < t < \infty$. In this sense, the NBU class of distributions is a natural class to consider in studying block replacement, as it is in studying age replacement.

By similar arguments, we may obtain stochastic comparisons under block replacement with different replacement intervals.

4.5. Theorem. $N_B(t, kT) \overset{\text{st}}{\geq} N_B(t, T)$ for $t \geq 0$, $T \geq 0$, $k = 1, 2, \ldots, \Leftrightarrow F$ is NBU.

Next we study the effect of varying the replacement interval T under age replacement.

4.6. Theorem. $N_A(t, T) \uparrow st$ in $T \geq 0$ for each $t \geq 0 \Leftrightarrow F$ is IFR.

PROOF. $\Leftarrow$ For fixed $T > 0$, $\{N_A(t, T); t \geq 0\}$ is a renewal counting process with underlying distribution

$$S_T(t) = 1 - [\bar{F}(T)]^n \bar{F}(t - nT) \tag{4.1}$$

where, for $t \geq 0$, integer n is determined from $nT \leq t < (n + 1)T$. By differentiating $S_T(t)$ with respect to T, it is easy to verify that $S_T(t)$ is increasing in $T \geq 0$ for fixed $t \geq 0$. Hence $P[N_A(t, T) \geq n] = S_T^{(n)}(t)$ is increasing in $T \geq 0$ for fixed $t \geq 0$.

$\Rightarrow$ $\bar{S}_T(t) = P[N_A(t, T) = 0]$ is decreasing in $T \geq 0$ for each $t \geq 0$. Again differentiating $S_T(t)$ with respect to T, it follows that F is IFR. ||

Thus under an age replacement policy with an IFR failure distribution, the number of failures observed in any time interval $[0, t]$ increases stochastically as the replacement interval T increases. A weaker comparison is possible for the broader NBU class, as shown in Theorem 4.7.

4.7. Theorem. $N_A(t, kT) \overset{\text{st}}{\geq} N_A(t, T)$ for $t \geq 0$, $T \geq 0$, $k = 1, 2, \ldots \Leftrightarrow F$ is NBU.

PROOF. $\Leftarrow$ For $nT \leq t \leq (n + 1)T$, $\bar{S}_T(t) - \bar{S}_{kT}(t) = \bar{F}^n(T)\bar{F}(t - nT) - \bar{F}^{[n/k]}(kT)\bar{F}(t - [n/k]kT)$. Since F is NBU, then $\bar{F}^{[n/k]}(kT) \leq [\bar{F}(T)]^{[n/k]k}$, and $\bar{F}(t - [n/k]kT) \leq \bar{F}(T)^{n-k[n/k]}\bar{F}(t - nT)$. Thus $\bar{S}_T(t) \geq \bar{S}_{kT}(t)$. Hence $P[N_A(t, kT) \geq n] = S_{kT}^{(n)}(t) \geq S_T^{(n)}(t) = P[N_A(t, T) \geq n]$ for $t \geq 0$, $T \geq 0$. It follows that $N_A(t, kT) \overset{\text{st}}{\geq} N_A(t, T)$.

$\Rightarrow$ Let $k = 2$. Then for $T \leq t \leq 2T$, $\bar{F}(T)\bar{F}(t - T) = P[N_A(t, T) = 0] \geq P[N_A(t, 2T) = 0] = \bar{F}(t)$. Since t, T are arbitrary nonnegative numbers, F is NBU. ||

Theorem 4.7 is the analog for age replacement of Theorem 4.5.

Comparisons between age and block replacement policies are made in B-P (1965). For convenience, we recapitulate the results here, omitting proofs. We define $R_A(t, T)(R_B(t, T))$ to be the number of *removals* of both failed and unfailed components during $[0, t]$ under an age (block) replacement policy with replacement interval T.

4.8. Theorem. For all $t > 0$, $T > 0$: (a) $R_A(t, T) \overset{\text{st}}{\leq} R_B(t, T)$; (b) if F is IFR, then $N_A(t, T) \overset{\text{st}}{\geq} N_B(t, T)$.

Thus for every underlying life distribution, block replacement leads to more removals (stochastically) than does age replacement. If the life distribution is actually IFR, then block replacement reduces (stochastically) the number of failures experienced.

5. PRESERVATION OF LIFE DISTRIBUTION CLASSES UNDER RELIABILITY OPERATIONS

In Section 4 of Chapter 4 we classified the reliability operations under which the IFR, DFR, IFRA, and DFRA classes of life distributions are preserved. In this section we perform a similar classification for the NBU, NWU, NBUE, and NWUE classes. As in Section 4 of Chapter 4 we consider the reliability operations of

(a) Formation of coherent systems,
(b) Addition of life lengths,
(c) Mixture of distributions.

Coherent Systems

To show that the NBU property is preserved under the formation of coherent systems, we will find it helpful to define and use the properties of the hazard transform of a coherent system.

Definition. For a coherent system of n components, let F_i be the life distribution of the ith component, $i = 1, \ldots, n$. Then the *hazard function*, R_i (or simply, hazard R_i), of component i is $R_i(t) = -\log \bar{F}_i(t)$.

Let R be the system hazard. Then the function η expressing system hazard in terms of component hazards $\mathbf{R} = (R_1, \ldots, R_n)$ is called the *hazard transform:*

$$R = \eta(\mathbf{R}).$$

If the system reliability function is h, then a more explicit expression for η is

$$\eta(\boldsymbol{\rho}) = -\log h(e^{-\rho_1}, \ldots, e^{-\rho_n}) \qquad \text{for} \quad 0 \leq \rho_i < \infty, \quad i = 1, \ldots, n.$$

It is easy to verify that η is increasing in $\rho_1, \ldots, \rho_n$, and that $\eta(\mathbf{0}) = 0$, $\eta(\infty) = \infty$. (See Exercise 10.)

5.1. Theorem. If each component of a coherent system has an NBU life distribution, then the system has an NBU life distribution.

PROOF. Let R_i be the ith component hazard and R the system hazard. Using Exercise 2 of Section 2, $R_i(s + t) \geq R_i(s) + R_i(t)$. Since the hazard transform η is increasing, then

$$\eta[\mathbf{R}(s + t)] \geq \eta[\mathbf{R}(s) + \mathbf{R}(t)]. \tag{5.1}$$

Since η is superadditive by Exercise 10, then

$$\eta[\mathbf{R}(s) + \mathbf{R}(t)] \geq \eta[\mathbf{R}(s)] + \eta[\mathbf{R}(t)]. \tag{5.2}$$

From (5.1) and (5.2) we deduce

$$R(s + t) \equiv \eta[\mathbf{R}(s + t)] \geq \eta[\mathbf{R}(s)] + \eta[\mathbf{R}(t)] \equiv R(s) + R(t).$$

Thus system survival probability $\bar{F}(s + t) = e^{-R(s+t)} \leq e^{-R(s)}e^{-R(t)} = \bar{F}(s)\bar{F}(t)$. Hence system life is NBU. ||

5.2. NBUE Not Preserved. To show that the NBUE class need not be preserved under the formation of coherent systems, consider a series system of two independent components, each having life distribution

$$F(x) = \begin{cases} 0 & \text{for } 0 \leq x < 1, \\ \frac{1}{2} & \text{for } 1 \leq x < 3, \\ 1 & \text{for } 3 \leq x. \end{cases}$$

It is easy to verify that F is NBUE, but not NBU. However, the mean life of a new system is $\int_0^3 \bar{F}^2(x)\,dx = \frac{3}{2}$, whereas the mean life of a system of age 1 is 2. Thus system life is *not* NBUE.

5.3. NWU, NWUE Not Preserved. To show that neither the NWU nor the NWUE classes are preserved under the formation of coherent systems, consider a parallel system of two independent components each having life distribution $1 - e^{-t}$, which is NWU, and hence also NWUE. Then system life distribution is given by

$$F(t) = (1 - e^{-t})^2,$$

so that system failure rate is

$$r(t) = 1 - \frac{1}{2e^t - 1},$$

a strictly increasing function. Thus system life is neither NWU nor NWUE.

Convolutions

Next we show that the sum of independent NBU random variables is itself NBU.

5.4. Theorem. Let F_1, F_2 be NBU distributions. Then the convolution

$$F(t) = \int_0^t F_1(t - x)\, dF_2(x)$$

is NBU.

PROOF.

$$\bar{F}(x + y) = \int_0^x \bar{F}_2(x + y - z)\, dF_1(z) + \int_0^\infty \bar{F}_2(y - z)\, d_z F_1(x + z).$$

But

$$\int_0^x \bar{F}_2(x + y - z)\, dF_1(z) \le \bar{F}_2(y) \int_0^x \bar{F}_2(x - z)\, dF_1(z)$$

$$= \bar{F}_2(y)[\bar{F}(x) - \bar{F}_1(x)].$$

Also, integrating by parts,

$$\int_0^\infty \bar{F}_2(y - z)\, d_z F_1(x + z) = \bar{F}_2(y)\bar{F}_1(x) + \int_0^\infty \bar{F}_1(x + z)[-d_z F_2(y - z)]$$

$$\le \bar{F}_2(y)\bar{F}_1(x) + \bar{F}_1(x) \int_0^\infty \bar{F}_1(z)[-d_z F_2(y - z)]$$

$$= \bar{F}_2(y)\bar{F}_1(x) + \bar{F}_1(x)[\bar{F}(y) - \bar{F}_2(y)].$$

Thus

$$\bar{F}(x + y) \le \bar{F}_2(y)\bar{F}(x) + \bar{F}_1(x)\bar{F}(y) - \bar{F}_1(x)\bar{F}_2(y)$$

$$= \bar{F}(x)\bar{F}(y) - [\bar{F}(x) - \bar{F}_1(x)][\bar{F}(y) - \bar{F}_2(y)] \le \bar{F}(x)\bar{F}(y). \quad \|$$

Similarly, we may show that the sum of independent NBUE random variables is also NBUE.

5.5. Theorem. Let F_1, F_2 be NBUE. Then the convolution

$$F(t) = \int_0^t F_1(t - x)\, dF_2(x)$$

is NBUE.

PROOF.

$$\int_0^\infty \overline{F}(t+x)\,dx = \int_0^\infty \int_0^t \overline{F}_1(t+x-u)\,dF_2(u)\,dx + \int_0^\infty \int_t^\infty \overline{F}_1(t+x-u)\,dF_2(u)\,dx.$$

But for $u \leq t$,

$$\int_0^\infty \overline{F}_1(t-u+x)\,dx \leq \mu_1 \overline{F}_1(t-u)$$

by (2.2), where μ_i is the mean of F_i, $i = 1, 2$. Thus

$$\int_0^\infty \int_0^t \overline{F}_1(t+x-u)\,dF_2(u)\,dx \leq \mu_1 \int_0^t \overline{F}_1(t-u)\,dF_2(u) = \mu_1[\overline{F}(t) - \overline{F}_2(t)].$$

For $u > t$,

$$\int_0^\infty \overline{F}_1(t+x-u)\,dx = u - t + \mu_1,$$

so that

$$\int_0^\infty \int_t^\infty \overline{F}_1(t+x-u)\,dF_2(u)\,dx = \int_t^\infty (u-t+\mu_1)\,dF_2(u).$$

Let $w = u - t$. Then

$$\int_t^\infty (u-t+\mu_1)\,dF_2(u = \mu_1\overline{F}_2(t) + \int_0^\infty w\,dF_2(w+t) = \mu_1\overline{F}_2(t) + \int_0^\infty \overline{F}_2(w+t)\,dw \leq \mu_1\overline{F}_2(t) + \mu_2\overline{F}_2(t).$$

Thus

$$\int_0^\infty \overline{F}(t+x)\,dx \leq \mu_1\overline{F}(t) + \mu_2\overline{F}_2(t) \leq (\mu_1+\mu_2)\overline{F}(t). \;\|$$

5.6. NWU, NWUE Not Preserved. To show that neither the NWU nor the NWUE classes are preserved under convolution, let $F(t) = 1 - e^{-t}$, which is both NWU and NWUE. Then $F^{(2)}(t) = 1 - (1+t)e^{-t}$, a gamma distribution of order 2, which has strictly increasing failure rate, and thus is neither NWU nor NWUE.

Mixtures

Recall that if $\mathscr{F} = \{F_\alpha: \alpha \in \mathscr{A}\}$ is a family of probability distributions and G is a probability distribution, then $F(t) = \int F_\alpha(t)\, dG(\alpha)$ is a *mixture* of probability distributions from $\mathscr{F}$.

As we will see below, none of the four classes—NBU, NBUE, NWU, NWUE—is preserved under mixtures. However, we can demonstrate preservation of a subclass of the NWU and the NWUE classes under mixtures.

5.7. Theorem. Suppose F is the mixture of F_α, $\alpha \in \mathscr{A}$, with each F_α NWU (NWUE), and no two distinct F_α, $F_{\alpha'}$ crossing on $(0, \infty)$, Then F is NWU (NWUE).

PROOF. *NWU Case.* By the Chebyschev inequality for similarly ordered functions (Hardy, Littlewood, and Pólya, 1952, Theorem 4.3),

$$\bar{F}(s)\bar{F}(t) \equiv \int \bar{F}_\alpha(s)\, d\quad(\alpha) \int \bar{F}_\alpha(t)\, dG(\alpha) \leq \int \bar{F}_\alpha(s)\bar{F}_\alpha(t)\, d\quad(\alpha).$$

By the NWU property,

$$\int \bar{F}_\alpha(s)\bar{F}_\alpha(t)\, dG(\alpha) \leq \int \bar{F}_\alpha(s + t)\, dG(\alpha) \equiv \bar{F}(s + t).$$

Thus F is NWU.

NWUE Case. Similarly,

$$\mu\bar{F}(t) \equiv \int \mu_\alpha\, dG(\alpha) \int \bar{F}_\alpha(t)\, dG(\alpha) \leq \int \mu_\alpha \bar{F}_\alpha(t)\, dG(\alpha).$$

Using the NWUE defining property (2.2) first and the Fubini Theorem next, we have

$$\begin{aligned} \int \mu_\alpha \bar{F}_\alpha(t)\, dG(\alpha) &\leq \iint \bar{F}_\alpha(t + x)\, dx\, dG(\alpha) \\ &= \iint \bar{F}_\alpha(t + x)\, dG(\alpha)\, dx = \int \bar{F}(t + x)\, dx. \end{aligned}$$

Thus by (2.2), F is NWUE. ‖

5.8. NBU, NBUE Not Preserved. To see that neither the NBU nor the NBUE class is preserved under mixtures, note that a mixture of nonidentical exponential distributions has a strictly decreasing failure rate, and thus cannot be NBU nor NBUE.

5.9. NWU Not Preserved. To see that the NWU class is not closed under mixtures, consider the following example. Let $\bar{F}_\delta(x) = e^{-k\delta}$ for $(k-1)\delta < x \le k\delta$, $k = 1, 2, \ldots$. Then it is easy to verify that F is NWU.

Next we show that the mixture $F = \frac{1}{2}F_\delta + \frac{1}{2}F_\gamma$ does *not* satisfy the NWU property $\bar{F}(x+y) \ge \bar{F}(x)\bar{F}(y)$ for $0 < y < \delta < x < \gamma < 2\delta < x + y < 2\gamma$ (for example, take $y = 3$, $\delta = 4$, $x = 6$, $\gamma = 7$). Then $\bar{F}_\delta(x) = e^{-2\delta}$, $\bar{F}_\delta(y) = e^{-\delta}$, $\bar{F}_\delta(x+y) = e^{-3\delta}$; $\bar{F}_\gamma(x) = e^{-\gamma}$, $\bar{F}_\gamma(y) = e^{-\gamma}$, $\bar{F}_\gamma(x+y) = e^{-2\gamma}$. We compute $\bar{F}(x+y) - \bar{F}(x)\bar{F}(y) = \frac{1}{2}(e^{-3\delta} + e^{-2\gamma}) - \frac{1}{4}(e^{-2\delta} + e^{-\gamma}) \times (e^{-\delta} + e^{-\gamma}) = \frac{1}{4}(e^{-\gamma} - e^{-\delta})(e^{-\gamma} - e^{-2\delta}) < 0$, so that F is not NWU.

Table 5.1 presents a concise summary of the results of this section. We conjecture that NWUE is *not* preserved under arbitrary mixtures.

Table 5.1. Preservation of Life Distribution Classes under Reliability Operations

	Formation of Coherent Systems	*Convolutions*	*Arbitrary Mixtures*	*Mixtures of Distributions That Do Not Cross*
NBU	Preserved	Preserved	Not preserved	Not preserved
NBUE	Not preserved	Preserved	Not preserved	Not preserved
NWU	Not preserved	Not preserved	Not preserved	Preserved
NWUE	Not preserved	Not preserved	?	Preserved

EXERCISES

1. Are coherent systems of *like* NBUE (NWU) ((NWUE)) necessarily NBUE (NWU) ((NWUE))?

2. Show that *any* distribution on $[0, \infty)$ may be obtained by mixing distributions from an appropriate subclass of the IFR class.

*3. Let $\mu_t = \int_0^\infty x^t\, dF(x)$ and $\lambda_t = \mu_t/\Gamma(t+1)$. Show that if F is NBU (NWU), then

$$\lambda_{r+s} \le (\ge) \lambda_r \lambda_s$$

for $r \ge 0$, $s \ge 0$ (see Corollary 6.5 of Chapter 4).

*4. Let F be NBUE (NWUE). Show that

$$\lambda_{r+1} \le (\ge) \lambda_r \lambda_1 \qquad \text{for} \quad r \ge 0,$$

where λ_r is defined in Exercise 3.

*5. Let F be NBUE (NWUE) with mean μ. Show that

$$\int_x^\infty \bar{F}(t)\, dt \le (\ge)\, \mu e^{-x/\mu} \qquad \text{for} \quad x \ge 0.$$

*6. Let F_i be NBUE (NWUE) with mean μ_i, $i = 1, 2, \ldots, n$. Show that

(a) $$\int_0^x \prod_{i=1}^n \bar{F}_i(t)\, dt \geq (\leq) \left(\sum_{i=1}^n \frac{1}{\mu_i}\right)^{-1} \left[1 - \exp\left[-x \sum_{i=1}^n \frac{1}{\mu_i}\right]\right]$$

for x $\geq$ 0.

(b) $$\int_x^\infty \left[1 - \prod_{i=1}^n F_i(t)\right] dt \leq (\geq) \int_x^\infty \left[1 - \prod_{i=1}^n (1 - e^{-t/\mu_i})\right] dt$$

for $x \geq 0$.

Note that by setting $x = \infty$ in (a) we conclude that the mean life of a series system of independent components is greater (less) when the components are NBUE (NWUE) than when they are exponential. Similarly, by setting $x = 0$ in (b), we conclude that the mean life of a parallel system of independent components is less (greater) when the components are NBUE (NWUE) than when they are exponential.

*7. Let F be NBU with $\bar{F}(t) = \alpha$ for a fixed value of t. Then

$$\bar{F}(x) \begin{cases} \geq \alpha^{1/k} & \text{for } \dfrac{t}{k+1} < x \leq \dfrac{t}{k}, \quad k = 0, 1, 2, \ldots, \\ \leq \alpha^k & \text{for } kt \leq x < (k+1)t, \quad k = 0, 1, 2, \ldots. \end{cases}$$

After proving these bounds hold, show that they are sharp.

*8. Let F be NWU with $\bar{F}(t) = \alpha$ for a fixed value of t. Then

$$\bar{F}(x) \begin{cases} \leq \alpha^{1/(k+1)} & \text{for } \dfrac{t}{k+1} \leq x < \dfrac{t}{k}, \quad k = 0, 1, 2, \ldots, \\ \geq \alpha^{k+1} & \text{for } kt \leq x < (k+1)t, \quad k = 0, 1, 2, \ldots. \end{cases}$$

After proving these bounds hold, show that they are sharp.

*9. Let F be NBUE with mean μ. Then for $t \leq \mu$,

$$F(t) \leq \frac{t}{\mu}.$$

The bound is sharp.

*10. Show that the hazard transform $\eta(\mathbf{R}) = -\log h(e^{-R_1}, \ldots, e^{-R_n})$ of a coherent system with reliability function $h(p_1, \ldots, p_n)$ (a) is increasing in each argument, (b) satisfies $\eta(\mathbf{0}) = 0$, $\eta(\infty) = \infty$, (c) is continuous where finite, and (d) is superadditive; that is, $\eta(\mathbf{R}^{(1)} + \mathbf{R}^{(2)}) \geq \eta(\mathbf{R}^{(1)}) + \eta(\mathbf{R}^{(2)})$.

6. NOTES AND REFERENCES

Section 1. Cox (1962) discusses age replacement. Flehinger (1962) considers block replacement. Both policies are extensively treated in B-P (1965).

Section 2. The NBU, NWU, NBUE, and NWUE classes are systematically analyzed and their applications in maintenance theory described in Marshall-Proschan (1972). Many of the results concerning these classes presented in this chapter are from that paper.

Section 3. The presentation of the beginning of this section follows to some extent the format of Section 2, Chapter 3, of B-P (1965). The main change is that the unnecessarily strong IFR (DFR) hypothesis is replaced by the NBU (NWU) hypothesis in most cases.

The NBU bound of Theorem 3.2 is essentially due to E. Straub (1970). Theorem 3.15, Corollary 3.16, and Theorem 3.17 were also motivated by the work of E. Straub (1970).

Section 4. The stochastic comparisons of Theorems 4.1, 4.2, and 4.4 are obtained in Barlow-Proschan (1964) under the unnecessarily strong hypothesis that the underlying life distribution is IFR (alternately in Theorem 4.1, DFR). The necessary conditions for stochastic comparison in the three theorems were not obtained in the paper.

Section 5. Theorem 5.1 is presented in Esary-Marshall-Proschan (1970). Properties and applications of hazard transforms are developed in detail in this paper.

7

MAINTENANCE AND REPLACEMENT MODELS

1. INTRODUCTION

In this chapter we study some models arising in the maintenance of systems. In Section 2 we obtain results concerning the availability of components and systems of components. In Section 3 we consider various models in which a system is maintained by replacing failed parts by spares and by repairing failed parts so that they may enter the pool of spares. Finally, in Section 4 we show how to determine optimal allocations of spares for the various components of a system when the total expenditure for spares is required not to exceed a specified amount. A general allocation model is formulated, and an algorithm for its solution is presented. The algorithm is a generalization of one due to Kettelle (1962). The general allocation model may be applied in a variety of reliability problems, as well as in other problems of optimization subject to a single constraint arising in operations research.

The treatment of maintenance policies in this chapter is not intended to be complete or comprehensive. Rather, certain selected topics of current research interest are discussed in some depth.

2. AVAILABILITY THEORY

An important figure of merit for a system undergoing repair is the probability that the system is operating at a specified time t. Let $X(t) = 1$ if the system is operating at time t, and 0 if it is not.

2.1. Definitions. (a) The *availability* $A(t)$ *at time* t is given by

$$A(t) = P[X(t) = 1] = EX(t). \tag{2.1}$$

(b) The *limiting availability* (or just *availability*) A is given by

$$A = \lim_{t\to\infty} A(t), \tag{2.2}$$

when it exists.

(c) The *average availability* in $[0, T]$ is $(1/T)\int_0^T A(t)\,dt$.

(d) The *limiting average availability* $A_{\text{av}} = \lim_{T\to\infty} (1/T)\int_0^T A(t)\,dt$.

Remarks. (a) Note that if repair is *not* permitted, then $A(t)$ reduces to system reliability, the probability that the system operates without failure during $[0, t]$.

(b) Note that the average availability in $[0, T]$ is also the expected proportion of time the system is operating during $[0, T]$. To see this, let $U(T)$ denote the total amount of operating time during $[0, T]$, that is, $U(T) = \int_0^T X(t)\,dt$. Then

$$\frac{1}{T} EU(T) = \frac{1}{T} E\int_0^T X(t)\,dt = \frac{1}{T}\int_0^T EX(t)\,dt = \frac{1}{T}\int_0^T A(t)\,dt,$$

by Fubini's Theorem, when any of the indicated integrals exist.

Next we show that limiting average availability is the same as limiting availability.

2.2. Lemma. If $\lim_{t\to\infty} A(t) = A$ exists, then

$$A_{\text{av}} = \lim_{T\to\infty} \frac{1}{T} E\int_0^T X(t)\,dt = A. \tag{2.3}$$

PROOF. Given $\varepsilon > 0$, there exists t_0 sufficiently large so that $|A(t) - A| < \varepsilon$ for all $t > t_0$. For $T > t_0$, it follows that

$$\frac{1}{T}\int_0^T A(t)\,dt \le \frac{1}{T}\int_0^{t_0} A(t)\,dt + \frac{1}{T}\int_{t_0}^T (A + \varepsilon)\,dt.$$

Thus, since t_0 is fixed, we have

$$\lim_{T\to\infty} \frac{1}{T}\int_0^T A(t)\,dt \le A + \varepsilon.$$

Similarly, we obtain

$$\lim_{T\to\infty} \frac{1}{T}\int_0^T A(t)\,dt \ge A - \varepsilon.$$

Since ε is arbitrary, the proof is complete.||

Next we obtain an expression for limiting availability in terms of mean functioning period and mean repair period. Let T_i be the duration of the ith functioning period, and D_i the down time for the ith repair or replacement. We do *not* assume T_i and D_i are independent, although we do assume the sequence $\{T_i + D_i\}_{i=1,2,\ldots}$ to be mutually independent.

Let F denote the common distribution of T_i, $i = 1, 2, \ldots$, and H the common distribution of $T_i + D_i$, $i = 1, 2, \ldots$. Let $M_H(t) = \sum_{n=1}^{\infty} H^{(n)}(t)$ be the renewal function corresponding to underlying distribution H (see Chapter 6, Section 3). Then it is not hard to show that availability may be expressed as

$$A(t) = \bar{F}(t) + \int_0^t \bar{F}(t - u)\, dM_H(u).$$

By the Key Renewal Theorem (Theorem 3.8 of Chapter 6 and the remark following),

$$A = \lim_{t\to\infty} A(t) = \frac{ET}{ET + ED} \tag{2.4}$$

when the distribution H is nonlattice.

Note that the limiting availability given in (2.4) depends only on the mean time to failure and the mean time to replace. Formula (2.4) is basic in availability theory.

Thus far we have obtained basic availability results for a component or a system treated as a single unit. Next we study system availability taking into account two alternative disciplines for component failure and repair.

System Availability: Independent Component Performance Processes, Model 1

First we consider a coherent system ϕ of n component "positions." We suppose the component in position i is replaced at failure, generating an alternating renewal process (see B-P, 1965, pp. 74–83). Let T_{ij} represent the (random) length of the jth operating period, with distribution F_i, having mean μ_i, $0 < \mu_i < \infty$, and D_{ij} the (random) length of the jth replacement period, with distribution G_i, having mean ν_i, $0 \leq \nu_i < \infty$, for $j = 1, 2, \ldots$; $i = 1, \ldots, n$. We assume the double array $\{T_{ij} + D_{ij}\}_{j=1,2,\ldots;\, i=1,\ldots,n}$ to be mutually independent random variables; however, we permit any pair T_{ij}, D_{ij} to be dependent. Figure 7.2.1 depicts a sample realization of the alternating sequence of operating and replacement periods; the ordinate $X_i(t)$ is 1 if the ith component is operating at time t, and 0 if failed, $i = 1, \ldots, n$.

We assume in this first model that *the n component positions operate independently of one another*. Specifically, while replacement of a failed component is occurring in one position, the components in the remaining positions continue to operate.

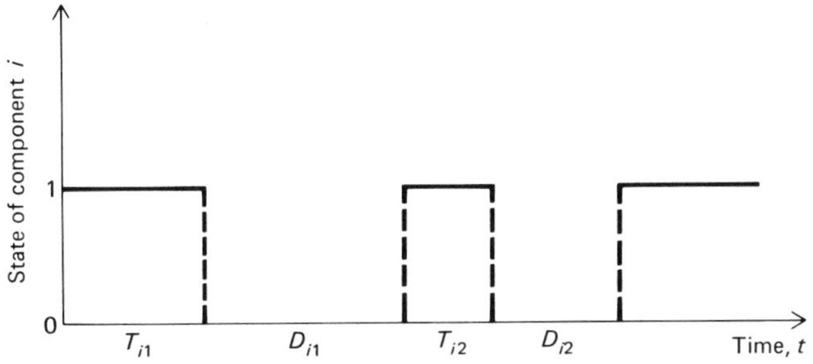

Figure 7.2.1.
Alternating failure and repair for component i.

We may now express the state $X(t)$ of the system in terms of the component states, $X_1(t), \ldots, X_n(t)$:

$$X(t) = \phi(X_1(t), \ldots, X_n(t)).$$

It follows that the availability $A(t)$ of the system at time t is given by

$$A(t) = EX(t) = h(EX_1(t), \ldots, EX_n(t)),$$

or, equivalently,

$$A(t) = h(A_1(t), \ldots, A_n(t)), \tag{2.5}$$

where h is the reliability function of structure ϕ (see Chapter 2, Section 1), and $A_i(t)$ is the availability at time t of component i, $i = 1, \ldots, n$. Since h is multilinear in its arguments, then system limiting availability $A = \lim_{t\to\infty} A(t)$ is given by

$$A = h(A_1, \ldots, A_n), \tag{2.6}$$

assuming component limiting availability $A_i = \lim_{t\to\infty} A_i(t)$ exists for $i = 1, \ldots, n$.

Assuming the distribution of $T_{i1} + D_{i1}$ is nonlattice, $i = 1, \ldots, n$, it follows from (2.4) that

$$A = h\left(\frac{\mu_1}{\mu_1 + \nu_1}, \ldots, \frac{\mu_n}{\mu_n + \nu_n}\right). \tag{2.7}$$

2.3. Example. A system consists of a computer in series with two electrical power generators in parallel, as shown in Figure 7.2.2.

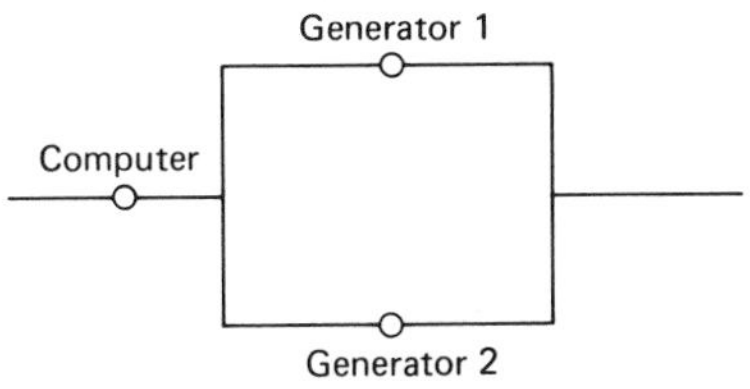

Figure 7.2.2. Computer system.

Assume that component life lengths and repair periods are independent, and that nonfailed components continue to operate while a failed component is undergoing replacement. The data necessary to compute limiting system availability are shown in Table 2.1.

Table 2.1. Component Data

Component	*Mean Life (hours)*	*Mean Time to Repair (hours)*	*Availability*
Computer	1000	1	.999
Generator 1	98	2	.98
Generator 2	96	4	.96

From (2.7), we compute limiting system availability as

$$A = .999\ (1 - .02 \times .04) = .998.$$

Series System Availability: Functioning Components Suspend Operation during Repair, Model 2

The second model differs from the first model in two respects:

(a) The system is assumed *series*, not an arbitrary coherent system. Thus system failure coincides with component failure.
(b) While a failed component is undergoing replacement, all other components remain in "suspended animation." When replacement of the failed component is completed, the remaining components resume operation. At that instant, they are not "as good as new," but rather as good as they were when the system stopped operating.

We make the additional assumption that two or more components cannot fail at the same time. This will be true, for example, if all failure distributions are continuous.

For this model, we will obtain almost sure convergence results for system availability, fractional down time of each component position, number of component failures per unit of time, mean time between failures of any component position or of the system, and so on. These results will be seen to depend only on the mean life length μ_i $(0 < \mu_i < \infty)$ and replacement period ν_i $(0 \leq \nu_i < \infty)$, $i = 1, \ldots, n$, and *not* on the actual life and replacement distributions F_i, G_i, $i = 1, \ldots, n$. It is remarkable that these asymptotic results turn out to be mathematically very simple in spite of the fact that the age distribution of components in the system quickly becomes stochastically very complicated. In fact, if $\xi(t)$ is defined to be i if the ith

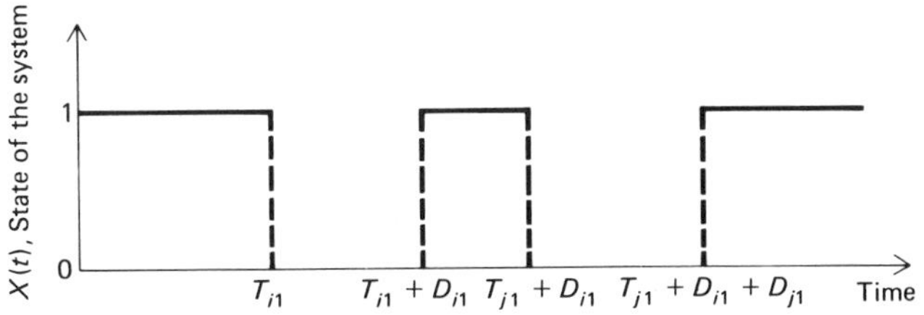

Figure 7.2.3.
Sample realization of system state.

component is down at time t, and 0 if all components are operating at time t, then the process $\{\xi(t), 0 \le t < \infty\}$ has no regeneration points.

A typical failure-repair history for such a series system might appear as in Figure 7.2.3.

In Figure 7.2.3 component i fails at time T_{i1} and the system is down D_{i1} hours. The system resumes operation at time $T_{i1} + D_{i1}$, continuing until component j fails at time $T_{j1} + D_{i1}$. Replacement of component j is completed at time $T_{j1} + D_{i1} + D_{j1}$, at which time the system resumes operation, and so on.

In Figure 7.2.3, we depicted a sample realization in terms of *total* time elapsed. An alternate mode of representation of outcomes is very useful in terms of system operating time (or up time) $U(t)$ cumulated by time t. Figure 7.2.4 depicts the same sample realization as in Figure 7.2.3, except that some additional history is included. Note that system down time,

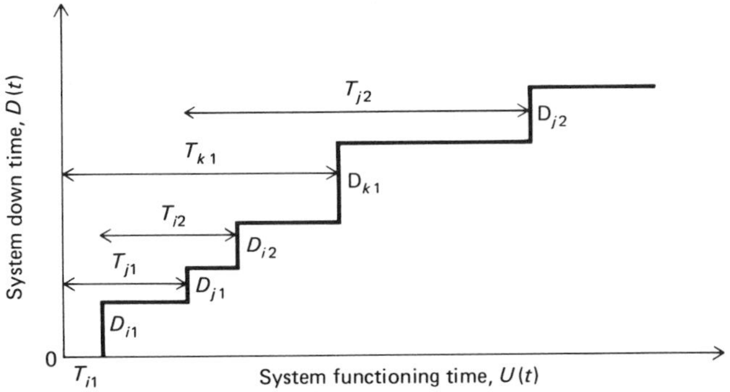

Figure 7.2.4.
Failure-repair history in terms of system up time.

$D(t)$, cumulated by time t is obtainable from $U(t) + D(t) = t$.

Let $\tilde{N}_i(t, \omega)$ denote the number of failures of component i during real time t corresponding to the particular sample outcome identified by the point ω in the sample space. Then

$$\tilde{N}_i(t, \omega) \equiv N_i[U(t, \omega), \omega], \tag{2.8}$$

where $\{N_i(t); t \geq 0\}$ is the renewal counting process associated with $\{T_{ir}\}_{r=1}^{\infty}$; that is, if $S_{i0} = 0$, and $S_{in} = \sum_{r=1}^{n} T_{ir}$ for $n \geq 1$, then $N_i(t) = \sup\{n \mid S_{in} \leq t\}$. It is important to note that (2.8) holds for each sample realization. However, it is *not* true that $\tilde{N}_i(t) \overset{\text{st}}{=} N_i(U(t))$, where the $U(\cdot)$ and $N_i(\cdot)$ are determined from independent processes $\{U(t), t \geq 0\}$, $\{N_i(x), x \geq 0\}$. For example, if $n = 1$ (only one component present), $D_1 > 1$ a.s., and the underlying failure distribution F_1 is strictly increasing on $[0, 1]$, then $P[\tilde{N}_1(1) > 1] = 0$, whereas $P[N_1(U(1)) > 1] > 0$. We will often omit ω in using the identity (2.8).

In the rest of this section we assume all components are new at $t = 0$ for definiteness. However, the limiting results are true regardless of the initial conditions.

2.4. Lemma. $\lim_{t\to\infty} [N_i(U(t, \omega), \omega)/U(t, \omega)] = 1/\mu_i$, except for a set of sample outcomes ω having probability 0, $i = 1, \ldots, n$.

PROOF. $U(t, \omega) \to \infty$ a.s. as $t \to \infty$. Also $N_i(x, \omega)/x \to (1/\mu_i)$ a.s. as $x \to \infty$ by Theorem 3.6 of Chapter 6. The result follows. ||

Note that we have specified that the sample point ω is common to the random variables involved.

We may now derive the expression for limiting average system availability A_{av}.

2.5. Theorem.

$$\lim_{t\to\infty} \frac{U(t)}{t} = \left(1 + \sum_{j=1}^{n} \frac{\nu_j}{\mu_j}\right)^{-1} \quad \text{a.s.} \tag{2.9}$$

PROOF. Note that

$$\sum_{i=1}^{n} \sum_{r=1}^{\tilde{N}_i(t)-1} D_{ir} \leq D(t) \leq \sum_{i=1}^{n} \sum_{r=1}^{\tilde{N}_i(t)} D_{ir}.$$

The left-hand inequality results from the fact that none of the final down time beginning before time t is included; the right-hand inequality results from the fact that the final down time beginning before time t may actually extend beyond time t.

Since $U(t) + D(t) \equiv t$,

$$\frac{U(t)}{t} = \left[1 + \frac{D(t)}{U(t)}\right]^{-1} \geq \left[1 + \sum_{i=1}^{n} \frac{1}{\tilde{N}_i(t)} \sum_{r=1}^{\tilde{N}_i(t)} D_{ir} \frac{N_i(U(t))}{U(t)}\right]^{-1}.$$

By the strong law of large numbers, $[1/\tilde{N}_i(t)] \sum_{r=1}^{\tilde{N}_i(t)} D_{ir} \overset{\text{a.s.}}{\to} \nu_i$ as $t \to \infty$. By Lemma 2.4, $N_i(U(t))/U(t) \overset{\text{a.s.}}{\to} 1/\mu_i$ as $t \to \infty$. Hence

$$\lim_{t\to\infty} \frac{U(t)}{t} \geq \left[1 + \sum_{i=1}^{n} \frac{\nu_i}{\mu_i}\right]^{-1} \quad \text{a.s.}$$

The reverse inequality is proved similarly. ||

Note that the limit given in (2.9) is a function only of component mean life lengths and replacement periods, and does *not* require a knowledge of the actual life and repair distributions.

The final result is given in Corollary 2.6.

2.6. Corollary.

$$A_{av} = \lim_{t\to\infty} \frac{EU(t)}{t} = \left(1 + \sum_{j=1}^{n} \frac{\nu_j}{\mu_j}\right)^{-1}. \tag{2.10}$$

PROOF. Since $U(t)/t \leq 1$ for all $t \geq 0$, and $U(t)/t \overset{\text{a.s.}}{\to} (1 + \sum_{j=1}^{n} (\nu_j/\mu_j))^{-1}$, it follows by the Lebesgue dominated convergence theorem that $EU(t)/t \to (1 + \sum_{j=1}^{n} (\nu_j/\mu_j))^{-1}$ as $t \to \infty$. $A_{av} = \lim_{t\to\infty} [EU(t)/t]$ by Remark (b) after Definition 2.1. ‖

Remark. Note that we have obtained limiting results for *average availability*. If limiting availability $A = \lim_{t\to\infty} P[\xi(t) = 0]$ exists, then $A = A_{av}$ by Lemma 2.2. However, the $\lim_{t\to\infty} P[\xi(t) = 0]$ need *not* always exist. Sufficient conditions for the existence of this limit are, for example, that any F_i is nonlattice and F_j is exponential for $j \neq i$.

In a similar fashion, we obtain asymptotic results for the total down time $D_i(t)$ resulting from failures in component position i during $[0, t]$. Define $D_{i,av} \overset{\text{a.s.}}{=} \lim_{t\to\infty} [D_i(t)/t]$ and $D_{av} \overset{\text{a.s.}}{=} \lim_{t\to\infty} [D(t)/t]$. Then we have the following corollary.

2.7. Corollary. With probability 1,

(a) $$D_{i,av} = \frac{\nu_i}{\mu_i} A_{av} \quad \text{for} \quad i = 1, \ldots, n, \tag{2.11a}$$

(b) $$D_{av} = A_{av} \sum_{i=1}^{n} \frac{\nu_i}{\mu_i}. \tag{2.11b}$$

PROOF. (a) Note that as in Theorem 2.5,

$$\sum_{r=1}^{\tilde{N}_i(t)-1} D_{ir} \leq D_i(t) \leq \sum_{r=1}^{\tilde{N}_i(t)} D_{ir}.$$

Thus

$$\frac{D_i(t)}{t} \leq \frac{1}{\tilde{N}_i(t)} \sum_{r=1}^{\tilde{N}_i(t)} D_{ir} \cdot \frac{N_i(U(t))}{U(t)} \cdot \frac{U(t)}{t}.$$

It follows that

$$\lim_{t\to\infty} \frac{D_i(t)}{t} \leq \frac{\nu_i}{\mu_i} A_{av},$$

using the strong law of large numbers on the first factor, Lemma 2.4 on the second, and Theorem 2.5 on the third.

The reverse inequality is proved similarly.

(b) Summing over i in (2.11a) yields (2.11b). ||

Remarks. (a) Using Lemma 2.2 and the method of proof of Corollary 2.6, we can show that if $\lim_{t\to\infty} P[\xi(t) = i]$ exists, then

$$\lim_{t\to\infty} P[\xi(t) = i] = D_{i,\text{av}}. \tag{2.12}$$

(b) A heuristic justification of (2.11a) may be devised as follows. The product $A_{\text{av}} \cdot (1/\mu_i)\, dt$ may be viewed as approximately the joint stationary probability of the system being in the functioning state and then entering the down state in the next dt units of time due to failure of component i. Similarly, the product $D_{i,\text{av}} \cdot (1/\nu_i)\, dt$ may be viewed as approximately the joint stationary probability of the system being in the down state due to failure of component i and then component i being repaired in the next dt units of time. Thus (2.11a) equates the stationary rates into and out of these states.

Next we obtain asymptotic results for the number $\tilde{N}_i(t)$ of failures in component position i during $[0, t]$.

2.8. Corollary.

(a)
$$\lim_{t\to\infty} \frac{\tilde{N}_i(t)}{t} = \frac{A_{\text{av}}}{\mu_i} \text{ a.s.} \quad \text{for } i = 1, \ldots, n, \tag{2.13a}$$

(b)
$$\lim_{t\to\infty} \frac{E\tilde{N}_i(t)}{t} = \frac{A_{\text{av}}}{\mu_i} \quad \text{for } i = 1, \ldots, n. \tag{2.13b}$$

PROOF. (a) $\tilde{N}_i(t)/t = [N_i(U(t))/U(t)][U(t)/t]$ by (2.8). But $N_i(U(t))/U(t) \overset{\text{a.s}}{\to} 1/\mu_i$ by Lemma 2.4, while $U(t)/t \overset{\text{a.s.}}{\to} A_{\text{av}}$ by (2.9) and (2.10).

(b) $\tilde{N}_i(t)/t = N_i(U(t))/t \le N_i(t)/t$. By the elementary renewal theorem (see Remark following Theorem 3.6 of Chapter 6), $EN_i(t)/t \to 1/\mu_i$. Also $EN_i(t) < \infty$ for each $t \ge 0$, as stated in Chapter 6, Section 3. Thus there exists M such that $\sup_t (EN_i(t)/t) < M$. The conclusion of (b) now follows from the Lebesgue dominated convergence theorem and from (a) just above. ||

Remarks. (a) A heuristic justification for (2.13a) may be given as follows. Cross-multiplying, (2.13a) asserts that the total functioning time during $[0, t]$ [approximately $\tilde{N}_i(t) \cdot \mu_i$] is the same as the total time multiplied by the fractional up time (approximately $t \cdot A_{\text{av}}$).

(b) Obviously $\tilde{N}(t)/t$ converges a.s. and in expectation to $A_{av} \sum_1^n (1/\mu_i)$, where $\tilde{N}(t) = \sum_1^n \tilde{N}_i(t)$, the total number of system failures during $[0, t]$.

Mean Time between System Failures

After any repair, the distribution of the time until the next system failure depends on the system history up until that moment. However, as we show next, the average length of system functioning periods during $[0, t]$ will converge to a limit—say, μ. Similarly, the average length of all replacement periods (system down times) during $[0, t]$ will converge to a limit—say, ν.

2.9. Theorem. (a) The average of system up times converges a.s. to $\mu = (\sum_1^n (1/\mu_i))^{-1}$.

(b) The average of system down times converges a.s. to $\nu = \mu \sum_{i=1}^n \nu_i/\mu_i$.

PROOF. (a) The average of system functioning periods in $[0, t]$ differs from $U(t)/\tilde{N}(t)$ by a term that converges to 0 a.s. as $t \to \infty$, as shown in previous proofs. By Theorem 2.5 and Corollary 2.6, $\lim_{t\to\infty} (U(t)/t) \overset{\text{a.s.}}{=} A_{av}$. By Corollary 2.8, $\lim_{t\to\infty} (\tilde{N}_i(t)/t) \overset{\text{a.s.}}{=} A_{av}/\mu_i$. Thus it follows that

$$\lim_{t\to\infty} \frac{U(t)}{\tilde{N}(t)} \overset{\text{a.s.}}{=} \frac{A_{av}}{A_{av} \sum_1^n \mu_i^{-1}} = \left(\sum_1^n \mu_i^{-1}\right)^{-1}.$$

The desired conclusion follows immediately.

(b) The average of system down times during $[0, t]$ will differ from $\sum_{i=1}^n \sum_{r=1}^{\tilde{N}_i(t)} D_{ir}/\tilde{N}(t)$ by a term that converges to 0 a.s. as $t \to \infty$. Rewriting the double sum as $\sum_{i=1}^n [\tilde{N}_i(t)/\tilde{N}(t)][1/\tilde{N}_i(t)] \sum_{r=1}^{\tilde{N}_i(t)} D_{ir}$, we note that $\tilde{N}_i(t)/\tilde{N}(t) \overset{\text{a.s.}}{\to} \mu_i^{-1}/\sum_{j=1}^n \mu_j^{-1}$ by Corollary 2.8, while $[1/\tilde{N}_i(t)] \sum_{r=1}^{\tilde{N}_i(t)} D_{ir} \overset{\text{a.s.}}{\to} \nu_i$ by the strong law of large numbers. It follows that the average system down time $\overset{\text{a.s.}}{\to} \sum_{i=1}^n (\nu_i/\mu_i)/\sum_{i=1}^n \mu_i^{-1} = \mu \sum_{i=1}^n \nu_i/\mu_i$. ‖

Remarks. (a) In the special case that failure distributions are exponential, then $(\sum_1^n \mu_i^{-1})^{-1}$ is the expected duration of each system functioning period regardless of system past history, because of the memoryless property of the exponential. (See Chapter 3, Section 2.) Theorem 2.9(a) tells us that for *arbitrary* failure distributions, the average of system up times converges a.s. to this same value.

(b) For a one-unit system with mean life μ and mean repair time ν, (2.4) states that the limiting average availability is $\mu/(\mu + \nu)$. For the present series system model the limiting average system availability is, by (2.9) and Theorem 2.9,

$$A_{av} = \left(1 + \sum_{i=1}^n \frac{\nu_i}{\mu_i}\right)^{-1} = \frac{\mu}{\mu + \mu \sum_{i=1}^n (\nu_i/\mu_i)} = \frac{\mu}{\mu + \nu}, \tag{2.14}$$

where the μ and ν of (2.14) are defined explicitly in Theorem 2.9, and represent respectively the average system up and down times. Clearly, (2.14) is the series system analog of the one-unit system result given in (2.4).

2.10. Numerical Example. Consider a system of four components in series operating as described in Model 2.

Table 2.2. Mean Times to Failure and Repair for Subsystems

Component Index, i	*Type*	*Mean Life, μ_i (hours)*	*Mean Repair Time, ν_i (hours)*
1	Power supply	50	.1
2	Analog equipment	100	.2
3	Digital equipment	1,000	1.0
4	Mechanical	10,000	20.0

From (2.9) we compute the long-run fraction of the time that the system is operating, $A_{\text{av}} = (1 + \sum_{j=1}^{4} \nu_j/\mu_j)^{-1} = 1.007^{-1} = .993$.

By (2.11a) the long-run fraction of the time that the system is down due to failure of each of the four component types may be computed as:

$$D_{1,\text{av}} = \frac{\nu_1}{\mu_1} A_{\text{av}} = \frac{.1}{50} \times .993 = .002,$$

$$D_{2,\text{av}} = \frac{\nu_2}{\mu_2} A_{\text{av}} = \frac{.2}{100} \times .993 = .002,$$

$$D_{3,\text{av}} = \frac{\nu_3}{\mu_3} A_{\text{av}} = \frac{1}{1000} \times .993 = .001,$$

$$D_{4,\text{av}} = \frac{\nu_4}{\mu_4} A_{\text{av}} = \frac{20}{10{,}000} \times .993 = .002.$$

By (2.13a), the long-run average number of failures per hour of each of the component types may be computed as:

$$\lim_{t\to\infty} \frac{\tilde{N}_1(t)}{t} \overset{\text{a.s.}}{=} \frac{A_{\text{av}}}{\mu_1} = \frac{.993}{50} = .020,$$

$$\lim_{t\to\infty} \frac{\tilde{N}_2(t)}{t} \overset{\text{a.s.}}{=} \frac{A_{\text{av}}}{\mu_2} = \frac{.993}{100} = .010,$$

$$\lim_{t\to\infty} \frac{\tilde{N}_3(t)}{t} \overset{\text{a.s.}}{=} \frac{A_{\text{av}}}{\mu_3} = \frac{.993}{1000} = .001,$$

$$\lim_{t\to\infty} \frac{\tilde{N}_4(t)}{t} \overset{\text{a.s.}}{=} \frac{A_{\text{av}}}{\mu_4} = \frac{.993}{10{,}000} = .0001.$$

By Theorem 2.9(a), the long-run average of system up times is $\mu = (\sum_1^n \mu_i^{-1})^{-1} = (.02 + .01 + .001 + .0001)^{-1} = 32.15$ hours. By Theorem 2.9(b), the long-run average of system down times is $\nu = \mu \sum_{i=1}^n (\nu_i/\mu_i) = 32.15 \times .007 = .225$ hours.

EXERCISES

1. A computer has exponential distribution, $F(t) = 1 - e^{-t/\mu}$, of time between failures. Repair time is also exponential with distribution $G(t) = 1 - e^{-t/\nu}$.
   (a) Compute the availability at time t_0 and average availability over $[0, t_0]$.
   (b) Show that both availability and average availability are decreasing in t_0.
   (c) Using this fact, obtain a simple conservative lower bound for availability and average availability for any time t_0.
2. In Model 2 suppose component i having life distribution F_i is replaced at operating age T_i, or at failure, whichever comes first. The expected replacement period is ν_i, whether the replacement is planned or as a result of failure, $i = 1, \ldots, n$. Compute limiting average availability.
3. Give an example in which the limiting availability of (2.2) does *not* exist.
4. Give an example to show that (2.4) need not hold if the $\{T_i + D_i\}_{i=1,2,\ldots}$ are *not* mutually independent.
5. Under Model 1 compute the limiting availability of a 2-out-of-3 system of independent components each having mean life 100 hours and mean repair period 2 hours.
6. Consider a series system of independent components in Model 1. Show that system *un*availability is approximately the sum of component unavailabilities, when mean lives are large and mean repair periods are small.
7. In Model 2 assume that components are stochastically alike. Suppose that as the number n of components in the series system increases, the mean repair time ν is decreased, so that $\nu = \nu_0/n$. Show that limiting average system availability A_{av} remains constant as n varies.
8. Explain heuristically why the limiting average of system up times $\mu = (\sum_1^n \mu_i^{-1})^{-1}$ [Theorem 2.9(a)] depends only on the $\mu_1, \ldots, \mu_n$, while the limiting average of system down times $\nu = \mu \sum_{i=1}^n \nu_i/\mu_i$ [Theorem 2.9(b)] depends on both $\mu_1, \ldots, \mu_n$ and $\nu_1, \ldots, \nu_n$.

3. MAINTENANCE THROUGH SPARES AND REPAIR

In the models of Section 2, a replacement for a failed component was always available. In the models of the present section, we assume that the number of spares available is limited; the system now fails when no spare is available to replace a failed component. In addition, a repair facility is present to repair failed components. We consider the following model first.

Model 3

A one-unit system is supported by a single spare and a repair facility. When the operating unit fails, it is replaced by the spare in negligible time; at the same instant, the failed unit is sent to the repair facility. This instant is a regeneration point. The system fails when the operating unit fails and no spare is available to replace it (that is, repair has not yet been completed on the previously failed unit). Assume operating units have life distribution F with mean μ, while repair periods are governed by distribution G with mean ν, with operating periods and repair periods mutually independent. The maintenance cycle is portrayed in Figure 7.3.1. Finally, assume that at time $t = 0$, the operating unit and spare are new.

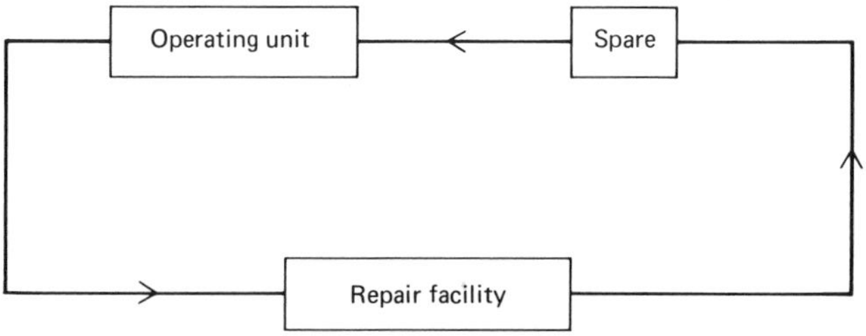

Figure 7.3.1.
Operating unit supported by spare and repair facility.

We may now compute various reliability quantities of interest in terms of F and G. We first express the time, T_1, until first system failure as

$$T_1 = X_1' + X_2 + X_3 + \cdots + X_N, \tag{3.1}$$

where $X_1', X_2, X_3, \ldots$ are independently distributed according to F, and N denotes the (random) number of operating unit failures culminating in a system failure. N has geometric-like frequency function given by

$$P[N = k + 1] = \alpha^{k-1}(1 - \alpha) \qquad \text{for} \quad k = 1, 2, \ldots, \tag{3.2}$$

where

$$\alpha = \int_0^\infty G(t)\, dF(t). \tag{3.3}$$

Note that α represents the probability that a failed unit is repaired (and thus becomes available as a spare) before the operating unit fails.

Since N is a stopping time (that is, the event $[N = n]$ is independent of $X_{n+1}, X_{n+2}, \ldots$ for each $n = 2, 3, \ldots$), it follows from a minor modification of Theorem 3.12 of Chapter 6 that

$$ET_1 = EX_1' + EX_2 + \cdots + EX_N = \mu EN.$$

Thus the mean time until first system failure is given by

$$ET_1 = \mu\left(1 + \frac{1}{1 - \alpha}\right). \tag{3.4}$$

In a similar fashion we may compute the mean time ET until system failure, measuring time from a regeneration point, that is, an instant when the operating unit fails and enters the repair facility, while simultaneously the spare unit is put into active operation. Reasoning as above, we obtain

$$ET = \frac{\mu}{1 - \alpha}. \tag{3.5}$$

Note that ET_1 in (3.4) exceeds ET in (3.5) by the amount μ, the mean time until the system reaches the first regeneration point starting at time 0.

Next, we may verify directly that the mean ED of system down times is given by

$$ED = \int_0^\infty \int_0^\infty \frac{\bar{G}(t + x)}{\bar{G}(x)}\, dt\, dF(x). \tag{3.6}$$

Equation (3.6) may be verified if we note that system failure occurs when the operating unit fails before the unit undergoing repair is returned to service as a spare (following completion of repair). Failure of the operating unit occurs x hours after it begins operating with probability measure $dF(x)$, while the probability that the unit undergoing repair requires at least an *additional* t hours (during which the system is down) for completion of repair is $\bar{G}(t + x)/\bar{G}(x)$. Taking into account all outcomes for x and t we obtain the mean, ED, of system down times given in (3.6).

To obtain the limiting system availability A, we use (2.4), obtaining

$$A = \frac{\mu(1 - \alpha)^{-1}}{\mu(1 - \alpha)^{-1} + ED}, \tag{3.7}$$

where ED is given in (3.6). Note that limiting system availability now depends on failure distribution F and repair distribution G, and not just on their means.

In (3.4) we gave a formula for *mean* time until first system failure. Now we compute the *distribution* H of time until first system failure in terms of Laplace transforms. Let H_1 be the distribution of time to first system failure, given that the operating unit has just failed and been replaced by the spare; that is, H_1 is the distribution of time elapsed between the first regeneration point and the first system failure. Then we may express H in terms of H_1 by

$$H(t) = \int_0^t H_1(t - x)\, dF(x). \tag{3.8}$$

We express H_1 in terms of F and G through the integral equation

$$\bar{H}_1(t) = \bar{F}(t) + \int_0^t \bar{H}_1(t - x)G(x)\, dF(x). \tag{3.9}$$

To verify (3.9), note that $\bar{F}(t)$ represents the probability that the operating unit does not fail in $[0, t]$, $G(x)\, dF(x)$ the probability measure corresponding to the event that the operating unit fails at time x *and* a spare is available to replace it, and $\bar{H}_1(t - x)$ the conditional probability that system failure does not occur in the remaining $t - x$ hours.

To solve for H_1 from (3.9) in terms of Laplace-Stieltjes transforms, let $H_1^*(s) = \int_0^\infty e^{-st}\, dH_1(t)$, $F^*(s) = \int_0^\infty e^{-st}\, dF(t)$, and $K^*(s) = \int_0^\infty e^{-st}G(t)\, dF(t)$. From (3.9) it follows immediately that

$$H_1^*(s) = F^*(s) + H_1^*(s)K^*(s) - K^*(s),$$

so that

$$H_1^*(s) = \frac{F^*(s) - K^*(s)}{1 - K^*(s)}.$$

From (3.8) we conclude that

$$H^*(s) = F^*(s)\, \frac{F^*(s) - K^*(s)}{1 - K^*(s)}. \tag{3.10}$$

By inverting (3.10), $H(t)$ can be obtained. Also, we may use (3.10) to check that the mean time to first failure is as given in (3.4); simply differentiate both sides of (3.10) and set $s = 0$.

In Model 3, one operating unit is supported by one spare and one repair facility. In Model 4 below we generalize to allow for any number of operating units, spares, and repair facilities. However, since the process of failures and repairs corresponding to *general* failure distribution F and *general* repair distribution G has no regeneration points, the mathematical analysis would become very complicated. We confine ourselves to the case of exponential failure and exponential repair, and use the method of balance equations to get stationary probabilities.

Model 4

A system consists of m operating units, n spares, and s service facilities. Each operating unit has life distribution $F(t) = 1 - e^{-\lambda t}$; unit repair time has distribution $G(t) = 1 - e^{-\gamma t}$. A unit in spare status cannot fail. All life lengths and repair times are mutually independent. As in model 3, when an operating unit fails, it is immediately replaced by a spare, if available; the failed unit is immediately sent to the repair facility. Repair begins immediately unless all s service facilities are occupied.

Let $X(t)$ denote the number of units down at time t, that is, the number of units undergoing repair or awaiting repair. Then $\{X(t), 0 \le t < \infty\}$ is a *continuous time Markov process*; that is, for time points $0 \le t_1 < t_2 < \cdots < t_k$ and nonnegative integers $x_1, x_2, \ldots, x_k$,

$$\begin{aligned} P[X(t_k) &= x_k \mid X(t_1) = x_1, \ldots, X(t_{k-1}) = x_{k-1}] \\ &= P[X(t_k) = x_k \mid X(t_{k-1}) = x_{k-1}]. \end{aligned}$$

More specifically, $\{X(t), 0 \le t < \infty\}$ is a *birth and death process*, that is, a continuous parameter Markov chain with state space $\{0, 1, 2, \ldots\}$ and homogeneous transition probabilities, and which changes only through transitions from a state to its immediate neighbors. See Parzen (1962), p. 278 ff. For birth and death processes, it is not difficult to solve for the stationary probability measure $\pi_i = \lim_{t\to\infty} P[X(t) = i]$.

We illustrate the procedure for the simplest case in which one operating unit is supported by one spare and one repair facility, as in Model 3 above. Let $P_i(t) = P[X(t) = i]$, $i = 0, 1, 2$. Using the exponential property, we write

$$P_0(t + h) = P_0(t)[1 - \lambda h + o(h)] + P_1(t)[\gamma h + o(h)]$$

$$P_1(t + h) = P_1(t)[1 - \lambda h - \gamma h + o(h)] + P_0(t)[\lambda h + o(h)] + P_2(t)[\gamma h + o(h)],$$

where, as usual, $o(h)$ denotes $\lim_{h\downarrow 0} [o(h)/h] = 0$. Subtracting, dividing by h, and taking limits as $h \to 0$, we obtain the differential equations:

$$\begin{aligned} P_0'(t) &= -\lambda P_0(t) + \gamma P_1(t), \\ P_1'(t) &= \lambda P_0(t) - (\lambda + \gamma)P_1(t) + \gamma P_2(t). \end{aligned}$$

It can be shown that $\lim_{t\to\infty} P_i'(t) = 0$ and that $\pi_i \overset{\text{def}}{=} \lim_{t\to\infty} P_i(t)$ exists. It follows that

$$\begin{aligned} -\lambda\pi_0 + \gamma\pi_1 \qquad\qquad &= 0, \\ \lambda\pi_0 - (\lambda + \gamma)\pi_1 + \gamma\pi_2 &= 0, \\ \pi_0 + \pi_1 + \pi_2 &= 1. \end{aligned} \tag{3.11}$$

Solving by Cramer's rule, we obtain

$$\pi_0 = \frac{1}{D}\begin{vmatrix} 0 & \gamma & 0 \\ 0 & -(\lambda + \gamma) & \gamma \\ 1 & 1 & 1 \end{vmatrix},$$

$$\pi_1 = \frac{1}{D}\begin{vmatrix} -\lambda & 0 & 0 \\ \lambda & 0 & \gamma \\ 1 & 1 & 1 \end{vmatrix},$$

$$\pi_2 = \frac{1}{D}\begin{vmatrix} -\lambda & \gamma & 0 \\ \lambda & -(\lambda + \gamma) & 0 \\ 1 & 1 & 1 \end{vmatrix},$$

where

$$D = \begin{vmatrix} -\lambda & \gamma & 0 \\ \lambda & -(\lambda + \gamma) & \gamma \\ 1 & 1 & 1 \end{vmatrix}.$$

Expanding determinants, we obtain

$$\begin{aligned} \pi_0 &= \left(1 + \frac{\lambda}{\gamma} + \frac{\lambda^2}{\gamma^2}\right)^{-1}, \\ \pi_1 &= \frac{\lambda}{\gamma}\left(1 + \frac{\lambda}{\gamma} + \frac{\lambda^2}{\gamma^2}\right)^{-1}, \\ \pi_2 &= \frac{\lambda^2}{\gamma^2}\left(1 + \frac{\lambda}{\gamma} + \frac{\lambda^2}{\gamma^2}\right)^{-1}. \end{aligned} \tag{3.12}$$

The first two equations of (3.11) are *balance equations* for the stationary situation. The first equation of (3.11) states that in the stationary situation, the rate $\pi_0\lambda$ at which transitions occur *out* of state 0 is equal to the rate $\pi_1\gamma$ at which transitions occur *into* state 0. The second equation states that the rate $\pi_1(\lambda + \gamma)$ at which transitions occur *out* of state 1 is equal to the rate $\pi_0\lambda + \pi_2\gamma$ at which transitions occur *into* state 1. The third equation of (3.11) simply states that the stationary probabilities sum to 1.

Since limiting availability $A = \pi_0 + \pi_1$, we have

$$A = \left(1 + \frac{\lambda}{\gamma}\right)\left(1 + \frac{\lambda}{\gamma} + \frac{\lambda^2}{\gamma^2}\right)^{-1}. \tag{3.13}$$

Note from (3.12) and (3.13) that the limiting values π_0, π_1, π_2, and A depend only on the ratio λ/γ, and not on λ and γ separately.

A similar procedure can be used for obtaining the steady-state probabilities of any continuous time Markov process with a finite state space S. Associated with each pair of states (i, j) is an exponential rate $\lambda_{ij} \geq 0$ at which a transition occurs from state i to state j. Since in the steady state the rate at which a transition occurs *into* state j must equal the rate at which a transition occurs *out* of state j, we obtain the balance equations:

$$\pi_j \sum_{\substack{i \in S \\ i \neq j}} \lambda_{ji} = \sum_{\substack{i \in S \\ i \neq j}} \pi_i \lambda_{ij}. \tag{3.14}$$

The left-hand side represents the steady-state rate of transition out of state j while the right-hand side represents the steady-state rate of transition into state j. For a detailed discussion, see Cox and Smith (1961).

A Spares Inventory Model: Model 5

In the models considered thus far, we have been interested in computing the reliability of a system supplied by spares. In the present model we focus on the operation of supplying spares; this will lead us in Section 4 below to a consideration of optimum allocations of spares of a variety of part types when the total expenditure for spares is constrained.

At time 0 a maintenance depot has n spares available of a given unit (that is, part type). Customers bring failed units to the depot at times governed by a Poisson process with rate λ. In exchange for a failed unit, the customer receives a spare, if available. Repair of the failed unit begins immediately since unlimited repair facilities are available. Repair is completed in a time governed by distribution G; repair times are independent of each other and of customer arrival times. The repaired unit becomes part of the spares inventory.

The distribution of $N(t)$, the number of units undergoing repair at time t, is Poisson, as stated in Theorem 3.1.

3.1. Theorem. In Model 5, for $j = 0, 1, 2, \ldots,$

$$P[N(t) = j] = \frac{[\lambda \int_0^t \bar{G}(u)\, du]^j}{j!} \exp\left[-\lambda \int_0^t \bar{G}(u)\, du\right], \tag{3.15}$$

and

$$\lim_{t\to\infty} P[N(t) = j] = \frac{(\lambda\nu)^j}{j!} e^{-\lambda\nu}, \tag{3.16}$$

where $\nu = \int_0^\infty \bar{G}(u)\, du$, the mean repair time.

PROOF. By hypothesis, the probability of m demands for spares during $[0, t]$ is $[(\lambda t)^m/m!]e^{-\lambda t}$. Given that m demands for spares are made during $[0, t]$, the times at which demands arise during $[0, t]$ have the same joint distribution as a sample of size m from a uniform distribution on $[0, t]$ (Theorem 3.7, Chapter 3). Let $P(t)$ be the probability that a "typical" failed unit brought in during $[0, t]$ is still undergoing repair at time t. Then

$$P(t) = \int_0^t \bar{G}(t - x)\, \frac{dx}{t} = \frac{1}{t} \int_0^t \bar{G}(u)\, du.$$

It follows by unconditioning on m that

$$\begin{aligned} P[N(t) = j] &= \sum_{m=j}^{\infty} \frac{(\lambda t)^m}{m!} e^{-\lambda t} \left\{\binom{m}{j} [P(t)]^j [1 - P(t)]^{m-j}\right\} \\ &= e^{-\lambda t} \frac{[\lambda t P(t)]^j}{j!} e^{\lambda t[1-P(t)]} = \frac{[\lambda t P(t)]^j}{j!} e^{-\lambda t P(t)}, \end{aligned}$$

yielding (3.15).

Equation (3.16) follows by letting $t \to \infty$ in (3.15). ‖

From (3.16) it is easy to see that the expected number of customers waiting for spares at time t is given by

$$E \max\{[N(t) - n], 0\}$$
$$= \sum_{j=n+1}^{\infty} (j - n) \frac{[\lambda \int_0^t \bar{G}(u)\, du]^j}{j!} \exp\left[-\lambda \int_0^t \bar{G}(u)\, du\right]. \quad (3.17)$$

Letting $t \to \infty$, we obtain the asymptotic expected number of customers waiting for spares:

$$\lim_{t\to\infty} E \max\{[N(t) - n], 0\} = \sum_{j=n+1}^{\infty} (j - n) \frac{(\lambda \nu)^j}{j!} e^{-\lambda \nu}. \quad (3.18)$$

Suppose we extend Model 5 so that instead of a single type of unit, there are now k different part types. At time 0, the depot has n_i spares of part type i, $i = 1, \ldots, k$. For each part type the depot operates as in Model 5. One measure of the degree of performance of the maintenance depot is the *fill rate* $R(\mathbf{n})$, that is, the expected long-run fraction of demands for spares filled without delay. It is intuitively clear that the fill rate $R(\mathbf{n})$ is given by

$$R(\mathbf{n}) = \sum_{i=1}^{k} \lambda_i \sum_{j=0}^{n_i - 1} \frac{(\lambda_i \nu_i)^j}{j!} e^{-\lambda_i \nu_i} \left(\sum_{i=1}^{k} \lambda_i\right)^{-1}, \quad (3.19)$$

where $\sum_{j=0}^{-1} \equiv 0$. Note that the denominator represents the long-run expected total number of demands per unit of time, while the numerator represents the long-run expected total number of demands filled without delay due to shortage of spares, per unit of time.

EXERCISES

1. Prove (3.5) in detail.
2. Use (3.10) to check that the mean time to first failure in Model 3 is as given in (3.4).
3. In Model 4 suppose we have $m = 1$ operating unit, $n = 2$ spares, and $s = 1$ service facility. Solve for the stationary probability π_i of state i, $i = 0, 1, 2, 3$. Does the solution depend on λ and γ separately, or just their ratio?
4. In Model 5 show that if repair distribution 1 is stochastically larger than repair distribution 2 [that is, $\bar{G}_1(u) \geq \bar{G}_2(u)$, $0 \leq u < \infty$], then the corresponding distributions for the number of units undergoing repair are similarly ordered stochastically $\{$that is, $P[N_1(t) \geq j] \geq P[N_2(t) \geq j]$, $0 \leq j < \infty\}$.
5. In Model 5 show that $N(t)$ is stochastically increasing in t. Using this fact, find a lower bound for $P[N(t) \leq j]$ valid for all repair distributions G, in terms of λ and ν.

6. Using (3.19) show that the fill rate is a decreasing function of λ_i and of ν_i, $i = 1, \ldots, k$. Why is this to be expected intuitively?
7. Let $\lambda_i \nu_i < 1$, $i = 1, \ldots, k$ (that is, mean time between arrivals is greater than mean time between repairs). Then the fill rate $R(\mathbf{n})$ is a concave function.

4. OPTIMAL SPARE PARTS ALLOCATION

A commonly occurring system maintenance problem is to determine the "best" spare parts kit possible for a given amount of money. Of course, we must specify the criterion by which we judge the value of the kit, before we can arrive at the best one. Some alternate criteria, or figures of merit, for judging the value of a spares kit are:

(a) Mean time until a shortage is experienced.
(b) Probability of *no* shortage in a specified period $[0, t]$.
(c) The fill rate, that is, the expected fraction of demands that are satisfied without delay during an indefinitely long period of operation.
(d) The expected long-run total delay experienced per unit of time in filling spares demands.

In this section we first show in detail how to determine the spares kit which maximizes long-run fill rate [criterion (c)] without violating a spares budget constraint. The algorithm used (Kettelle, 1962) may be used equally well for optimizing each of the other three figures of merit, (a), (b), (d), subject to a budget constraint. In fact, we generalize the Kettelle algorithm to solve a quite broad class of optimization problems in which a constraint is imposed on the resources available.

4.1. Fill Rate Model. Assume, as in Model 5 of Section 3:

(a) Demands for spares of type i at a maintenance depot are governed by a Poisson process with demand rate λ_i.
(b) Enough repair facilities are available at the depot so that repair of a failed unit is initiated as soon as it is received.
(c) The mean time to repair a failed unit of type i is ν_i.
(d) The cost of purchasing n_i units of type i is $n_i c_i$.
(e) There are k part types; that is, $i = 1, \ldots, k$.

The depot manager wishes to maximize fill rate $R(\mathbf{n})$ given in (3.18) by appropriate choice of spares allocation $\mathbf{n} = (n_1, \ldots, n_k)$ satisfying the budget constraint,

$$\sum_{i=1}^{k} n_i c_i \le c_0; \tag{4.1}$$

that is, the total cost of the initial supply of spares must not exceed the budget c_0 available.

To explain the Kettelle (1962) algorithm for constructing optimal spares allocations subject to a budget constraint, we must define certain concepts.

4.2. Definitions. (a) Allocation $\mathbf{n}$ is said to *dominate* allocation $\mathbf{n}'$ if either

(i) $R(\mathbf{n}) > R(\mathbf{n}')$ while $\sum_1^k n_i c_i \le \sum_1^k n_i' c_i$, or
(ii) $R(\mathbf{n}) = R(\mathbf{n}')$ while $\sum_1^k n_i c_i < \sum_1^k n_i' c_i$.

We write $\mathbf{n} \overset{\mathrm{D}}{>} \mathbf{n}'$.

Clearly, if $\mathbf{n} \overset{\mathrm{D}}{>} \mathbf{n}'$, then $\mathbf{n}$ is a preferable allocation; it achieves at least as high a fill rate as $\mathbf{n}'$, at no greater cost.

(b) $\mathbf{n}$ is an *undominated allocation* if there does not exist $\mathbf{n}'$ such that $\mathbf{n}' \overset{\mathrm{D}}{>} \mathbf{n}$.

If $\mathbf{n}$ is an undominated allocation, greater fill rate can be achieved only by spending more money for spares.

(c) A *complete sequence of undominated allocations* [ending in allocation $\mathbf{n}^{(s)}$, say] is a sequence of undominated allocations $\mathbf{n}^{(1)}, \ldots, \mathbf{n}^{(s)}$ such that

(i) $R(\mathbf{n}^{(1)}) \le R(\mathbf{n}^{(2)}) \le \cdots \le R(\mathbf{n}^{(s)})$,
(ii) $\sum_1^k c_i n_i^{(1)} \le \sum_1^k c_i n_i^{(2)} \le \cdots \le \sum_1^k c_i n_i^{(s)}$, and
(iii) $\mathbf{n}$ undominated and yielding a fill rate-cost pair distinct from those of $\mathbf{n}^{(1)}, \ldots, \mathbf{n}^{(s)} \Rightarrow R(\mathbf{n}) \ge R(\mathbf{n}^{(s)})$ and $\sum_1^k c_i n_i \ge \sum_1^k c_i n_i^{(s)}$.

If $\mathbf{n}^{(1)}, \ldots, \mathbf{n}^{(s)}$ constitute a complete sequence of undominated allocations, then no additional undominated allocations may be interpolated, except possibly those yielding identical fill rate-cost pairs.

The simplest and clearest way to explain the Kettelle algorithm is to illustrate its operation in a small numerical example. Suppose we wish to maximize fill rate for four parts, given a budget $c_0 = 1500$ (in dollars), with demand, repair, and cost data as shown in Table 4.1.

Table 4.1. Input Data for Spares Allocation

Part Type, i	*Demand Rate (per hour), λ_i*	*Mean Time to Repair Failed Part, v_i*	*Dollar Cost of Part, c_i*	*Computed Value: $\lambda_i v_i$*
1	.01	100	200	1.0
2	.02	150	100	3.0
3	.03	60	300	1.8
4	.01	200	250	2.0

Application of Kettelle Algorithm

The basic principle is that we generate complete sequences of undominated allocations taking into account one additional part type at each successive stage, combining appropriately the undominated allocations obtained at earlier stages of the calculation.

To simplify computation throughout, we compute fill rate numerator $\sum_{i=1}^{k} \lambda_i R_i(n_i)$, where

$$R_i(n) = \sum_{j=0}^{n-1} \frac{(\lambda_i \nu_i)^j}{j!} e^{-\lambda_i \nu_i} \qquad \text{for} \quad i = 1, \ldots, k; \tag{4.2}$$

that is, we drop the denominator $\sum_1^k \lambda_i$ in (3.18), since it is constant throughout.

The steps taken in carrying out the Kettelle algorithm are the following.

1. Obtain a complete sequence of undominated allocations considering part types 1 and 2 only. To obtain such a complete sequence, set up Table 4.2. In the n_1th column and n_2th row ($n_1 = 0, 1, 2, \ldots; n_2 = 0, 1, 2, \ldots$) of the body of Table 4.2 are entered in vertical succession (a) the pair: the number n_1 of spares of type 1 and the number n_2 of spares of type 2; (b) the total cost of these spares; and (c) the corresponding fill rate numerator

$$.01 \sum_{j=0}^{n_1-1} \frac{(1.0)^j}{j!} e^{-1.0} + .02 \sum_{j=0}^{n_2-1} \frac{(3.0)^j}{j!} e^{-3.0}.$$

Note that entries are made only if the total cost $c_1 n_1 + c_2 n_2 \leq c_0 = 1500$, the budget available.

2. Start with (0, 0), the first undominated spares allocation. The fill rate numerator is 0, of course.

3. The second undominated spares allocation is the *cheapest cost* entry with fill rate numerator higher than 0. Since (0, 1) has lower cost, 100, then (1, 0) with cost 200, select (0, 1).

4. In general, the *next* undominated spares allocation is the cheapest cost entry with fill rate numerator higher than the one just obtained. If several entries of identical cost qualify, choose the one with highest fill rate numerator. If several entries of identical cost and identical highest fill rate numerator qualify, choose the one with lowest part type index. If the present entry is at the intersection of row i and column j, search for the next entry may be confined to the union of rows $1, 2, \ldots, i$, and columns $1, 2, \ldots, j$.

5. Computation at this stage stops just before exceeding the budget constraint $c_0 = 1500$.

6. For part types 3 and 4, obtain in a similar fashion a complete sequence of undominated allocations (see Table 4.3). For each additional pair of part types obtain in a similar fashion a complete sequence of undominated allocations.

Table 4.2. Undominated Allocations for Parts 1 and 2

PART TYPE 1

PART TYPE 2

Number of Spares *Cost* *Fill Rate Numerator*	*0* *0* *0*	*1* *200* *.00368*	*2* *400* *.00736*	*3* *600* *.00920*	*4* *800* *.00981*	*5* *1000* *.00996*	*6* *1200* *.00999*	*7* *1400* *.01000*
0 0 0	(0, 0) 0 0	(1, 0) 200 .00368	(2, 0) 400 .00736	(3, 0) 600 .00920	(4, 0) 800 .00981	(5, 0) 1000 .00996	(6, 0) 1200 .00999	(7, 0) 1400 .01000
1 100 .00100	(0, 1) 100 .00100	(1, 1) 300 .00468	(2, 1) 500 .00836	(3, 1) 700 .01020	(4, 1) 900 .01081	(5, 1) 1100 .01096	(6, 1) 1300 .01099	(7, 1) 1500 .01100
2 200 .00398	(0, 2) 200 .00398	(1, 2) 400 .00766	(2, 2) 600 .01134	(3, 2) 800 .01318	(4, 2) 1000 .01379	(5, 2) 1200 .01394	(6, 2) 1400 .01397	
3 300 .00846	(0, 3) 300 .00846	(1, 3) 500 .01214	(2, 3) 700 .01612	(3, 3) 900 .01766	(4, 3) 1100 .01827	(5, 3) 1300 .01842	(6, 3) 1500 .01845	
4 400 .01294	(0, 4) 400 .01294	(1, 4) 600 .01662	(2, 4) 800 .02030	(3, 4) 1000 .02214	(4, 4) 1200 .02275	(5, 4) 1400 .02290		
5 500 .01630	(0, 5) 500 .01630	(1, 5) 700 .01998	(2, 5) 900 .02366	(3, 5) 1100 .02550	(4, 5) 1300 .02611	(5, 5) 1500 .02626		
6 600	(0, 6) 600	(1, 6) 800	(2, 6) 1000	(3, 6) 1200	(4, 6) 1400			

7 700 .01932	(0, 7) 700 .01932	(1, 7) 900 .02300	(2, 7) 1100 .02668	(3, 7) 1300 .02852	(4, 7) 1500 .02913				
8 800 .01976	(0, 8) 800 .01976	(1, 8) 1000 .02344	(2, 8) 1200 .02712	(3, 8) 1400 .02896					
9 900 .01992	(0, 9) 900 .01992	(1, 9) 1100 .02350	(2, 9) 1300 .02728	(3, 9) 1500 .02912					
10 1000 .01998	(0, 10) 1000 .01998	(1, 10) 1200 .02366	(2, 10) 1400 .02734						
11 1100 .02000	(0, 11) 1100 .02000	(1, 11) 1000 .02368	(2, 11) 1500 .02736						
12 1200 .02000	(0, 12) 1200 .02000	(1, 12) 1400 .02368							
13 1300 .02000	(0, 13) 1300 .02000	(1, 13) 1500 .02368							
14 1400 .02000	(0, 14) 1400 .02000								
15 1500 .02000	(0, 15) 1500 .02000								

Table 4.3. Undominated Allocations for Parts 3 and 4

PART TYPE 3

PART TYPE 4	*Number of Spares / Cost / Fill Rate Numerator*	*0 / 0 / 0*	*1 / 300 / .00495*	*2 / 600 / .01389*	*3 / 900 / .02193*	*4 / 1200 / .02673*	*5 / 1500 / .02892*		
	0 / 0 / 0	(0, 0) / 0 / 0	(1, 0) / 300 / .00495	(2, 0) / 600 / .01389	(3, 0) / 900 / .02193	(4, 0) / 1200 / .02673	(5, 0) / 1500 / .02892		
	1 / 250 / .00135	(0, 1) / 250 / .00135	(1, 1) / 550 / .00630	(2, 1) / 850 / .01524	(3, 1) / 1150 / .02328	(4, 1) / 1450 / .02808			
	2 / 500 / .00406	(0, 2) / 500 / .00406	(1, 2) / 800 / .00901	(2, 2) / 1100 / .01795	(3, 2) / 1400 / .02599				
	3 / 750 / .00677	(0, 3) / 750 / .00677	(1, 3) / 1050 / .01172	(2, 3) / 1350 / .02066					
	4 / 1000 / .00857	(0, 4) / 1000 / .00857	(1, 4) / 1300 / .01352						
	5 / 1250 / .00947	(0, 5) / 1250 / .00947							
	6 / 1500 / .00983	(0, 6) / 1500 / .00983							

7. Next, obtain a complete sequence of undominated spares allocations for part types 1, 2, 3, and 4 by using the undominated allocations of part types 1, 2 and the undominated allocations of part types 3, 4, as shown in Table 4.4. Follow the same principle as in step 4. The movements may be considerably more irregular as the number of parts involved becomes larger. In the present example involving only four part types, the resulting sequence of undominated allocations of Table 4.4 constitutes a complete sequence of undominated spares allocations that do not exceed the budget constraint $c_0 = 1500$.

8. In a similar fashion, combine part types 5, 6, 7, 8, and so on, until all part types have been combined. It is not necessary that groups of part types being combined be of equal size, although experience shows that wherever possible it is preferable to use groups of equal size.

The combination of spares $n_1 = 0$, $n_2 = 6$, $n_3 = 3$, $n_4 = 0$ attained finally in Table 4.4 for an expenditure of $\sum_1^4 n_i c_i = 1500$ represents the solution to the problem of achieving maximum fill rate without exceeding the budget $c_0 = 1500$ available for spares. The corresponding fill rate is

$$R(\mathbf{n}) = \sum_{i=1}^{4} \lambda_i \sum_{j=0}^{n_i-1} \frac{(\lambda_i v_i)^j}{j!} e^{-\lambda_i v_i} \left(\sum_{i=1}^{4} \lambda_i \right)^{-1}$$

$$= \frac{.01(0) + .02 \sum_{j=0}^{5} [(3.0)^j/j!]\, e^{-3.0} + .03 \sum_{j=0}^{2} [(1.8)^j/j!]\, e^{-1.8} + .01(0)}{.01 + .02 + .03 + .01} = .575.$$

Note further that each undominated spares allocation in Table 4.4 is the solution to *some* optimization problem. For example, the spares combination $n_1 = 0$, $n_2 = 5$, $n_3 = 3$, $n_4 = 0$ (just to the left of the solution of the original problem) maximizes fill rate when the budget available is \$1400, rather than \$1500, as in the original problem. Figure 7.4.1 shows optimal fill rate attained for each possible budget between 0 and \$1500.

Before we verify that the Kettelle algorithm does yield optimal spares kits in the present problem, we consider a more general model. We will show that the Kettelle algorithm furnishes a solution to this general optimization model. It then becomes easy to verify that the Fill Rate Model is just a special case.

4.3. General Optimization Model. Let $x_1, \ldots, x_k$ represent the decision variables; $x_i \in S_i = \{x_i^{(1)} < x_i^{(2)} < \cdots\}$, $i = 1, \ldots, k$, in the following. Let $\mathbf{x}_i = (x_1, \ldots, x_i)$, $i = 1, \ldots, k$. Let $f_1, \ldots, f_n$ be strictly increasing functions, with $y_1(x_1) = f_1(x_1)$, $y_2(\mathbf{x}_2) = f_2(y_1, x_2)$, $y_3(\mathbf{x}_3) = f_3(y_2, x_3), \ldots$, $y_k(\mathbf{x}_k) = f_k(y_{k-1}, x_k)$. Similarly, let $g_1, \ldots, g_k$ be strictly increasing functions, with $z_1(x_1) = g_1(x_1)$, $z_2(\mathbf{x}_2) = g_2(z_1, x_2)$, $z_3(\mathbf{x}_3) = g_3(z_2, x_3), \ldots, z_k(\mathbf{x}_k) = g_k(z_{k-1}, x_k)$. We drop the argument of the y_i and z_i, where convenient.

Table 4.4. Undominated Allocations Using Parts 1, 2, 3, and 4

PART TYPES 1, 2

PART TYPES 3, 4	*Spares* *Cost* *FRN*	*(0,0)* *0* *0*	*(0,1)* *100* *.00100*	*(0,2)* *200* *.00398*	*(0,3)* *300* *.00846*	*(0,4)* *400* *.01294*	*(0,5)* *500* *.01630*	*(0,6* *60* *.018*
	(0,0) 0 0	(0,0,0,0) 0 0	0,1,0,0 100 .00100	0,2,0,0 200 .00398	0,3,0,0 300 .00846	0,4,0,0 400 .01294	0,5,0,0 500 .01630	0,6,(60 .018
	(0,1) 250 .00135	0,0,0,1 250 .00135	0,1,0,1 350 .00235	0,2,0,1 450 .00533	0,3,0,1 550 .00981	0,4,0,1 650 .01429	0,5,0,1 750 .01765	0,6,(85 .019
	(1,0) 300 .00495	0,0,1,0 300 .00495	0,1,1,0 400 .00595	0,2,1,0 500 .00893	0,3,1,0 600 .01341	0,4,1,0 700 .01789	0,5,1,0 800 .02125	0,6,1 90 .023
	(1,1) 550 .00630	0,0,1,1 550 .00630	0,1,1,1 650 .00730	0,2,1,1 750 .01028	0,3,1,1 850 .01476	0,4,1,1 950 .01924	0,5,1,1 1050 .02260	0,6,1 115 .024
	(2,0) 600 .01389	0,0,2,0 600 .01389	0,1,2,0 700 .01489	0,2,2,0 800 .01787	0,3,2,0 900 .02235	0,4,2,0 1000 .02683	0,5,2,0 1100 .03019	0,6,2 120 .032
	(2,1) 850 .01524	0,0,2,1 850 .01524	0,1,2,1 950 .01624	0,2,2,1 1050 .01922	0,3,2,1 1150 .02370	0,4,2,1 1250 .02818	0,5,2,1 1350 .03154	0,6,2 145 .033
	(3,0) 900 .02193	0,0,3,0 900 .02193	0,1,3,0 1000 .02293	0,2,3,0 1100 .02591	0,3,3,0 1200 .03039	0,4,3,0 1300 .03487	0,5,3,0 1400 .03823	0,6,3 150 .040
	(3,1) 1150 .02328	0,0,3,1 1150 .02328	0,1,3,1 1250 .02428	0,2,3,1 1350 .02726	0,3,3,1 1450 .03174			
	(4,0) 1200 .02673	0,0,4,0 1200 .02673	0,1,4,0 1300 .02773	0,2,4,0 1400 .03071	0,3,4,0 1500 .03519			
	(4,1) 1450 .02808	0,0,4,1 1450 .02808						
	(5,0) 1500 .02892	0,0,5,0 1500 .02892						

5) 0 98	*(1,6)* *800* *.02200*	*(2,5)* *900* *.02366*	*(2,6)* *1000* *.02568*	*(2,7)* *1100* *.02668*	*(3,6)* *1200* *.02752*	*(3,7)* *1300* *.02852*	*(3,8)* *1400* *.02896*	*(4,7)* *1500* *.02913*
0,0 0 8 98	1,6,0,0 800 9 .02200	2,5,0,0 900 10 .02366	2,6,0,0 1000 .02568	2,7,0,0 1100 .02668	3,6,0,0 1200 .02752	3,7,0,0 1300 .02852	3,8,0,0 1400 .02896	4,7,0,0 1500 .02913
0,1 0 33	1,6,0,1 1050 .02335	2,5,0,1 1150 .02501	2,6,0,1 1250 .02703	2,7,0,1 1350 .02803	3,6,0,1 1450 .02887			
1,0 00 93	1,6,1,0 1100 .02695	2,5,1,0 1200 .02861	2,6,1,0 1300 .03063	2,7,1,0 1400 .03163	3,6,1,0 1500 .03247			
1,1 50 28	1,6,1,1 1350 .02830	2,5,1,1 1450 .02996						
2,0 00 87	1,6,2,0 1400 .03589	2,5,2,0 1500 .03755						

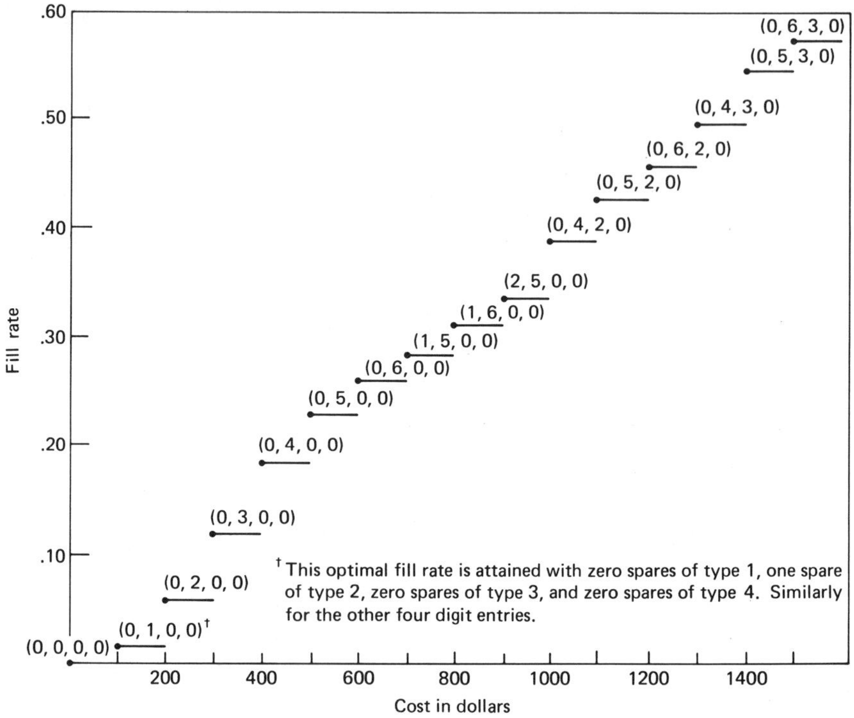

Figure 7.4.1.
Best fill rate attainable for various spares budgets.

It is suggestive to call $\mathbf{x}_i$ an "allocation" of order i, y_i a "payoff" of order i, and z_i a "cost" of order i ($i = 1, \ldots, k$), even though the model is more general than these terms would indicate.

Problem 1. Maximize payoff $y_k(\mathbf{x}_k)$ subject to the cost constraint $z_k(\mathbf{x}_k) \leq c$.

In order to solve Problem 1, we need to define concepts generalizing those of Definition 4.2.

4.4. Definitions. (a) Allocation $\mathbf{x}_i$ *dominates* allocation $\mathbf{x}_i'$ if either

(i) $y_i(\mathbf{x}_i) > y_i(\mathbf{x}_i')$ and $z_i(\mathbf{x}_i') \leq z_i(\mathbf{x}_i')$, or
(ii) $y_i(\mathbf{x}_i) = y_i(\mathbf{x}_i')$ and $z_i(\mathbf{x}_i') < z_i(\mathbf{x}_i')$.

We write $\mathbf{x}_i \overset{\text{D}}{>} \mathbf{x}_i'$. We also say the corresponding payoff-cost pair (y_i, z_i) dominates (y_i', z_i'), and write $(y_i, z_i) \overset{\text{D}}{>} (y_i', z_i')$.

(b) $\mathbf{x}_i$ is an *undominated allocation* of order i if there exists no $\mathbf{x}_i'$ such that $\mathbf{x}_i' \overset{\text{D}}{>} \mathbf{x}_i$. We also say the corresponding payoff-cost pair (y_i, z_i) is undominated.

(c) A *complete sequence of undominated allocations* of order i [ending in $\mathbf{x}_i^{(s)}$, say] is a sequence of undominated allocations $\mathbf{x}_i^{(1)}, \ldots, \mathbf{x}_i^{(s)}$ each of order i, such that

(i) $y_i(\mathbf{x}_i^{(1)}) \leq \cdots \leq y_i(\mathbf{x}_i^{(s)})$,
(ii) $z_i(\mathbf{x}_i^{(1)}) \leq \cdots \leq z_i(\mathbf{x}_i^{(s)})$, and
(iii) $\mathbf{x}_i$ undominated and yielding a payoff-cost pair distinct from those of $\mathbf{x}_i^{(1)}, \ldots, \mathbf{x}_i^{(s)} \Rightarrow y_i(\mathbf{x}_i) \geq y_i(\mathbf{x}_i^{(s)})$ and $z_i(\mathbf{x}_i) \geq z_i(\mathbf{x}_i^{(s)})$.

We also call the corresponding sequence $(y_i^{(1)}, z_i^{(1)}), \ldots, (y_i^{(s)}, z_i^{(s)})$ a complete undominated sequence of payoff-cost pairs of order i.

To solve Problem 1, we use the Generalized Kettelle Algorithm. The basic ideas are similar to those used in the special case in which fill rate was maximized subject to a constraint on total cost for spares.

Generalized Kettelle Algorithm (GKA)

First generate the longest complete sequence of undominated allocations of order 1 not exceeding the cost constraint c_0. Next generate the longest complete sequence of undominated allocations of order 2 based on the first two coordinates (x_1, x_2) not exceeding the cost constraint. The corresponding complete sequence of payoff-cost pairs, $(y_2^{(1)}, z_2^{(1)}), (y_2^{(2)}, z_2^{(2)}), \ldots, (y_2^{(s_2)}, z_2^{(s_2)})$ is used to generate the longest complete sequence of undominated allocations of order 3 not exceeding the cost constraint, with corresponding payoff-cost pairs $(y_3^{(1)}, z_3^{(1)}), (y_3^{(2)}, z_3^{(2)}), \ldots, (y_3^{(s_3)}, z_3^{(s_3)})$. Continue in this fashion, incorporating at each stage of the recursive calculation one additional x coordinate, until at the last (kth) stage, arrive at the longest complete sequence of undominated payoff-cost pairs, $(y_k^{(1)}, z_k^{(1)})$, $(y_k^{(2)}, z_k^{(2)}), \ldots, (y_k^{(s_k)}, z_k^{(s_k)})$, not exceeding the cost constraint.

Specifically, proceed as follows:

1. At stage 1 of the calculation, compute $y_1^{(i)} = f_1(x_1^{(i)})$, $z_1^{(i)} = g_1(x_1^{(i)})$ for $i = 1, \ldots, s_1$, where $z_1^{(s_1)} \leq c$, while $z_1^{(s_1+1)} > c$. Trivially, $(y_1^{(1)}, z_1^{(1)}), \ldots, (y_1^{(s_1)}, z_1^{(s_1)})$ constitute a complete sequence of undominated payoff-cost pairs of order 1, not violating the cost constraint.

2. At stage 2, compute $y_{2ij} = f_2(y_1^{(i)}, x_2^{(j)})$, $z_{2ij} = g_2(z_1^{(i)}, x_2^{(j)})$ for $i = 1, 2, \ldots; j = 1, 2, \ldots$ such that $z_{2ij} \leq c$. Enter the payoff-cost pair (y_{2ij}, z_{2ij}) in row i, column j of a table of payoff-cost pairs.

3. The lowest cost undominated pair $(y_2^{(1)}, z_2^{(1)})$ in the table is clearly (y_{211}, z_{211}).

4. To determine $(y_2^{(2)}, z_2^{(2)})$, find (y_{2ij}, z_{2ij}) such that $y_{2ij} > y_2^{(1)}$, with z_{2ij} minimum among entries satisfying this inequality. If several entries of identical cost qualify, choose the highest payoff y_{2ij}. If several entries of identical cost and identical highest payoff qualify, choose one at random, since all are equally cost-effective. The payoff-cost entry so chosen is $(y_2^{(2)}, z_2^{(2)})$.

5. Continue in this fashion; that is, having found $(y_2^{(r)}, z_2^{(r)})$, then $(y_2^{(r+1)}, z_2^{(r+1)})$ is the pair (y_{2ij}, z_{2ij}) such that $y_{2ij} > y_2^{(r)}$ with z_{2ij} minimum among pairs satisfying this inequality. In case of ties, follow the rules for breaking ties described in 4 above.

6. Stop at $(y_2^{(s_2)}, z_2^{(s_2)})$, where s_2 is determined so that $z_2^{(s_2)} \leq c$, while $z_2^{(s_2+1)} > c$.

7. The sequence $(y_2^{(1)}, z_2^{(1)}), \ldots, (y_2^{(s_2)}, z_2^{(s_2)})$ obtained this way constitutes a complete sequence of undominated payoff-cost pairs of order 2 not exceeding the cost limit c.

8. At stage 3, obtain a complete sequence of undominated payoff-cost pairs not violating the cost constraint, by proceeding in a similar fashion, using the payoff-cost pairs (y_{3ij}, z_{3ij}), where $y_{3ij} = f_3(y_2^{(i)}, z_3^{(j)})$, $z_{3ij} = g_3(z_2^{(i)}, x_3^{(j)})$ for $i = 1, 2, \ldots; j = 1, 2, \ldots$ such that $z_{3ij} \leq c$.

9. Continue in this fashion, until at the kth stage, arrive at the complete sequence of undominated cost-payoff pairs $(y_k^{(1)}, z_k^{(1)}), \ldots, (y_k^{(s_k)}, z_k^{(s_k)})$, where $z_k^{(s_k)} \leq c$, while $z_k^{(s_k+1)} > c$.

10. Then $y_k^{(s_k)}$ is the maximum payoff achievable under the cost constraint; corresponding cost is $z_k^{(s_k)}$. The allocation $\mathbf{x}^{(s_k)}$ yielding the payoff-cost pair $(y_k^{(s_k)}, z_k^{(s_k)})$ is the solution to Problem 1 above.

Next, we will prove that the payoff-cost pair $(y_k^{(s_k)}, z_k^{(s_k)})$ obtained following the Generalized Kettelle Algorithm (GKA) above does indeed represent the solution to Problem 1.

4.5. Lemma. (a) Every payoff-cost pair of order 2 obtained by the GKA is undominated.

(b) Every undominated payoff-cost pair of order 2 may be obtained by the GKA.

PROOF. (a) By construction $(y_2^{(1)}, z_2^{(1)})$ is undominated. Suppose then for $i > 1$, $(y_2^{(i)}, z_2^{(i)})$, obtained by the GKA, were dominated by (y_2, z_2), say. Let j be the largest integer such that $z_2^{(j)} < z_2$. Then $z_2^{(j)} < z_2 \leq z_2^{(i)}$, and so $j < i$. It follows that $y_2 \geq y_2^{(i)} \geq y_2^{(j+1)}$. Since $z_2^{(j)} < z_2 \leq z_2^{(j+1)}$ and $y_2 \geq y_2^{(j+1)}$, then (y_2, z_2) must be the successor to $(y_2^{(j)}, z_2^{(j)})$ obtained by the GKA. Since by construction, no entry in the GKA table is dominated by any other entry in the table, the desired conclusion follows.

(b) Let (y_2, z_2) be an undominated payoff-cost pair of order 2. Let the table entry $(y_2^{(i)}, z_2^{(i)})$ have the largest value i such that $z_2^{(i)} < z_2$. Then $y_2 > y_2^{(i)}$, since otherwise $(y_2^{(i)}, z_2^{(i)}) \overset{D}{>} (y_2, z_2)$. It follows that (y_2, z_2) must be the successor to $(y_2^{(i)}, z_2^{(i)})$.

Another basic result we will need in demonstrating that the GKA solves Problem 1 is given in the following lemma.

4.6. Lemma. Let $(y_2, z_2) \overset{D}{>} (y_2', z_2')$, $y_3 = f_3(y_2, x_3)$, $z_3 = g_3(z_2, x_3)$, $y_3' = f_3(y_2', x_3)$, and $z_3' = g_3(z_2', x_3)$. Then $(y_3, z_3) \overset{D}{>} (y_3', z_3')$.

PROOF. By hypothesis, either

(a) $y_2 > y_2'$ and $z_2 \leq z_2'$, or
(b) $y_2 = y_2'$ and $z_2 < z_2'$.

In case (a), $y_3 > y_3'$ since f_3 is strictly increasing. Also $z_3 \leq z_3'$ since g_3 is strictly increasing. Thus $(y_3, z_3) \overset{D}{>} (y_3', z_3')$.

In case (b), $y_3 = y_3'$, while $z_3 < z_3$ since g_3 is strictly increasing.

In both cases, $(y_3, z_3) \overset{D}{>} (y_3', z_3')$. ‖

Remark. Lemma 4.6 shows that at each stage of the recursive calculation, we need only retain the undominated payoff-cost pairs, since a *dominated* payoff-cost pair of order i will yield only dominated payoff-cost pairs of order $i + 1$.

We may now show the optimality of the GKA.

4.7. Theorem. The payoff-cost pair $(y_k^{(s_k)}, z_k^{(s_k)})$ obtained using the GKA is the solution to Problem 1. That is, $y_k^{(s_k)}$ is the maximum payoff achievable with corresponding cost $z_k^{(s_k)} \leq c$, the cost constraint.

PROOF. By repeated applications of Lemma 4.5 and 4.6, we see that the sequence $(y_n^{(1)}, z_n^{(1)}), \ldots, (y_k^{(s_k)}, z_k^{(s_k)})$ of payoff-cost pairs obtained at the last stage of the GKA calculation consists of *all* the undominated payoff-cost pairs of order k satisfying the cost constraint. Since $y_k^{(s_k)}$ achieves highest payoff among all the undominated pairs not exceeding the cost constraint, $(y_k^{(s_k)}, z_k^{(s_k)})$ constitutes the solution to Problem 1. ‖

Next we present some examples of the General Optimization Model formulated in 4.3, and solved by the Generalized Kettelle Algorithm.

Example 1. A series system of k subsystems is being designed. For subsystem i, the designer can choose any one of a finite or infinite set of possibilities with respective reliability-cost combinations $(r_{i1}, c_{i1}), (r_{i2}, c_{i2}), \ldots$ for $i = 1, \ldots, n$. What choice j_i shall the designer make for subsystem i, $i = 1, \ldots, k$, so as to maximize system reliability $\prod_{i=1}^{k} r_{ij_i}$ subject to the cost constraint $\sum_{i=1}^{k} c_{ij_i} \leq c$?

In the setting of the General Optimization Model of 4.3, $j_1, \ldots, j_k$ represent the decision variables, $f_1(j_1) \equiv r_{1j_1}$, $g_1(j_1) = c_{1j_1}$, $f_i(r, j_i) = rr_{ij_i}$, and $g_i(c, c_{ij_i}) = c + c_{ij_i}$, $i = 2, \ldots, k$.

Special cases of Example 1 that arise frequently in reliability are the following:

Example 1(a): Parallel Redundancy. Let $r_{ij} = 1 - q_i^j$, where q_i denotes the unreliability of a single unit of type i, and r_{ij} represents the reliability of the ith subsystem consisting of j units of type i in parallel. Let $c_{ij} = jc_i$, where c_i is the cost of a single unit of type i, and c_{ij} is the cost of j units of type i.

In this model, the designer must choose the level of redundancy for the ith subsystem, consisting of like units in parallel, $i = 1, \ldots, k$, when the total budget for redundancy is subject to a constraint. See B-P (1965), Chapter 6, for a detailed discussion with numerical examples.

Example 1(b): Spares Redundancy. In this model a system consists of k "component positions" in series. When the component occupying position i fails, it is immediately replaced by a spare component of the same type, as long as such a spare is available. A component of type i costs c_i and has life distribution F_i, $i = 1, \ldots, k$, with all life lengths mutually independent. Component position i is required to function for a period of length t_i, $i = 1, \ldots, k$.

The problem is to determine the number n_i of spares of type i, $i = 1, \ldots, k$, so as to minimize the probability of system shutdown due to shortage of spares, without exceeding a constraint c on the total expenditure $\sum_{i=1}^{k} n_i c_i$ for spares.

In this model $r_{ij} = \bar{F}^{(j+1)}(t_i)$, the survival probability corresponding to the $j + 1$st convolution of F_i.

For a detailed discussion with numerical examples, see B-P (1965), Chapter 6.

Example 1(c): Spares Redundancy for k-out-of-m Subsystems. A system consists of n subsystems in series. Subsystem i functions if at least k_i of the m_i units in the subsystem function. Subsystem i is required to function for a period of length t_i. A unit of type i costs c_i and has life distribution F_i; all life lengths are mutually independent. A number n_i spares of type i are provided. Replacement of a failed part is immediate, so long as a spare is available.

The problem is to determine the number n_i of spares of type i, $i = 1, \ldots, k$, so as to minimize the probability of system shutdown due to shortage of spares, without exceeding a constraint c on the total expenditure for spares.

In this model r_{ij} is somewhat complicated, although expressible in closed form. See Exercise 1.

Examples 1(a), 1(b), and 1(c) are special cases of Example 1, in that the figure of merit being maximized is the probability of series system survival.

In the following examples of the General Optimization Model of 4.3, other figures of merit are selected for optimization.

Example 2: Fill Rate. The fill rate model described in 4.1 is a special case of the General Optimization Model. The decision variables are the numbers of spares of the k different types, $(n_1, \ldots, n_k)$. The functions f_i and g_i of the General Model of 4.3 are given by

$$f_1(n_1) = \lambda_1 \sum_{j=0}^{n_1-1} \frac{(\lambda_1 v_1)^j}{j!} e^{-\lambda_1 v_1},$$

$$f_i(R, n_i) = R + \lambda_i \sum_{j=0}^{n_i-1} \frac{(\lambda_i v_i)^j}{j!} e^{-\lambda_j v_j},$$

$$g_1(n_1) = c_1 n_1,$$

$$g_i(C, n_i) = C + c_i n_i,$$

so that all f_i, g_i are strictly increasing functions.

It follows by Theorem 4.7 that the Kettelle Algorithm yields the solution to the problem of maximization of fill rate subject to the cost constraint on spares.

Example 3: Minimizing Expected Total Weighted Shortage. For k component types, assume n_i spares of type i are initially available, each costing c_i. During a fixed period, the probability is p_{ij} that j spares of type i are called for. The penalty per shortage of type i is w_i. The expected total weighted shortage is given by

$$S(\mathbf{n}) = \sum_{i=1}^{k} w_i \sum_{j=n_i}^{\infty} p_{ij}(j - n_i).$$

We wish to minimize $S(\mathbf{n})$ subject to the usual constraint that total cost $\sum_{i=1}^{k} c_i n_i \le c$.

It is left to Exercise 2 to verify that this model is a special case of the General Model of 4.3.

Remark. In the examples displayed above, the cost for spares of each component type is linear in the number of spares of that type. It is apparent, however, that all that is really required to apply the GKA is that the cost be strictly increasing in the number of spares of each type.

It is clear that a great many other models requiring optimization subject to a constraint arising in reliability and more generally in operations research are special cases of the General Optimization Model of 4.3. Such problems may be solved by the Generalized Kettelle Algorithm.

EXERCISES

1. An l-out-of-m system is required to operate for t hours. Failed components are immediately replaced from an initial stock of n spares so long as spares are available. Components and spares have mutually independent life lengths with common distribution F. Express the probability that the system survives during $[0, t]$ without shutdown due to shortage of spares.
2. Verify that the expected total weighted shortage model of Example 3 is a special case of the General Model of 4.1.
3. Formulate a model in which the expected time until system shutdown due to shortage of spares is to be minimized subject to a cost constraint. Does the Generalized Kettelle Algorithm provide a solution?
4. Formulate an alternative model in which the expected time until system failure is to be minimized subject to a cost constraint. In this model redundant units operate in parallel (active redundancy), rather than in a standby capacity as spares. Does the GKA provide a solution?
5. Show that the GKA may be used to solve the problem of maximizing (a) $\sum_{i=1}^{k} u_i(x_i)$, or, alternately, (b) $\prod_{i=1}^{n} u_i(x)$, subject to $\sum_{i=1}^{k} v_i(x_i) \leq c$, where each u_i and v_i is strictly increasing, and, additionally, in (b) each $u_i > 0$.
6. Using the same notation as in 4.3, suppose we wish to maximize $h(y_k(\mathbf{x}_k), z_k(\mathbf{x}_k))$, where $h(u, v)$ is increasing in u and decreasing in v, subject to the constraint that $z_k(\mathbf{x}_k) \leq c$. Show that the GKA provides the solution.

   An example of such a function $h(u, v)$ is $h(u, v) = g(u) - v$, where $g(u)$ may be interpreted as the utility corresponding to the figure of merit u, and v the corresponding expenditure required to achieve it. Thus $h(y_k(\mathbf{x}_k), z_k(\mathbf{x}_k))$ may be interpreted as the net return corresponding to the choice of decision variables $(x_1, \ldots, x_k)$.

5. NOTES AND REFERENCES

Section 2. The treatment of Model 1 is based on a forthcoming paper by Esary and Proschan, while the discussion of Model 2 is based on Barlow and Proschan (1973). Model 2 is also considered by Bazovsky, MacFarlane, and Wunderman (1962), who give a heuristic proof for Theorem 2.5. A discussion of Model 2 assuming exponential failure and exponential repair is presented in *Handbook of Reliability Engineering* (1968).

Section 3. Models 3 and 4 are discussed in Gnedenko, Belyayev, and Solovyev (1969) and Barlow (1962).

Section 4. A comprehensive discussion of optimal spares provisioning and, more generally, optimal redundancy allocation is presented in B-P (1965), Chapter 6. The fill rate model and the numerical illustration of 4.1 were developed in a consulting application at the Boeing Scientific Research Laboratories.

The General Optimization Model of 4.3 has not been considered in this degree of generality before. The associated Generalized Kettelle Algorithm providing a family of solutions in this optimization model, although a direct generalization of the Kettelle (1962) algorithm, is new.

8

LIMITING LIFE DISTRIBUTIONS FOR COHERENT SYSTEMS

In Chapter 4 we characterized the smallest class of life distributions, containing the exponential distributions, which is closed under the formation of coherent systems and the taking of limits in distribution. This is the class of IFRA distributions. In the present chapter, we obtain more restricted classes of life distributions by studying other limiting operations corresponding to various reliability models.

In Sections 1, 2, and 3 we give a complete answer to the following question: *Given a series or parallel system of like components whose number is increasing without bound, what are the possible types of limiting distributions for system lifetime?* Finally, in Section 4 we obtain the limiting life distribution for a series system of unlike components, with components replaced at failure.

1. LIMIT DISTRIBUTIONS FOR SERIES AND PARALLEL SYSTEMS

As we have seen, the series (parallel) system is basic in reliability theory and practice. It is therefore of some interest to determine the type of life distribution that a series (parallel) system of like components can have as the number of components increases without bound. (In Section 4 we describe results obtained in the case of unlike components.) Of course, the series (parallel) system life length corresponds to the smallest (largest) order statistic in a random sample, so that the results we obtain below are of interest both in reliability theory and in order statistics theory.

For greater generality, unless otherwise specified, we consider $T_1, \ldots, T_n$ to be independent, identically distributed observations from a common distribution F, where the support of F is not necessarily restricted to $[0, \infty)$. Set $\xi_n = \min(T_1, \ldots, T_n)$. (In the reliability context, the T_i might denote the life length of the ith component of a series system, and ξ_n the system life length.) Obviously, as $n \to \infty$, ξ_n converges a.s. to the left-hand limit of the support of F. Thus to avoid trivialities, we normalize ξ_n by scale parameter a_n and translation parameter b_n, and seek the limiting distribution of the sequence $(\xi_n - b_n)/a_n$, where $a_n > 0$, b_n are suitably chosen normalizing constants. Also to avoid trivialities, we rule out degenerate distributions for the limiting distributions, since these can simply be obtained by taking a_n sufficiently large so that $\xi_n/a_n \xrightarrow{P} 0$ and choosing b_n appropriately. Intuitively, in the reliability context we may view a_n as the unit in which time is measured.

Before proceeding with the formal theory, it may be instructive to consider some special component life distributions F, and to calculate the corresponding norming constants a_n and b_n, and limiting series system life distributions. Note that in general

$$P[\xi_n > x] = \bar{F}^n(x),$$

and

$$P\left[\frac{\xi_n - b_n}{a_n} > x\right] = P[\xi_n > a_n x + b_n] = \bar{F}^n(a_n x + b_n).$$

1.1. Example. Let $F(x) = 1 - e^{-x^\alpha}$ for $x \geq 0$, where parameter $\alpha > 0$, that is, a Weibull distribution with shape parameter α (see Chapter 3, Section 5). Then $\bar{F}^n(a_n x + b_n) = e^{-n(a_n x + b_n)^\alpha} = e^{-x^\alpha}$, if we choose $a_n = n^{-1/\alpha}$ and $b_n = 0$ for $n = 1, 2, \ldots$. Thus we conclude that the class of Weibull distributions is closed under the formation of series systems of identical components.

1.2. Example. Let $F(x) = x$ for $0 \leq x \leq 1$. Choosing $a_n = 1/n$, $b_n = 0$, we have $\lim_{n\to\infty} \bar{F}^n(a_n x + b_n) = \lim_{n\to\infty}(1 - (x/n))^n = e^{-x}$. Thus limiting series system life distribution is exponential when identical component life length is uniform.

1.3. Example. Let $F(x) = e^{-(1-x)/x}$ for $0 \leq x \leq 1$. Choosing $a_n = (\log n)^{-2}$ and $b_n = (\log n)/(1 + \log n)$, we may verify (Exercise 2) that

$$\lim_{n\to\infty} \bar{F}^n(a_n x + b_n) = 1 - e^{-e^x} \quad \text{for} \quad -\infty < x < \infty.$$

A general method for finding minimum extreme value distributions is to use the transformation

$$U_n = nF(\xi_n)$$

so that

$$P[U_n > y] = P[nF(\xi_n) > y] = P\left[\xi_n > F^{-1}\left(\frac{y}{n}\right)\right] = \left[1 - \frac{y}{n}\right]^n.$$

Hence

$$\lim_{n\to\infty} P[U_n > y] = e^{-y}.$$

If F can be inverted, even asymptotically as $n \to \infty$, it is possible to determine a_n, b_n, and the limiting extreme value distribution. However, this method does not prove, as we will show in Section 2, that only three extreme value limit distributions are possible. (See Cramér, 1951, pp. 370–378, for an application of this method.)

1.4. Example. Let

$$F(x) = \int_{-\infty}^{x} \frac{e^{-u^2/2}}{\sqrt{2\pi}}\, du.$$

It is shown in Cramér (1951), pp. 374–376, that

$$\xi_n = -\sqrt{2 \log n} + \frac{\log \log n + \log 4\pi}{2\sqrt{2 \log n}} + \frac{\log U_n}{\sqrt{2 \log n}} + 0\left(\frac{1}{\log n}\right).$$

Let

$$b_n = -\sqrt{2 \log n} + \frac{\log \log n + \log 4\pi}{2\sqrt{2 \log n}}$$

and $a_n = (2 \log n)^{-1/2}$. Then

$$\lim_{n\to\infty} P\left[\frac{\xi_n - b_n}{a_n} \leq x\right]$$

$$= \lim_{n\to\infty} P[\log U_n \leq x]$$

$$= \lim_{n\to\infty} P[U_n \leq e^x] = 1 - e^{-e^x} \qquad \text{for} \quad -\infty < x < \infty.$$

From the approximation for ξ_n above it is clear that the convergence is like $(\log n)^{-1}$, which is less rapid than any power of n. Fisher and Tippet (1928) present tables illustrating the slowness of convergence in this case.

The extreme value limiting distributions derived below differ from each other in *type*.

1.5. Definition. Distributions G and H are of the same *type* if and only if there exist constants $A > 0$, B such that

$$G(Ax + B) = H(x) \qquad \text{for all} \quad x.$$

Khinchin proved the following fundamental lemma (see Feller, 1966, p. 246, for a proof).

1.6. Fundamental Lemma. If $\{F_n\}$ is a sequence of distribution functions such that for all x,

(a) $F_n(a_n x + b_n) \to G(x)$,
(b) $F_n(a_n^* x + b_n^*) \to G^*(x)$, where G and G^* are nondegenerate distribution functions, then G and G^* are of the same type, that is, for some $A > 0$, B, $G(x) = G^*(Ax + B)$ for all x.

It is also clear that if we can obtain $G(x)$ as a limit distribution, then we can obtain all distributions of the same type by appropriately modifying a_n and b_n. It follows from Lemma 1.6 and from this last assertion that in seeking all possible limiting life distributions for ξ_n, we need only consider limit distributions that differ in type.

In Section 2 we will derive the basic result, due to Fisher and Tippet (1928) and Gnedenko (1943), that the only possible limiting types of distributions for $\xi_n = \min(T_1, \ldots, T_n)$, where the $T_1, \ldots, T_n$ are a random sample from arbitrary distribution F, are:

$$W_1(x) = 1 - e^{-x^\alpha} \quad \text{for} \quad x \geq 0, \quad \text{where } \alpha > 0. \tag{1.1}$$

$$W_2(x) = 1 - e^{-(-x)^{-\alpha}} \quad \text{for} \quad x \leq 0, \quad \text{where } \alpha > 0. \tag{1.2}$$

$$\Lambda(x) = 1 - e^{-e^x} \quad \text{for} \quad -\infty < x < \infty. \tag{1.3}$$

$W_1(x)$ is the Weibull distribution discussed in Chapter 3, Section 5. $W_2(x)$ is of little interest in reliability theory or application since it is confined to the negative axis, and since it cannot arise as a limit distribution for the minimum of nonnegative random variables (for example, life lengths), as we will see in Section 2. Even though $\Lambda(x)$ permits negative values of the random variable, it is of some interest in the reliability context, since it *can* arise as a limiting distribution of the minimum of life lengths. For appropriate choice of A and B, $\Lambda(Ax + B)$ provides a close approximation to the Makeham distribution

$$F(t) = 1 - e^{-[\alpha t + (\beta/\gamma)(e^{\gamma t} - 1)]},$$

with failure rate

$$r(t) = \alpha + \beta e^{\gamma t}, \qquad t \geq 0.$$

This distribution is widely used in life insurance and mortality studies. Perhaps a theoretical explanation for its success lies in its close resemblance to Λ.

As before, let $T_1, \ldots, T_n$ be a random sample from distribution F, not necessarily restricted to the positive axis. Let $\eta_n = \max(T_1, \ldots, T_n)$. (In the reliability context, η_n might represent the life length of a parallel system of independent components with common life distribution F.) Note that $\eta_n = \max(T_1, \ldots, T_n) = -\min(-T_1, \ldots, -T_n)$. It follows that if $G(x)$ is a type of limit distribution for ξ_n, then $\bar{G}(-x)$ is a type of limit distribution for η_n. Thus the possible limit distributions for $(\eta_n - b_n)/a_n$ are of the following types:

$$W_1^*(x) = e^{-(-x)^\alpha} \quad \text{for} \quad x \le 0, \quad \text{where } \alpha > 0. \tag{1.1*}$$

$$W_2^*(x) = e^{-x^{-\alpha}} \quad \text{for} \quad x \ge 0, \quad \text{where } \alpha > 0. \tag{1.2*}$$

$$\Lambda^*(x) = e^{-e^{-x}} \quad \text{for} \quad -\infty < x < \infty. \tag{1.3*}$$

W_1^* is of little interest in reliability theory or application since it is confined to the negative axis, and since it cannot arise as a limit distribution for the maximum of nonnegative random variables (for example, life lengths).

The six extreme value distributions presented in (1.1), (1.2), (1.3), (1.1*), (1.2*), and (1.3*) are certainly appropriate for describing measurements obtained by instruments measuring extreme values. For example, Gumbel (1958) claims that Λ^* holds for highest temperatures, flood levels, and wind velocities; similarly, Λ is useful as a model describing the minima of temperatures and pressures. The Weibull distribution, W_1, has been fitted to static and dynamic breaking strengths of materials (Weibull, 1939). Many applications of the extreme value distributions may be found in Gumbel (1958). Other applications are mentioned briefly in David (1970), along with references.

1.7. Example: Air Pollution Measurements. Various air pollutants such as oxides of sulphur, carbon monoxide, oxides of nitrogen, ozone, and particulate matter are monitored at specified intervals, with pollutant concentrations $X_1, X_2, \ldots$ punched into a computer tape. Larsen (1969) claims on the basis of empirical evidence that air pollution concentrations can be considered as random variables from a lognormal distribution. Since state and national air quality standards require that the maximum value over, say, a 24-hour period do not exceed a specified standard, it is of interest and importance to determine the distribution of $\eta_n = \max(X_1, \ldots, X_n)$. As we will see in Section 3, η_n, properly normalized, has approximately the extreme value distribution $\Lambda^*(x)$ given in (1.3*). From this knowledge, the probability of exceeding an air pollution standard can be computed.

EXERCISES

1. Let $\xi_n = \min(T_1, \ldots, T_n)$, where $T_1, \ldots, T_n$ are i.i.d. random variables. Show how to choose a_n so that $\xi_n/a_n \to 0$ in probability.
2. Verify the limit claimed in Example 1.3.

3. Let $X_1 \sim W_1(x)$ in (1.1) and $X_1^* \sim W_1^*(x)$ in (1.1*). Express X_1^* as stochastically equivalent to a function of X_1. Similarly, relate (a) W_2 in (1.2) and W_2^* in (1.2*), and (b) Λ in (1.3) and Λ^* in (1.3*).

4. To find the limiting distribution of η_n use the transformation

$$V_n = n\bar{F}(\eta_n),$$

so that

$$\lim_{n\to\infty} P[V_n > y] = e^{-y}.$$

Use this transformation and Example 1.4 to find b_n and a_n so that

$$\lim_{n\to\infty} P\left[\frac{\eta_n - b_n}{a_n} \le x\right] = e^{-e^{-x}} \qquad \text{for} \quad -\infty < x < \infty$$

when $F(x) = \int_{-\infty}^{x} (e^{-u^2/2}/\sqrt{2\pi})\, du$.

5. Let

$$F(x) = \int_0^x \frac{x^{\lambda-1}e^{-x/\theta}}{\theta^\lambda \Gamma(\lambda)}\, dx.$$

Show that

$$\lim_{n\to\infty} P\left[\frac{\eta_n - b_n}{a_n} \le x\right] = e^{-e^{-x}} \qquad \text{for} \quad -\infty < x < \infty,$$

where $a_n = \theta$ and

$$b_n = \theta \log\left(\frac{n}{\theta^{\lambda-1}\Gamma(\lambda)}\right) + \theta(\lambda - 1)\log(\theta \log n).$$

(See Gurland, 1955.)

2. DERIVATION OF EXTREME VALUE DISTRIBUTIONS

Minimum Stable Distributions

A distribution function G is *minimum stable* if for all positive integers k, there exist constants $\alpha_k > 0$, β_k such that

$$\bar{G}^k(\alpha_k x + \beta_k) = \bar{G}(x) \qquad \text{for all} \quad x. \tag{2.1}$$

To show that W_1, W_2, and Λ given in (1.1), (1.2), and (1.3) are the only possible types of limit distributions for sample minimum ξ_n, we will prove first that (a) all possible limit distributions for ξ_n are minimum stable, and second that (b) W_1, W_2, and Λ are the only possible minimal stable distributions.

2.1. Theorem. G is a limiting distribution for ξ_n if and only if G is minimum stable.

PROOF. First suppose that $\bar{F}^n(a_n x + b_n) \to \bar{G}(x)$ as $n \to \infty$. Then

$$\bar{F}^{nk}(a_{nk}x + b_{nk}) \equiv \{\bar{F}^n(a_{nk}x + b_{nk})\}^k \to \bar{G}(x),$$

that is, $\bar{F}^n(a_{nk}x + b_{nk}) \to \bar{G}^{1/k}(x)$ as $n \to \infty$. By the Fundamental Lemma 1.6, it follows that $\bar{G}$ and $\bar{G}^{1/k}$ are survival probabilities of the same type. Thus by definition, G is minimum stable.

Conversely, suppose G is minimum stable. Then there exist constants $\alpha_k > 0$, β_k such that (2.1) holds. Thus G is a limit distribution for sample minima. ||

Possible Sequences for Minimum Stable Distributions

Next we will show that for minimum stable G, the norming constants α_k in (2.1) are either all <1, all $=1$, or all >1. To this end we first establish Lemma 2.2.

2.2. Lemma. Let α_k and β_k be the norming constants satisfying (2.1) for minimum stable distribution G. Then for all positive integers j and k,

$$\alpha_{jk} = \alpha_j \alpha_k \tag{2.2}$$

and

$$\beta_{jk} = \beta_k + \alpha_k \beta_j = \beta_j + \alpha_j \beta_k. \tag{2.3}$$

PROOF. From (2.1) we have $\bar{G}^k(x) = \bar{G}((x - \beta_k)/\alpha_k)$ for some α_k, β_k, all $k = 1, 2, \ldots$, and all x. Thus

$$\begin{aligned}\bar{G}^{jk}(x) &= \bar{G}^j\left(\frac{x - \beta_k}{\alpha_k}\right)\\ &= \bar{G}\left(\frac{[(x - \beta_k)/\alpha_k] - \beta_j}{\alpha_j}\right)\\ &= \bar{G}\left(\frac{[(x - \beta_j)/\alpha_j] - \beta_k}{\alpha_k}\right)\end{aligned}$$

for all x. It follows that

$$\frac{x - \beta_k - \alpha_k\beta_j}{\alpha_j\alpha_k} = \frac{x - \beta_j - \alpha_j\beta_k}{\alpha_j\alpha_k} = \frac{x - \beta_{jk}}{\alpha_{jk}}$$

for all x, so that (2.2) and (2.3) must hold. ||

We immediately deduce the following lemma.

2.3. Lemma. If $\alpha_j = 1$ for some $j > 1$ in (2.1), then $\alpha_j = 1$ for all j.

PROOF. Suppose $\alpha_{j_0} = 1$ for some $j_0 > 1$ in (2.1). Then by (2.3), $\beta_k + \alpha_k\beta_{j_0} = \beta_{j_0} + \beta_k$ for all $k \geq 1$. Thus either $\beta_{j_0} = 0$ or $\alpha_k = 1$ for all $k \geq 1$. If $\beta_{j_0} = 0$, then $\bar{G}^{j_0}(x) = \bar{G}(x)$ for all x. But this is impossible since G is nondegenerate. Hence $\alpha_k = 1$ for all k. ||

Note that $\Lambda(x) = 1 - e^{-e^x}$ satisfies (2.1) with $\alpha_k = 1$ and $\beta_k = -\log k$ for $k \geq 1$.

Next we show that if in (2.1) $\alpha_j < 1$ for some $j > 1$, then the minimum stable distribution G has its support bounded from below.

2.4. Lemma. Let $\alpha_j < 1$ for some $j > 1$ in (2.1). Then

(a) There exists x_0 such that $G(x_0) = 0$ and $G(x) > 0$ for all $x > x_0$;

(b) $\beta_k/(1 - \alpha_k) = x_0$ for all $k > 1$.

PROOF. Suppose $\alpha_{j_0} < 1$ for some $j_0 > 1$ in (2.1). Let $x^* = \beta_{j_0}/(1 - \alpha_{j_0})$. Note that $x < x^*$ implies that $x > (x - \beta_{j_0})/\alpha_{j_0}$. Hence for $x < x^*$,

$$\bar{G}(x) \leq \bar{G}\left(\frac{x - \beta_{j_0}}{\alpha_{j_0}}\right) = \bar{G}^{j_0}(x).$$

Thus for $x < x^*$, $\bar{G}(x) = 0$ or 1. Since $\bar{G}$ is decreasing and nondegenerate, it follows that $\bar{G}(x) = 1$ for $x < x^*$. Hence there exists $x_0 \geq x^*$ such that $\bar{G}(x_0) = 1$ and $\bar{G}(x) < 1$ for all $x > x_0$. Therefore $\bar{G}^k(x_0) = \bar{G}((x_0 - \beta_k)/\alpha_k) = 1$, while $\bar{G}^k(x_0 + \varepsilon) = \bar{G}([(x_0 - \beta_k)/\alpha_k] + [\varepsilon/\alpha_k]) < 1$ for arbitrary $\varepsilon > 0$, $k > 1$. This implies that $(x_0 - \beta_k)/\alpha_k = x_0$ for $k > 1$. Letting $k = j_0$, we conclude from the definition of x^* that $x_0 = x^*$. The result of (b) follows immediately. ||

In a similar fashion we may show the following lemma.

2.5. Lemma. Let $\alpha_j > 1$ for some $j > 1$ in (2.1). Then

(a) There exists x_0 such that $G(x_0) = 1$ and $G(x) < 1$ for all $x < x_0$;

(b) $\beta_k/(1 - \alpha_k) = x_0$ for all $k > 1$.

The proof is left to Exercise 3.

From Lemmas 2.3, 2.4, and 2.5, we immediately deduce the desired result.

2.6. Lemma. Let α_j, $j = 1, 2, \ldots$, be the norming constants of (2.1). Then either

(a) $\alpha_j < 1$ for all $j > 1$, or

(b) $\alpha_j = 1$ for all $j \geq 1$, or
(c) $\alpha_j > 1$ for all $j > 1$.

The proof is left to Exercise 4.

From Example 2.1 we see that W_1 satisfies the conditions of Lemma 2.4. Similarly, we may show that W_2 satisfies the hypothesis of Lemma 2.5 (see Exercise 5). Without loss of generality, we can take $x_0 = 0$ in Lemmas 2.4 and 2.5. Thus the problem of finding all minimum stable distributions reduces to the solution of the following functional equations:

$$\bar{G}^n(\alpha_n x) = \bar{G}(x), \qquad \text{with} \quad G(x) = 0 \quad \text{for } x \leq 0, \tag{2.4}$$

$$\bar{G}^n(\alpha_n x) = \bar{G}(x), \qquad \text{with} \quad G(x) = 1 \quad \text{for } x \geq 0, \tag{2.5}$$

$$\bar{G}^n(x + \beta_n) = \bar{G}(x). \tag{2.6}$$

To solve these functional equations, we use the notion of regular variation introduced by Karamata (1930); see Feller, II (1966), p. 268 ff.

2.7. Lemma. Let U be a positive monotone function on $(0, \infty)$ such that

$$\frac{U(\gamma x)}{U(\gamma)} \to \psi(x) \qquad \text{as} \quad \gamma \to \infty, \tag{2.7}$$

where $\psi(x)$ is finite and positive in some interval. Then $\psi(x) = x^\rho$, where $-\infty < \rho < \infty$.

PROOF. Since

$$\frac{U(\gamma x_1 x_2)}{U(\gamma)} = \frac{U(\gamma x_1 x_2)}{U(\gamma x_2)} \cdot \frac{U(\gamma x_2)}{U(\gamma)},$$

it follows that $\psi(x_1 x_2) = \psi(x_1)\psi(x_2)$, where finite. Letting $x_i = e^{y_i}$, $i = 1, 2$, and $\phi(y) = \psi(e^y)$, we obtain the functional equation:

$$\phi(y_1 + y_2) = \phi(y_1)\phi(y_2).$$

The only monotone solution is known to be

$$\phi(y) = e^{\rho y} \qquad \text{for} \quad -\infty < \rho < \infty.$$

(See Theorem 2.2, Chapter 3, for a closely related result.) Thus $\psi(x) = \phi(\log x) = e^{\rho \log x} = x^\rho$ for $-\infty < \rho < \infty$. ||

2.8. Definitions. (a) Let L be a positive function defined on $(0, \infty)$. Then *L varies slowly at infinity* if for each $x > 0$,

$$\frac{L(\gamma x)}{L(\gamma)} \to 1 \qquad \text{as} \quad \gamma \to \infty. \tag{2.8}$$

(b) Let $U(x) = x^\rho L(x)$ for $x > 0$, $-\infty < \rho < \infty$, where L varies slowly. Then U is said to *vary regularly with exponent* ρ.

Note that neither L nor U is required to be monotone in Definition 2.8. Note also that the notions of slow variation and regular variation do not depend on the behavior of the function in finite intervals. A useful criterion for regular variation is provided by the following lemma.

2.9. Lemma. Suppose that (a) $\lambda_{n+1}/\lambda_n \to 1$ and $c_n \to \infty$ as $n \to \infty$, and (b) U is a positive monotone function such that

$$\lim_{n\to\infty} \lambda_n U(c_n x) = \chi(x) \le \infty \tag{2.9}$$

exists on a dense set, with χ finite and positive in some interval. Then U varies regularly.

PROOF. By a change of scale, we may assume that $\chi(1) = 1$, and that (2.9) holds for $x = 1$. For given γ, define n_γ as the smallest integer such that $c_{n_\gamma+1} > \gamma$. Then $c_{n_\gamma} \le \gamma < c_{n_\gamma+1}$. Thus for an increasing U,

$$\frac{U(c_{n_\gamma}x)}{U(c_{n_\gamma+1})} \le \frac{U(\gamma x)}{U(\gamma)} \le \frac{U(c_{n_\gamma+1}x)}{U(c_{n_\gamma})}. \tag{2.10}$$

Since $\lambda_n U(c_n) \to 1$ as $n \to \infty$, then from the first inequality in (2.10),

$$\chi(x) = \lim_{\gamma\to\infty} \frac{\lambda_{n_\gamma} U(c_{n_\gamma}x)}{\lambda_{n_\gamma+1}U(c_{n_\gamma+1})} \cdot \frac{\lambda_{n_\gamma+1}}{\lambda_{n_\gamma}} \le \lim_{\gamma\to\infty} \frac{U(\gamma x)}{U(\gamma)},$$

while from the second inequality,

$$\chi(x) = \lim_{n\to\infty} \frac{\lambda_{n_\gamma+1}U(c_{n_\gamma+1}x)}{\lambda_{n_\gamma}U(c_{n_\gamma})} \cdot \frac{\lambda_{n_\gamma}}{\lambda_{n_\gamma+1}} \ge \lim_{\gamma\to\infty} \frac{U(\gamma x)}{U(x)}.$$

It follows that $\chi(x) = \lim_{\gamma\to\infty} U(\gamma x)/U(\gamma)$. Thus by Lemma 2.7 and Definition 2.8, U varies regularly.

For decreasing U, the inequalities in (2.10) are reversed, but the succeeding arguments are similar. ‖

Now we may prove our main result concerning minimum-stable distributions.

2.10. Theorem. Let distribution G be minimum-stable. Then it must be one of the following three types:

(a) $W_1(x) = 1 - e^{-x^\alpha}$ for $x \ge 0$, where $\alpha > 0$.
(b) $W_2(x) = 1 - e^{-(-x)^{-\alpha}}$ for $x \le 0$, where $\alpha > 0$.
(c) $\Lambda(x) = 1 - e^{-e^x}$ for $-\infty < x < \infty$.

PROOF. As established after Lemma 2.6, we need examine only three cases.

Case (*a*). From (2.4), where $\alpha_n < 1$ for $n > 1$, we obtain $\bar{G}(x) = \bar{G}^{1/n}(x/\alpha_n)$, so that $\lim_{n\to\infty} [-(1/n) \log \bar{G}(x/\alpha_n)] = -\log \bar{G}(x)$. Since $\alpha_n < 1$ for all $n > 1$, from (2.2) we deduce that $\alpha_{n^k} = \alpha_n{}^k \to 0$ as $n \to \infty$. Applying Lemmas 2.7 and 2.9 with $c_n = 1/\alpha_n$, $\lambda_n = 1/n$, and $U(x) = -\log \bar{G}(x)$ for $x > 0$, we conclude that $-\log \bar{G}(x)$ is of the form cx^ρ. Since $\bar{G}(0) = 1$ and $\bar{G}(x)$ is decreasing, it follows that $\rho > 0$ and $c > 0$. Hence G is of the same type as W_1.

Case (*b*). The proof is left to Exercise 6.

Case (*c*). From (2.6) we see that $\beta_n \le 0$. From (2.3) and the fact that $\alpha_n = 1$ for $n \ge 1$, we conclude that either $\beta_n \equiv 0$ or $\lim_{n\to\infty} \beta_n = -\infty$. But $\beta_n \equiv 0$ implies that G is degenerate, a case that we have ruled out. Therefore $\beta_n \to -\infty$ as $n \to \infty$.

Next make the transformation $z = e^x$, $\alpha_n = e^{\beta_n}$, and $\bar{H}(z) = \bar{G}(\log z)$ for $z > 0$. Then $\alpha_n \to 0$, while $\bar{H}^n(\alpha_n z) = \bar{H}(z)$. Applying the result of case (a) and then transforming variables, we obtain $\bar{G}(x) = e^{-e^{\rho x}}$, where $\rho > 0$. Thus G is of the same type as Λ. ||

Combining Theorems 2.1 and 2.8 we have Gnedenko's basic result.

2.11. Theorem. Let $\xi_n = \min(T_1, \ldots, T_n)$, where the $T_1, \ldots, T_n$ are a random sample from an arbitrary distribution. Then the only possible limiting types of distributions for ξ_n are W_1, W_2, and Λ, specified respectively in (1.1), (1.2), and (1.3).

In the reliability context, Theorem 2.11 states that the possible limit distributions for the life length of a series system of i.i.d. components cannot be of types other than W_1, W_2, or Λ [see (1.1), (1.2), and (1.3)]. In Section 3 we will see that W_2, with support on the negative axis, may be eliminated as a possibility, so that only types W_1 and Λ are possible series system limit distributions. Furthermore, since Λ assigns positive probability to *negative* outcomes for series system life length, it would seem more reasonable to model the life length of a large series system of identical components in terms of W_1, the Weibull distribution. This may constitute one justification of its popularity in reliability analysis.

Limit Distributions for Sample Maxima

From the results obtained above for possible limit distributions for sample minima, we may obtain analogous results for sample maxima, using the relationship: $\eta_n = \max(T_1, \ldots, T_n) = -\min(-T_1, \ldots, -T_n)$. The dual of Theorem 2.11 is the following.

2.11. Theorem. Let $\eta_n = \max(T_1, \ldots, T_n)$, where the $T_1, \ldots, T_n$ are a random sample from an arbitrary distribution. Then the only possible limiting types of distributions for η_n are W_1^*, W_2^*, and Λ^*, specified respectively in (1.1*), (1.2*), and (1.3*).

The proof is left to Exercise 7.

*k*th Order Statistics

Let $\xi_{k,n}$ be the kth smallest of n independent random variables. Smirnov (1952) shows that the only possible limiting distributions are (for fixed k)

$$\frac{1}{(k-1)!}\int_0^{x^\alpha} e^{-u}u^{k-1}\,du, \qquad x > 0, \quad \alpha > 0, \tag{2.11}$$

$$\frac{1}{(k-1)!}\int_0^{|x|^{-\alpha}} e^{-u}u^{k-1}\,du, \qquad x < 0, \quad \alpha > 0, \tag{2.12}$$

and

$$\frac{1}{(k-1)!}\int_0^{e^x} e^{-u}u^{k-1}\,du. \tag{2.13}$$

Smirnov (1952) and Chibisov (1964) find all possible limiting distributions when k as well as n is allowed to become infinite.

These results are clearly of interest in the study of k-out-of-n structures.

Compositions of Coherent Structures

Most of the classical problems in extreme value theory are contained in the following general question: given a coherent structure with a finite number of components and some procedure to increase the number of components without bound, what are the possible limiting distribution functions for the system lifetime for given component lifetime distributions? For example, the minimum of a set of random variables corresponds to the lifetime of a series structure, and the maximum corresponds to a parallel structure. The k-out-of-n structure generates order statistics, and series-parallel and parallel-series systems give rise to the minimax and maximin (Chernoff and Teicher, 1965) of double arrays of random variables.

We now describe results obtained when the method of "expanding" the structure is that of repeated composition. Thus, starting with a system of n components, each component is replaced by a similar system giving a system of n^2 components. Then each component in this system is replaced by a replica of the original system, giving a system of n^3 components, and so forth. The first three stages of composition for a three-component system are illustrated in Figure 8.2.1. This method of repeated composition was discussed in Chapter 2, Section 6, as a method of constructing more reliable safety systems out of 2-out-of-3 systems.

It is assumed that all components are independent and identical. In particular, let $p = P\{X_i = 1\}$ be the *reliability* of component i, where X_i

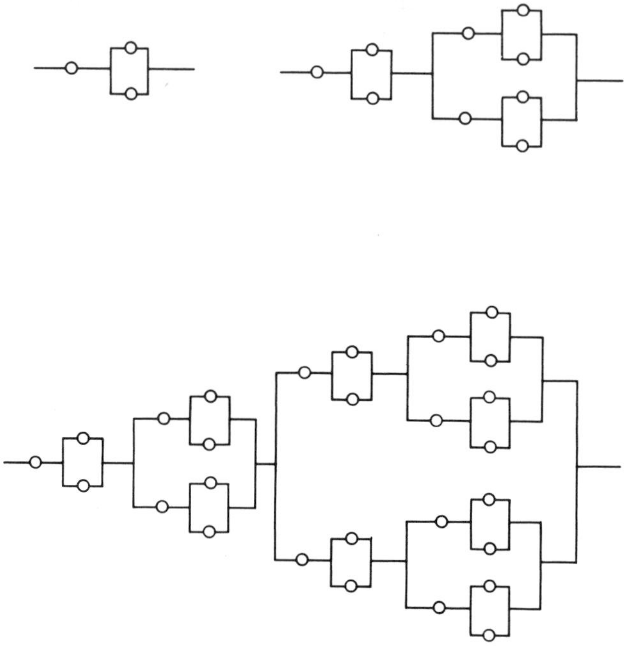

Figure 8.2.1.
Repeated compositions of a coherent structure.

is the binary random variable designating the state of component i. The reliability of the structure will be denoted by $h(p) = P\{\phi(\mathbf{X}) = 1\}$ where ϕ is the structure function. Now suppose that the components fail over time and have a common lifetime distribution F with binary indicator random variables $X_i(t)$ which are 1 or 0 according as component i is in a working or failed state respectively at time t. Then $P\{X_i(t) = 1\} = \bar{F}(t)$ and so the system has lifetime distribution $1 - h(\bar{F}(t))$. The system which results after n compositions has lifetime distribution $1 - h^{(n)}(\bar{F}(t))$, where $h^{(n)}$ denotes the n-fold composition of the function h with itself. It is assumed that $h(p) \not\equiv p$. Suppose that for suitable $\{a_n > 0\}$ and $\{b_n\}$

$$h^{(n)}(\bar{F}(a_n x + b_n)) \to \bar{G}(x). \tag{2.14}$$

Then we have the following lemma.

2.12. Lemma. If G is a limiting distribution as in (2.14), then $h(\bar{G}(x)) = \bar{G}(\alpha x + \beta)$ for some $\alpha > 0$, β.

Remark. This lemma generalizes the concept of *minimum stable* distribution introduced at the beginning of Section 2. The relation $h^{(n)}(\bar{G}(x)) = \bar{G}(\alpha_n x + \beta_n)$

for some $\alpha_n > 0$, β_n and for all n is no more general than the conclusion of Lemma 2.12, since it is implied by the lemma with $\alpha_n = \alpha^n$ and $\beta_n = (1 + \alpha + \cdots + \alpha^{n-1})\beta$.

PROOF. From (2.14)

$$h^{(n+1)}(\bar{F}(a_n x + b_n)) \to h(\bar{G}(x))$$

and

$$h^{(n+1)}(\bar{F}(a_{n+1} x + b_{n+1})) \to \bar{G}(x).$$

But by Fundamental Lemma 1.6 these two distributions must be of the same type, that is,

$$h(\bar{G}(x)) = \bar{G}(\alpha x + \beta). \;\|$$

If $\alpha \neq 1$, set $G^*(x) = G(x + [\beta/(1 - \alpha)])$, so that

$$h(\bar{G}^*(x)) = h\left(\bar{G}\left(x + \frac{\beta}{1-\alpha}\right)\right) = \bar{G}\left(\alpha x + \frac{\beta}{1-\alpha}\right) = \bar{G}^*(\alpha x). \quad (2.15)$$

If $\alpha = 1$, set $G^*(x) = G(\log x)$ and $\alpha^* = e^\beta$, so that

$$h(\bar{G}^*(x)) = h(\bar{G}(\log x)) = \bar{G}(\log x + \log \alpha^*) = \bar{G}^*(\alpha^* x). \quad (2.16)$$

2.13. Theorem. G is a limiting distribution as in (2.14) if and only if G is type equivalent to $\psi(x)$ or $\psi(e^x)$, where $1 - \psi(x)$ is a d.f. and

$$h(\psi(x)) = \psi(\alpha x) \quad \text{for all } x, \text{ where } \alpha > 0, \ \alpha \neq 1. \quad (2.17)$$

(Note that we are restricting attention to component life distributions which are necessarily 0 for negative argument.)

PROOF. The necessity part follows from (2.15), (2.16), and Lemma 2.12. For sufficiency, notice that each such distribution is in its own domain of attraction (see Definition 3.1) by choosing $a_n = \alpha^{-n}$, $b_n = 0$ or $a_n = 1$, $b_n = -n \log \alpha$ respectively. ‖

The class of limit distributions which can arise in this way is very large. In the case of composition of series structures, the class of limit distributions has been characterized by Mejzler (1965) and includes distributions other than (1.1), (1.2), (1.3) or (1.1*), (1.2*), (1.3*). For example, the function

$$\psi(x) = \exp\left[-x^3\left(5 + \sin\left(2\pi \frac{\log x}{\log 2}\right)\right)\right] \quad \text{for } x > 0$$

is the tail of a distribution and satisfies (2.17) for $h(p) = p^8$, $\alpha = 2$.

EXERCISES

1. (a) Let $U(x) = (\log x)^\alpha$ for $x > 1$, $-\infty < \alpha < \infty$. Then $U(x)$ varies regularly with exponent $\rho = 0$.
   (b) Let $U(x)$ be monotone, with $\lim_{x\to\infty} U(x) = a$, finite. Then the same conclusion holds.
   (c) Let $U(x) = (1 + x^2)^p$ for $x \geq 0$. Then $U(x)$ varies regularly with exponent $\rho = 2p$.
   (d) Does $U(x) = e^x$ vary regularly?
2. Verify that $W_1(x)$, $W_2(x)$, and $\Lambda(x)$ are minimum stable.
3. Prove Lemma 2.5.
4. Prove Lemma 2.6.
5. Prove that W_2 satisfies the hypothesis of Lemma 2.5.
6. Carry out the argument for case (b) of Theorem 2.10.
7. Prove Theorem 2.11*.
8. Let distribution F have failure rate function r, with $F(0) = 0$. For $a \leq 1$, let $\lim_{t\to\infty} t^a r(t) = L(a)$ exist (finite or infinite). Then for all $b \geq 0$,

$$\lim_{t\to\infty} \frac{\bar{F}(t + bt^a)}{\bar{F}(t)} = \begin{cases} e^{-bL(a)} & \text{for} \quad a < 1, \\ (1 + b)^{-L(a)} & \text{for} \quad a = 1. \end{cases}$$

9. Let $X_1, X_2, \ldots,$ be a sequence of independent random variables with distribution function F. Then $\xi_n = \max(X_1, \ldots, X_n)$ are said to satisfy the law of large numbers if there exist real numbers $\{A_n\}$ such that $P[|\xi_n - A_n| < \varepsilon] \to 1$ as $n \to \infty$ for every $\varepsilon > 0$. Gnedenko (1943) showed that this condition is equivalent to $\lim_{t\to\infty} [\bar{F}(t + \varepsilon)/\bar{F}(t)] = 0$. Using Exercise 8, show that this is equivalent to $\lim_{t\to\infty} r(t) = \infty$.
10. Distributions F_1, F_2 are said to have *proportional hazards* R_1, R_2 (where $R_i = -\log \bar{F}_i$) if there exists $a > 0$ such that $R_2(t) = aR_1(t)$ for all $t \geq 0$. Show that if F is a possible limit distribution for sample minima, then any distribution with a proportional hazard also is.

3. DOMAINS OF ATTRACTION

In Section 2, we derived the basic result that sample minima (maxima) could have only three possible types of limit distributions. In this section we characterize the underlying distributions from which the samples are taken for which the minima (maxima) converge to each of the three types. We may state our goal more precisely using the following definitions.

3.1. Definitions. A distribution F belongs to the *minimum domain of attraction* of a distribution G if there exist constants $\alpha_n > 0$, β_n such that $\bar{F}^n(\alpha_n x + \beta_n) \xrightarrow{L} \bar{G}(x)$. [Recall from Chapter 4, Section 2, that $F_n \xrightarrow{L} F$ means $\lim_{n\to\infty} F_n(x) = F(x)$ at each point of continuity of F.]

Similarly, F belongs to the *maximum domain of attraction* of G if there exist $\alpha_n > 0$, β_n such that $F^n(\alpha_n x + \beta_n) \xrightarrow{L} G(x)$.

Our goal then is to determine the minimum domains of attraction for W_1, W_2, and Λ, and the maximum domains of attraction for $W_1{}^*$, $W_2{}^*$, and Λ^*. We will need the following lemma.

3.2. Lemma. In order that

$$\bar{F}^n(\alpha_n x + \beta_n) \xrightarrow{L} \bar{G}(x), \tag{3.1}$$

it is necessary and sufficient that

$$nF(\alpha_n x + \beta_n) \to -\log \bar{G}(x) \tag{3.2}$$

for all continuity points x of G such that $\bar{G}(x) \neq 0$.

PROOF. Since $1 - y \leq -\log y \leq (1 - y)/y$, it follows that

$$\begin{aligned}\lim_{n\to\infty} nF(\alpha_n x + \beta_n) &\leq \mathrm{l\,m}_{n\to\infty} [-n \log \bar{F}(\alpha_n x + \beta_n)] \\ &\leq \lim_{n\to\infty} \frac{nF(\alpha_n x + \beta_n)}{\bar{F}(\alpha_n x + \beta_n)}.\end{aligned} \tag{3.3}$$

Assuming either (3.1) or (3.2), then $\lim_{n\to\infty} \bar{F}(\alpha_n x + \beta_n) = 1$, so that the upper and lower limits in (3.3) must be equal. It follows that

$$\lim_{n\to\infty} [-n \log \bar{F}(\alpha_n x + \beta_n)] = \lim_{n\to\infty} nF(\alpha_n x + \beta_n).$$

Thus (3.1) and (3.2) are equivalent. ||

Domain of Attraction of the Weibull Distribution

The following theorem gives necessary and sufficient conditions for a distribution to belong to the minimum domain of attraction of the Weibull distribution W_1.

3.3. Theorem. A distribution F belongs to the minimum domain of attraction of $W_1(x) = 1 - e^{-x^\alpha}$, $x \geq 0$, $\alpha > 0$, iff

(a) There exists x_0 such that $F(x_0) = 0$ and $F(x_0 + \varepsilon) > 0$ for all $\varepsilon > 0$;

(b) $\lim_{t\downarrow 0} [F(xt + x_0)/F(t + x_0)] = x^\alpha$ for $x > 0$, where $\alpha > 0$.

PROOF. Assume (a) and (b) hold. By the Fundamental Lemma 1.6, we may assume $x_0 = 0$. Define $a_n = \sup\{x \mid F(x^-) \leq 1/n \leq F(x^+)\}$. Then $a_n > 0$, and by (a), $a_n \to 0$. From the definition, $nF(a_n{}^-) \leq 1 \leq nF(a_n{}^+)$. Taking limits as $n \to \infty$,

$$\begin{aligned}\lim(n+1)F(a_{n+1}) &\\ &= \lim nF(a_{n+1}) \leq \lim nF(a_n{}^-) \leq 1 \leq \lim nF(a_n{}^+) \leq \lim nF(a_{n-1}) \\ &= \lim(n-1)F(a_{n-1}).\end{aligned}$$

Thus

$$\lim_{n \to \infty} nF(a_n) = 1. \tag{3.4}$$

Since $\lim_{t \downarrow 0} [F(xt)/F(t)] = x^\alpha$, $x > 0$, then $x^\alpha = \lim_{n \to \infty} [F(a_n x)/F(a_n)] = \lim_{n \to \infty} [nF(a_n x)/nF(a_n)] = \lim_{n \to \infty} nF(a_n x)$, by (3.4). By Lemma 3.2

$$\lim_{n \to \infty} \overline{F}^n(a_n x) = e^{-x^\alpha} \qquad \text{for} \quad x \geq 0.$$

This proves the sufficiency of conditions (a) and (b). For a proof of the necessity of conditions (a) and (b) we refer the reader to Gnedenko (1943). ‖

Although the distribution $W_1{}^*(x)$ defined in (1.1*) is not likely to be of interest in the reliability context since its domain is the negative axis, for completeness we state Gnedenko's characterization of the maximum domain of attraction of $W_1{}^*(x)$.

3.4. Corollary. F belongs to the maximum domain of attraction of $W_1{}^*(x) = e^{-(-x)^\alpha}$ for $x \leq 0$, where $\alpha > 0$, if and only if

(a) There exists x_0 such that $F(x_0) = 1$ and $F(x_0 - \varepsilon) < 1$ for all $\varepsilon > 0$;
(b) $\lim_{t \uparrow 0} [\overline{F}(x_0 + xt)/\overline{F}(x_0 + t)] = x^\alpha$ for $x > 0$, where $\alpha > 0$.

3.5. Example. Let F be the gamma distribution with density $f(y) = \lambda^\alpha y^{\alpha-1} e^{-\lambda y}/\Gamma(\alpha)$ for $x \geq 0$. To show that F belongs to the minimum domain of attraction of the Weibull, W_1, we use Theorem 3.3, verifying that:

(a) For $x_0 = 0$, $F(x_0) = 0$, and
(b) $\lim_{t \downarrow 0} [F(xt)/F(t)] = \lim_{t \downarrow 0} [xf(xt)/f(t)]$ [by l'Hospital's rule] $= x^\alpha$.

In a similar fashion Gnedenko shows the following theorem.

3.6. Theorem. (a) F belongs to the minimum domain of attraction of $W_2(x) = 1 - e^{-(-x)^{-\alpha}}$ for $x \leq 0$, where $\alpha > 0$, if and only if

$$\lim_{t \to -\infty} \frac{F(t)}{F(tx)} = x^{\alpha}$$

for every $x > 0$.

(b) F belongs to the maximum domain of attraction of $W_2^*(x) = e^{-x^{-\alpha}}$ for $x \geq 0$, where $\alpha > 0$, if and only if $\lim_{t \to \infty} [\bar{F}(t)/\bar{F}(tx)] = x^{\alpha}$ for every $x > 0$.

Note that $W_2(x)$, the distribution of a negative random variable, is not likely to be of direct interest in the reliability context.

Finally, we present sufficient conditions developed by von Mises (1939) for a distribution to belong to the maximum domain of attraction of $\Lambda^*(x) = e^{-e^{-x}}$. Let the distribution F have a differentiable density f, and let $r(x) = f(x)/\bar{F}(x)$ be the failure rate.

3.7. Theorem. (a) Let $F(x) < 1$ for all x, and $\lim_{x \to \infty} [d/dx][1/r(x)] = 0$. Then F belongs to the maximum domain of attraction of Λ^*. Furthermore, $\lim_{n \to \infty} F^n((x - b_n)/a_n) = e^{-e^{-x}}$, where b_n satisfies $\bar{F}(b_n) = 1/n$ and $a_n = 1/nf(b_n)$.

(b) Let there exist x_0 such that for all $\varepsilon > 0$, $G(x_0 - \varepsilon) < 1$, $G(x_0) = 1$, and let $\lim_{x \uparrow x_0} [d/dx][1/r(x)] = 0$. Then G belongs to the maximum domain of attraction of Λ^*.

The proof may be found in von Mises (1939) and in Marcus and Pinsky (1969).

In reliability applications we are often interested in the *minimum* domain of attraction for life distributions, especially when dealing with series systems. For this purpose we present the analog of Theorem 3.7 for minimum domains of attraction.

3.8. Theorem. Let there exist x_0 such that $F(x_0) = 0$ and $F(x_0 + \varepsilon) > 0$ for all $\varepsilon > 0$, and let $\lim_{x \downarrow x_0} [d/dx][1/\phi'(x)] = 0$, where $\phi(x) = -\log F(x)$. Then F belongs to the minimum domain of attraction of $\Lambda(x) = 1 - e^{-e^x}$.

Remark. Note that by Theorem 3.3 when F belongs to the minimum domain of attraction of W_1, the norming constants are $b_n = x_0$ and $a_n = F^{-1}(1/n)$ so that

$$\lim_{n \to \infty} P\left[\frac{\xi_n - x_0}{F^{-1}(1/n)} \leq x\right] = W_1(x).$$

A distribution need not belong to the domain of attraction of any limit law (other than the degenerate). For example, Gnedenko (1943) shows that the Poisson is not attracted to any limit law.

EXERCISES

1. Show that the Weibull distribution W_1 belongs to the maximum domain of attraction of Λ^*.

2. Show that the distribution $\Lambda(x) = 1 - e^{-e^x}$ belongs to the maximum domain of attraction of Λ^*.

3. Show that the normal distribution belongs to the maximum domain of attraction of Λ^*.

4. Show that the normal distribution belongs to the minimum domain of attraction of Λ. [Hint: Modify condition (a) of Theorem 3.7 to obtain the corresponding result for the minimum domain of attraction of Λ.]

5. Show that the lognormal distribution belongs to the maximum domain of attraction of Λ^*.

6. Let $X_1, X_2, \ldots,$ be i.i.d. random variables with distribution F. Let

$$\eta_{k,n} = \max\left\{\frac{1}{k}(X_1 + \cdots + X_k), \frac{1}{k}(X_{k+1} + \cdots + X_{2k}), \ldots, \frac{1}{k}(X_{n-k+1} + \cdots + X_n)\right\}.$$

Assume the k-fold convolution $F^{(k)}$ belongs to the maximum domain of attraction of Λ^*. Show that the norming constants for $\eta_{k,n}$ (k fixed) satisfy $\bar{F}^{(k)}(kb_{k,n}) = 1/n$ and $a_{k,n} = 1/nf^{(k)}(kb_{k,n})$. Hence

$$\lim_{n\to\infty} P\left[\frac{\eta_{k,n} - b_{k,n}}{a_{k,n}} \le x\right] = \Lambda^*(x).$$

(Hint: See Theorem 3.7.)

7. Assume F is NBU. Use the bounds on $F^{(k)}$ given in Theorem 3.2 of Chapter 6 to obtain bounds on $b_{k,n}$ of Exercise 6 above. Note that $b_{k,n}$ may be written as $(1/k)R_k^{-1}(\log n)$, where $R_k \equiv -\log \bar{F}^{(k)}$.

8. A life distribution F has failure rate $r(t)$ satisfying $\lim_{t\to\infty} tr(t) = \alpha > 0$. Prove that $\lim_{t\to\infty} [\bar{F}(t)/\bar{F}(xt)] = x^\alpha$ for all $x > 0$. From this it follows that the condition $\lim_{t\to\infty} tr(t) = \alpha > 0$ implies that F belongs to the maximum domain of attraction of $W_2^*(x) = e^{-x^{-\alpha}}$ for $x \ge 0$.

9. Let F belong to the minimum domain of attraction of $W_{1,\alpha}(x) = 1 - e^{-x^\alpha}$, $x \ge 0$, $\alpha > 0$. Then $G(x) = F^\beta(x)$, $\beta > 0$, belongs to the minimum domain of attraction of $W_{1,\beta\alpha}$.

10. Let F be a life distribution belonging to the minimum domain of attraction of $W_{1,\alpha}(x) = 1 - e^{-x^\alpha}$, $x \ge 0$, $\alpha > 0$. Let $G(x) = F^\beta(x^{1/\beta})$, where $\beta > 0$. Then G belongs to the same domain.

11. Let F belong to the minimum domain of attraction of the Weibull $W_{1,\alpha}$. Then the life distribution of any parallel system of independent components

each distributed according to F belongs to the domain of attraction of $W_{1,\alpha'}$, where α' is an integer multiple of α.

12. Show that the exponential distribution $1 - e^{-x}$ belongs to the maximum domain of attraction of $\Lambda^*(x) = e^{-e^{-x}}$.

4. LIMIT DISTRIBUTIONS OF SERIES SYSTEMS WITH COMPONENT RENEWAL

In previous sections of this chapter we found the limiting distributions of series and parallel systems as the number of components became very large; components were not replaced at failure. Now we assume component replacement at failure, and calculate the distribution of time between system failures as the number of components increases indefinitely. We will show that under fairly weak restrictions the limiting distribution of time between system failures is exponential even though component life distributions are not necessarily exponential.

We make the following assumptions about the system:

(a) The system consists of n component positions in series, each containing a component.
(b) Each failed component is immediately replaced.
(c) In the rth component position, the initial component has life distribution $\hat{F}_{nr}$ while all subsequent components have life distribution F_{nr}; $r = 1, \ldots, n$.
(d) All life lengths are mutually independent.

Let $X_{nr}(t)$ be the number of replacements occurring in the rth component position during $[0, t]$. As discussed in Cox (1962), pp. 27–28, $\{X_{nr}(t), 0 \leq t < \infty\}$ is a modified renewal counting process. Let $X_n(t) = \sum_{r=1}^{n} X_{nr}(t)$, the total number of replacements in all n component positions during $[0, t]$.

Our problem is to determine necessary and sufficient conditions on the $\hat{F}_{nr}$, F_{nr} so that $\{X_n(t), 0 \leq t < \infty\}$ will converge to a Poisson process (not necessarily homogeneous) as $n \to \infty$.

4.1. Definition. $\{X(t), 0 \leq t < \infty\}$ is a *nonhomogeneous Poisson process with mean value function* $\Lambda(t)$ if it has independent increments that assume only nonnegative integer values, and if for all $0 \leq t_1 < t_2$,

$$P[X(t_2) - X(t_1) = k] = \frac{[\Lambda(t_2) - \Lambda(t_1)]^k}{k!} e^{-[\Lambda(t_2) - \Lambda(t_1)]}.$$

This problem has a long history. Initially Palm (1943) gave a somewhat heuristic treatment. Khinchin (1956b) gave sufficient conditions for convergence of $\{X_n(t), 0 \le t < \infty\}$ to a Poisson process. Ososkov (1956) gave necessary and sufficient conditions for such convergence. Drenick (1960) gave sufficient conditions, and concentrated mainly on reliability applications. Our discussion is based on a paper by Grigelionis (1964).

Since we wish to study the limiting behavior of $X_n(t)$ as $n \to \infty$, we must insure that no one component dominates in the limiting process. To this end we will assume in the remainder of the section that for fixed $t > 0$,

$$\lim_{n\to\infty} \max_{1\le r\le n} \hat{F}_{nr}(t) = 0 = \lim_{n\to\infty} \max_{1\le r\le n} F_{nr}(t). \tag{4.1}$$

For example, we could have $\hat{F}_{nr}(t) = \hat{F}_r(t/n)$ and $F_{nr}(t) = F_r(t/n)$, with $\lim_{n\to\infty} \max_{1\le r\le n} F_r(t/n) = 0$, so that the F_r are restricted in their behavior at the origin.

Our main result may now be stated.

4.2. Theorem (Grigelionis). In order for $\{X_n(t), 0 \le t < \infty\}$ to converge in probability (see Billingsley, 1968) to a nonhomogeneous Poisson process with mean value function $\Lambda(t)$, it is necessary and sufficient that for each fixed $t > 0$,

$$\lim_{n\to\infty} \sum_{r=1}^{n} \hat{F}_{nr}(t) = \Lambda(t). \tag{4.2}$$

It is interesting to note that, aside from the requirement (4.1), no further conditions on the F_{nr} are imposed. Before proving Theorem 4.2, we present some examples.

4.3. Example. Let

$$\hat{F}_{nr}(t) = \frac{1}{\mu_{nr}} \int_0^t \bar{F}_{nr}(u)\, du, \tag{4.3}$$

where μ_{nr} is the mean of F_{nr}, for $r = 1, \ldots, n$, $n = 1, 2, \ldots$. As pointed out in Chapter 6, Section 3, $\hat{F}_{nr}(t)$ represents the probability of a failure in component position r during $[0, t]$, given that the renewal process started at time $-\infty$ with underlying distribution F_{nr}. Thus the assumption (4.3) is equivalent to assuming *the system is in the time stationary case* (that is, the system has been operating for an indefinitely long time before we begin to observe it). Let $\lim_{n\to\infty} \sum_{r=1}^{n} 1/\mu_{nr} = \Lambda < \infty$. Then

$$\sum_{r=1}^{n} \hat{F}_{nr}(t) = \sum_{r=1}^{n} \frac{t}{\mu_{nr}} - \sum_{r=1}^{n} \frac{1}{\mu_{nr}} \int_0^t F_{nr}(u)\, du \to \Lambda t$$

as $n \to \infty$, since $\max_{1\le r\le n} F_{nr}(t) \to 0$ as $n \to \infty$, by (4.1). It follows from Theorem 4.2 that $\{X_n(t), 0 \le t < \infty\} \to$ a homogeneous Poisson process with parameter Λ.

4.4. Example. Let $\hat{F}_{nr}(x) = F_{nr}(x) = F(\lambda_{nr}x)$, where $F(x) = \Lambda_1 x^\alpha + o(x^\alpha)$ for $x \downarrow 0$, and Λ_1, α, λ_{nr} are all >0, with $\max_{1 \le r \le n} \lambda_{nr} \to 0$ as $n \to \infty$. By Theorem 4.2, $\{X_n(t), 0 \le t < \infty\}$ converges to a Poisson process with mean value function $\Lambda(t) = \Lambda_1 \Lambda_2 t^\alpha$ if and only if $\lim_{n\to\infty} \sum_{r=1}^n \lambda_{nr}{}^\alpha = \Lambda_2$. Note that for $\alpha = 1$, the Poisson process is homogeneous.

A distribution F of the required form is the Weibull $F(t) = 1 - e^{-\Lambda_1 t^\alpha}$ for $t \ge 0$.

4.5. Example. Let $\hat{F}_{nr}(x) = F_{nr}(x) = F(x, \lambda_{nr}) = \int_0^x [\lambda_{nr}{}^\alpha u^{\alpha-1}/\Gamma(\alpha)] e^{-\lambda_{nr}u}\, du$, a gamma distribution with shape parameter $\alpha > 0$. By Theorem 4.2, if $\sum_{r=1}^n \lambda_{nr}{}^\alpha \to \Lambda$ as $n \to \infty$, then $\{X_n(t), 0 \le t < \infty\}$ converges to a Poisson process with mean value function $\Lambda(t) = \Lambda t^\alpha/\Gamma(\alpha + 1)$, as $n \to \infty$.

4.6. Example. Let $\hat{F}_{nr}(x) = F_{nr}(x) = \int_0^x \sigma_{nr}^{-1} \sqrt{2/\pi}\, e^{-u^2/2\sigma_{nr}{}^2}\, du$, a "folded-over" normal distribution. Suppose that $\sum_{r=1}^n \sigma_{nr}^{-1} \to \sigma^{-1} > 0$ as $n \to \infty$. Then $\sum_{r=1}^n \hat{F}_{nr}(x) = \sum_{r=1}^n \int_0^{x/\sigma_{nr}} \sqrt{2/\pi}\, e^{-u^2/2}\, du \to \sqrt{2/\pi} \cdot x/\sigma$, as $n \to \infty$. Thus by Theorem 4.2, $\{X_n(t), 0 \le t < \infty\}$ converges to a homogeneous Poisson process with event rate $\sqrt{2/\pi}/\sigma$.

We will need the following lemma in the proof of Theorem 4.2.

4.7. Lemma. $\{X_n(t), 0 \le t < \infty\}$ converges to a Poisson process with mean value function $\Lambda(t)$ as $n \to \infty$ if and only if

$$\lim_{n\to\infty} \sum_{r=1}^{n} PX[_{nr}(t_2) - X_{nr}(t_1) = 1] = \Lambda(t_2) - \Lambda(t_1), \tag{4.4}$$

for $0 \le t_1 < t_2$, and

$$\lim_{n\to\infty} \sum_{r=1}^{n} P[X_{nr}(t) \ge 2] = 0 \qquad \text{for} \quad t \ge 0. \tag{4.5}$$

This lemma is a special case of Lemma 4.8 below.

PROOF OF THEOREM 4.2. *Necessity.*

$$\begin{aligned}\lim_{n\to\infty} \sum_{r=1}^{n} \hat{F}_{nr}(t) &= \lim_{n\to\infty} \sum_{r=1}^{n} [P[X_{nr}(t) \ge 1] \\ &= \lim_{n\to\infty} \sum_{r=1}^{n} P[X_{nr}(t) \ge 2] + \lim_{n\to\infty} \sum_{r=1}^{n} P[X_{nr}(t) = 1].\end{aligned}$$

By (4.5), the first limit is 0. By (4.4), the second limit is $\Lambda(t)$. Equation (4.2) follows.

Sufficiency. Assuming $\lim_{n\to\infty} \sum_{r=1}^n \hat{F}_{nr}(t) = \Lambda(t)$, it will suffice by Lemma 4.7 to verify (4.4) and (4.5).

The renewal function $EX_{nr}(t)$ satisfies the renewal equation (3.5) of Chapter 6:

$$EX_{nr}(t) = \hat{F}_{nr}(t) + \int_0^t EX_{nr}(t-u)\, d\hat{F}_{nr}(u).$$

Summing, we have

$$\sum_{r=1}^{n} EX_{nr}(t) = \sum_{r=1}^{n} \hat{F}_{nr}(t) + \sum_{r=1}^{n} \int_0^t EX_{nr}(t-u)\, d\hat{F}_{nr}(u).$$

Since $\sum_{r=1}^{n} \int_0^t EX_{nr}(t-u)\, d\hat{F}_{nr}(u) \le \max_{1 \le r \le n} \hat{F}_{nr}(t) \sum_{r=1}^{n} EX_{nr}(t)$, it follows that

$$\sum_{r=1}^{n} EX_{nr}(t)[1 - \max_{1 \le r \le n} \hat{F}_{nr}(t)] \le \sum_{r=1}^{n} \hat{F}_{nr}(t).$$

By (4.1), we conclude that

$$\lim_{n\to\infty} \sum_{r=1}^{n} EX_{nr}(t) \le \lim_{n\to\infty} \sum_{r=1}^{n} \hat{F}_{nr}(t).$$

Since the reverse inequality is trivially true, we have established

$$\lim_{n\to\infty} \sum_{r=1}^{n} EX_{nr}(t) = \lim_{n\to\infty} \sum_{r=1}^{n} \hat{F}_{nr}(t).$$

Finally, by hypothesis (4.2), we conclude

$$\lim_{n\to\infty} \sum_{r=1}^{n} EX_{nr}(t) = \Lambda(t). \tag{4.6}$$

However, we still have not proven (4.4). To this end we consider the remaining life random variable in component position r at time t, with distribution $F_{t,nr}(\cdot)$. [See (3.9) of Chapter 6.] We may easily verify that for $0 \le t_1 < t_2$,

$$EX_{nr}(t_2) - EX_{nr}(t_1) = F_{t_1,nr}(t_2 - t_1)$$
$$+ \int_0^{t_2 - t_1} [EX_{nr}(t_2) - EX_{nr}(t_1 + z)]\, dF_{t_1,nr}(z).$$

Using (4.1), we have, as $n \to \infty$,

$$F_{t_1,nr}(t_2 - t_1) = EX_{nr}(t_2) - EX_{nr}(t_1) + o(1). \tag{4.7}$$

By definition,

$$\sum_{r=1}^{n} P[X_{nr}(t_2) - X_{nr}(t_1) = 1]$$
$$= \sum_{r=1}^{n} [F_{t_1,nr}(t_2 - t_1) - F_{t_1,nr} * F_{nr}(t_2 - t_1)]$$
$$= \sum_{r=1}^{n} F_{t_1,nr}(t_2 - t_1) + o(1) \qquad \text{as} \quad n \to \infty, \text{ by (4.1)}.$$

From (4.7)

$$\sum_{r=1}^{n} P[X_{nr}(t_2) - X_{nr}(t_1) = 1] = \sum_{r=1}^{n} [EX_{nr}(t_2) - EX_{nr}(t_1)] + o(1).$$

Finally, using (4.6), we conclude that (4.4) holds.

To complete the proof of sufficiency, it remains to show that (4.5) holds. Note that

$$\sum_{r=1}^{n} P[X_{nr}(t) \geq 2] = \sum_{r=1}^{n} \hat{F}_{nr} * F_{nr}(t) \leq \max_{1 \leq r \leq n} F_{nr}(t) \sum_{r=1}^{n} \hat{F}_{nr}(t).$$

By (4.1), lim $\sum_{r=1}^{n} P[X_{nr}(t) \geq 2] \to 0$ as $n \to \infty$, establishing (4.5). The sufficiency of Theorem 4.2 now follows from Lemma 4.7. ‖

We now state and prove a result about step processes which yields Lemma 4.7 as a special case. We call the random process $\{X(t), 0 \leq t < \infty\}$ a *step process* if $X(0) = 0$, and for $0 \leq t_1 < t_2$, the increments $X(t_2) - X(t_1)$ assume only nonnegative integer values. Thus $X(t)$ might be the number of random events that have occurred during $[0, t]$. In particular, the process $\{X_n(t), 0 \leq t < \infty\}$, representing the total number of component renewals, is a step process obtained by superposing (that is, adding) the n independent step processes $\{X_{nr}(t), 0 \leq t < \infty\}$, $1 \leq r \leq n$, representing the number of component renewals in the various component positions.

4.8. Lemma. Let $\{X_{nr}(t), 0 \leq t < \infty\}$, $1 \leq r \leq n$, be n independent step processes, satisfying

$$\lim_{n\to\infty} \max_{1 \leq r \leq n} P[X_{nr}(t) \geq 1] = 0, \qquad \text{for each} \quad t \geq 0. \tag{4.8}$$

Let $\{X_n(t), 0 \leq t < \infty\}$ be their superposition; that is, $X_n(t) = \sum_{r=1}^{n} X_{nr}(t)$. Then $\{X_n(t), 0 \leq t < \infty\}$ converges to a Poisson process with mean value function $\Lambda(t)$ as $n \to \infty$ if and only if, for $0 \leq t_1 < t_2$,

$$\lim_{n\to\infty} \sum_{r=1}^{n} P[X_{nr}(t_2) - X_{nr}(t_1) = 1] = \Lambda(t_2) - \Lambda(t_1), \tag{4.9}$$

and

$$\lim_{n\to\infty} \sum_{r=1}^{n} P[X_{nr}(t_2) - X_{nr}(t_1) \geq 2] = 0. \tag{4.10}$$

PROOF. *Necessity.* The necessity of conditions (4.9) and (4.10) follows from the well-known theory of sums of infinitesimally small random variables [that is, satisfying (4.8)]. See Gnedenko and Kolmogorov (1954), p. 132.

Sufficiency. Assuming (4.9) and (4.10), we will show (a) the asymptotic independence of the increments of the process $\{X_n(t), 0 \le t < \infty\}$, and (b) the convergence of the univariate distributions of the increments to the corresponding Poisson distributions. We will use characteristic functions.

Let $f_{nr}(\boldsymbol{\alpha}, \mathbf{t}) = E \exp [i \sum_{\nu=1}^{s} \alpha_\nu \Delta_\nu(X_{nr})]$, $f_n(\boldsymbol{\alpha}, \mathbf{t}) = E \exp [i \sum_{\nu=1}^{s} \alpha_\nu \Delta_\nu X_n]$, where $\boldsymbol{\alpha} = (\alpha_1, \ldots, \alpha_s)$ and $\mathbf{t} = (t_1, \ldots, t_s)$, $0 \equiv t_0 \le t_1 < \cdots < t_s$, $\Delta_\nu(X_{nr}) = X_{nr}(t_\nu) - X_{nr}(t_{\nu-1})$, and $\Delta_\nu(X_n) = X_n(t_\nu) - X_n(t_{\nu-1})$. To show convergence of the distributions of the increments $\Delta_\nu X_n$, it suffices to show convergence of their characteristic functions.

Since the processes $\{X_{nr}(t), 0 \le t < \infty\}$ are independent, we have

$$f_n(\boldsymbol{\alpha}, \mathbf{t}) = \prod_{r=1}^{n} f_{nr}(\boldsymbol{\alpha}, \mathbf{t}). \tag{4.11}$$

We may write

$$f_{nr}(\boldsymbol{\alpha}, \mathbf{t}) = \sum_{\mathbf{m}} P[\boldsymbol{\Delta}(X_{nr}) = \mathbf{m}] \exp\left[i \sum_{1}^{s} \alpha_\nu m_\nu\right],$$

where $\boldsymbol{\Delta}(X_{nr}) = (\Delta_1(X_{nr}), \ldots, \Delta_s(X_{nr}))$, and $\mathbf{m} = (m_1, \ldots, m_s)$, a vector of nonnegative integers. Expanding the right-hand side, we obtain

$$\begin{aligned} f_{nr}(\boldsymbol{\alpha}, \mathbf{t}) &= 1 + \sum_{\mathbf{m} \ne \mathbf{0}} P[\boldsymbol{\Delta}(X_{nr}) = \mathbf{m}] \left(\exp\left[i \sum_{1}^{s} \alpha_\nu m_\nu\right] - 1\right) \\ &= \exp\left\{\sum_{\mathbf{m} \ne \mathbf{0}} P[\boldsymbol{\Delta}(X_{nr}) = \mathbf{m}] \left(\exp\left[i \sum_{1}^{s} \alpha_\nu m_\nu\right] - 1\right)\right. \\ &\qquad \left. + \, o\left(\left(\sum_{\mathbf{m} \ne \mathbf{0}} P[\boldsymbol{\Delta}(X_{nr}) = \mathbf{m}]\right)^2\right)\right\} \end{aligned}$$

Now

$$\sum_{r=1}^{n} \left(\sum_{\mathbf{m} \ne \mathbf{0}} P[\boldsymbol{\Delta}(X_{nr}) = \mathbf{m}]\right)^2 \le \max_{1 \le r \le n} P[X_{nr}(t_s) \ge 1] \sum_{r=1}^{n} P[X_{nr}(t_s) = 1] \to 0$$

as $n \to \infty$, by (4.8) and (4.9).

Thus

$$\begin{aligned} \lim_{n \to \infty} f_n(\boldsymbol{\alpha}, \mathbf{t}) &= \lim_{n \to \infty} \prod_{r=1}^{n} f_{nr}(\boldsymbol{\alpha}, \mathbf{t}) \\ &= \exp\left\{\lim_{n \to \infty} \sum_{r=1}^{n} \sum_{\mathbf{m} \ne \mathbf{0}} P[\boldsymbol{\Delta}(X_{nr}) = \mathbf{m}] \left(\exp\left[i \sum_{\nu=1}^{s} \alpha_\nu m_\nu\right] - 1\right)\right\}. \end{aligned}$$

But

$$\lim_{n \to \infty} \sum_{r=1}^{n} \sum_{\mathbf{m} \ne \mathbf{0}} P[\boldsymbol{\Delta}(X_{nr}) = \mathbf{m}] = \lim_{n \to \infty} \sum_{r=1}^{n} \sum_{\nu=1}^{s} P[X_{nr}(t_\nu) - X_{nr}(t_{\nu-1}) = 1] + A,$$

where $A \leq \lim_{n\to\infty} \sum_{r=1}^{n} P[X_{nr}(t_s) \geq 2]$. This last limit is 0 by (4.10).
It follows that

$$\lim_{n\to\infty} f_n(\boldsymbol{\alpha}, \mathbf{t}) = \exp\left\{\sum_{\nu=1}^{s} [\Lambda(t_\nu) - \Lambda(t_{\nu-1})]\left(\exp\left[i \sum_{\nu=1}^{s} \alpha_\nu m_\nu\right] - 1\right)\right\},$$

yielding the sufficiency of Lemma 4.8. ‖

EXERCISES

1. Let $\{X(t), 0 \leq t < \infty\}$ be a nonhomogeneous Poisson process with mean value function $\Lambda(t)$. Then $\{X(\Lambda^{-1}(t)), 0 \leq t < \infty\}$ is a homogeneous Poisson process with event rate 1.
2. Let $F_{nr}(t) = F^n(t), 1 \leq r \leq n, n = 1, 2, \ldots$, where $F(t) < 1$ for $0 \leq t < \infty$. (a) Is (4.1) satisfied? (b) What happens to the condition (4.2)? Does Theorem 4.2 hold?
3. Let $\{\hat{F}_{nr}{}^{(i)}(t)\}$, $\{F_{nr}{}^{(i)}(t)\}$ satisfy the conditions of (4.1) and (4.2), with corresponding $\{X_n{}^{(i)}(t), 0 \leq t < \infty\}$ converging to a Poisson process with mean value function $\Lambda^{(i)}(t)$, $i = 1, \ldots, k$. If $\hat{F}_{nr} = \sum_{i=1}^{k} p_i \hat{F}_{nr}$ and $F_{nr} = \sum_{i=1}^{k} p_i F_{nr}$ are obtained as mixtures with positive weights, then the corresponding $\{X_n(t), 0 \leq t < \infty\}$ converges to a Poisson process with mean value function $\Lambda(t) = \sum_1^k p_i \Lambda^{(i)}(t)$.
4. Let $\{\hat{F}_{nr}{}^{(i)}\}$, $\{F_{nr}{}^{(i)}\}$ satisfy (4.1) and (4.2), $i = 1, 2$. Do the convolutions $\{\hat{F}_{nr}{}^{(1)} * \hat{F}_{nr}{}^{(2)}\}$, $\{F_{nr}{}^{(1)} * F_{nr}{}^{(2)}\}$ satisfy the same conditions?
5. Let $\{\hat{F}_{nr}\}$, $\{F_{nr}\}$ satisfy (4.1) and (4.2), and let $\hat{G}_{nr} = \hat{F}_{nr} * G$, $G_{nr} = F_{nr} * G$, where G is a life distribution, $r = 1, 2, \ldots, n$; $n = 1, 2, \ldots$. Then $\{\hat{F}_{nr}\}$, $\{G_{nr}\}$ satisfy (4.1) and (4.2). It follows that Theorem 4.2 holds for the $X_n(t)$ process corresponding to $\{\hat{G}_{nr}\}$, $\{G_{nr}\}$.
6. Assume (4.1) and (4.2) hold. Let each component position in the system be replaced by a pair of like component positions in parallel. Does (4.1) still hold? What is the $\lim_{n\to\infty} \sum_{r=1}^{n} \hat{F}_{nr}{}^2(t)$? Does Theorem 4.2 still hold?

5. NOTES AND REFERENCES

Section 1. The theory of extreme values has a long history, benefiting from contributions by a number of distinguished statisticians. The first results were obtained by Fréchet (1927) and by Fisher and Tippet (1928). The most complete results, summarizing and rounding out this series of investigations, were obtained by Gnedenko (1943).

Results for order statistics of fixed and increasing rank were obtained by Smirnov (1952), who completely characterized the limiting types and their domains of attraction. This has applications to k-out-of-n structures.

Generalizations of these results for the maximum of a sequence of observations have been made by Juncosa (1949), who dropped the assumption of a common distribution, and by Watson (1954), who proved that under slight restrictions the limiting distribution of the maximum in a stationary sequence of m-dependent random variables is the same as in the independent case. A very general model in terms of triangular arrays is treated exhaustively by Loève (1956). A bibliography and discussion of applications is contained in Gumbel (1958).

Harris (1970) considers extreme value theory in a replacement policy context. In his model, a system consists of n independent and identical components in series, with m standby spares available for instantaneous replacement of failed components. After spares are exhausted, the system fails at the next component failure.

His basic results are summarized in the following theorems, the first corresponding to the case in which m is of fixed size, the second corresponding to the case in which m increases to ∞ along with n.

5.1. Theorem (Harris). Let m be fixed as $n \to \infty$. Then the only possible limit distributions for system lifetime (appropriately normalized) are of the following two types:

$$G_{1m}(x) = \int_0^{x^\alpha} e^{-y} \frac{y^{m-1}}{(m-1)!}\, dy \qquad \text{for } x \geq 0, \text{ where } \alpha > 0.$$

$$G_{2m}(x) = \int_0^{e^x} e^{-y} \frac{y^{m-1}}{(m-1)!}\, dy \qquad \text{for } -\infty < x < \infty.$$

5.2. Theorem (Harris). Let $m = 0(n^\alpha)$, where $0 < \alpha < \frac{1}{2}$. Then the only possible limit distributions for system lifetime (appropriately normalized) are of the following two types:

$$G_3(x) = \Phi(x) \qquad \text{for } x \geq 0.$$

$$G_4(x) = \Phi(\beta \log x) \qquad \text{for } x > 0, \quad \text{where } \beta > 0;$$

Φ is the normal distribution with mean 0, variance 1.

Section 2. The use of regularly varying functions to obtain the three limit laws for minima (alternately, maxima) is due to Feller (1966, pp. 268–272). Section 2 is a revision of lecture notes provided by Albert W. Marshall. Recently, De Haan (1970) has written a monograph on regular variation and extreme value theory.

In Exercise 10 a result is stated relating distributions with proportional hazards and the possible limit distributions for sample minima. Further results are discussed in Sethuraman (1965) and David (1970). Limit distributions for compositions of coherent structures was discussed by Robert Harris in his Ph.D. thesis.

Section 3. Gnedenko (1943) discusses domains of attraction of the limit laws. However, he was not satisfied with his results for the limit law $\Lambda(x) = 1 - e^{-e^x}$, $-\infty < x < \infty$. More recently, Marcus and Pinsky (1969) have treated the problem of domains of attraction for Λ and Λ^* more satisfactorily using the notion of regular variation. Von Mises' (1939) condition (see Theorems 3.7 and 3.8) for a distribution to belong to the domain of attraction of Λ (or Λ^*) still seems to be the most useful. De Haan (1970) also obtains results on this problem.

Section 4. The Grigelionis result (Theorem 4.2) is analogous to a central limit theorem for reliability theory. It furnishes another important justification for the use of the exponential distribution. The basic mathematical ideas go back to Palm (1943) and Khinchin (1956a, b). However, the treatment by Grigelionis (1964) seems to be the most elegant to date.

APPENDIX

IMPLEMENTING COHERENT STRUCTURE THEORY FOR COMPLEX SYSTEMS

The engineer faced with analyzing a very complex structure or system needs a method for translating his system into the mathematical terminology of the previous chapters. In Chapter 1 coherent structures were characterized as two terminal networks (allowing for component replication). In this Appendix we consider an alternative representation of a coherent structure called an *event tree*. When system failure, rather than success, is stressed, the event tree is commonly called a *fault tree*. Constructing the event tree is often itself a useful exercise in understanding a system. Once the event tree is drawn, it is a relatively simple matter to transfer this information to a computer which can then utilize the preceding mathematical results. In particular, we can determine the minimal cut sets in the system, and order components according to their structural importance. If additional information is provided, such as component reliabilities or component failure rates, then one can compute system reliability and order components according to their *reliability importance* (see Chapter 2, Section 1). This type of analysis is very useful for engineering design.

Event trees are described in Section A.1. An algorithm for finding the min cut sets of an event tree is presented in Section A.2. Section A.3 points out the usefulness of dual event trees. The Boolean representation of event trees in Section A.4 provides the link with coherent structure theory. Numerical examples illustrating structural ordering of basic events and also probability importance ordering of basic events are given in Sections A.5 and A.6.

A.1. EVENT TREES

To construct event trees we employ a useful but different symbolism from that used previously. Component states or, more generally, *basic events* will be represented by circles and diamonds. A system event of major importance will be represented by a rectangle called the *top event*, appearing at the top

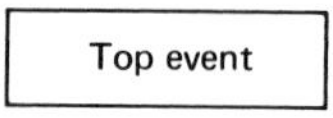

of the event tree. For example, this may be system success or system failure. Intermediate system or subsystem events will also be represented by rectangles. Immediately below each rectangle will be either an AND gate represented by

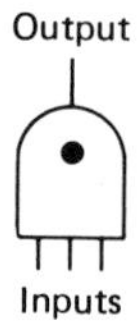

AND gate

or an OR gate represented by

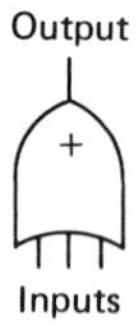

OR gate

The output event to an AND gate occurs if and only if all input events occur. It is helpful to put a dot (for set product or intersection) in the center of the AND gate. For example, to symbolize that if each of the events A, B, C, and D occur, then the event E will occur, the event tree analyst would draw a diagram such as that shown in Figure A.1.1.

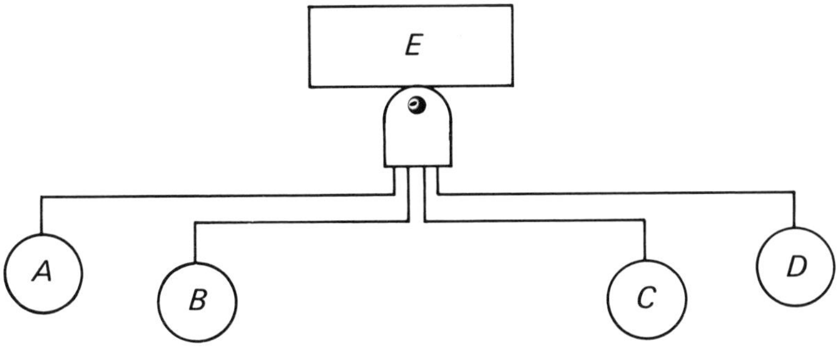

Figure A.1.1.
AND gate.

The output event to an OR gate occurs if one or more of the input events occur. It is helpful to put a plus (for set sum or union) in the center of the OR gate. For example, E occurs if one or more of the events A, B, C, or D occurs in Figure A.1.2.

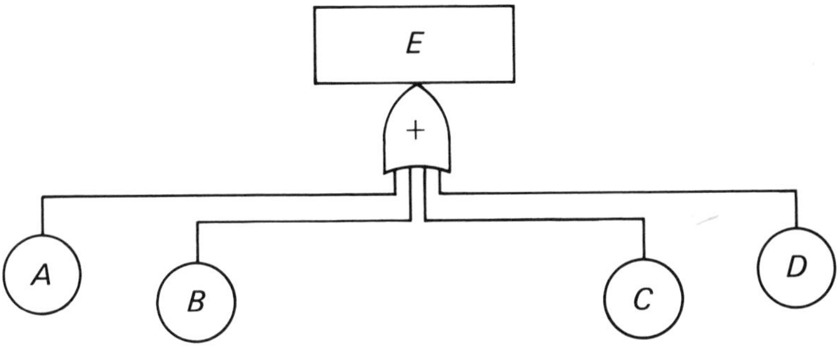

Figure A.1.2.
OR gate.

All possible (up to permutations of labels) event trees with three basic events (represented by circles) are shown here. Compare the corresponding two-terminal network representations in Figure 1.2.1 of Chapter 1.

Example 5 illustrates the repetition of basic events, which is permitted in an event tree.

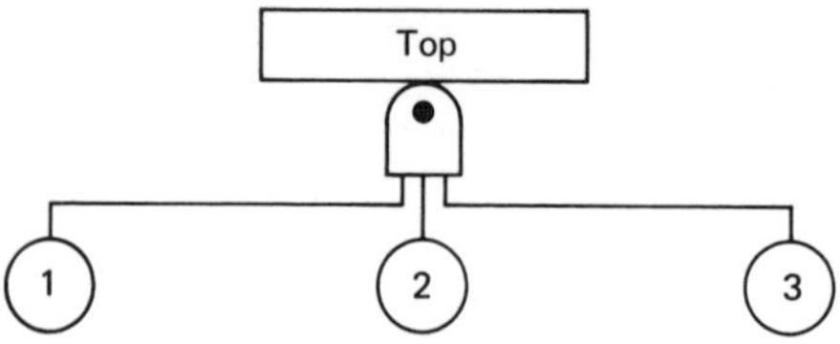

Example 1

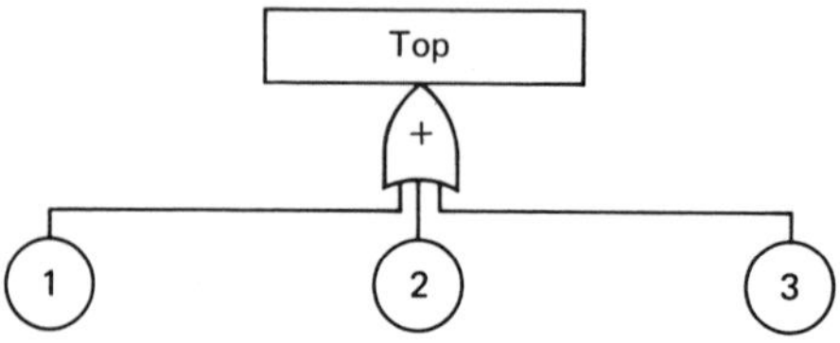

Example 2

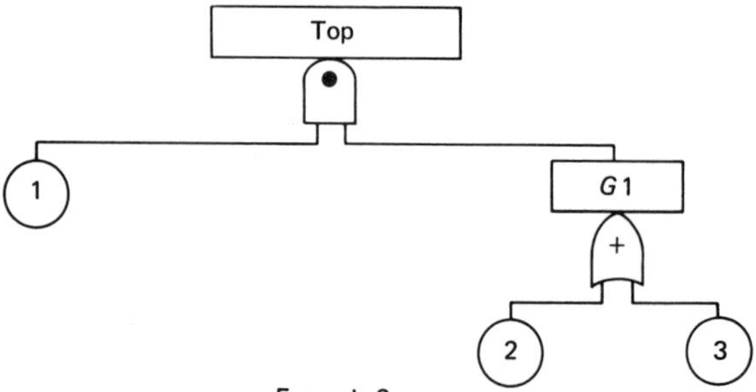

Example 3

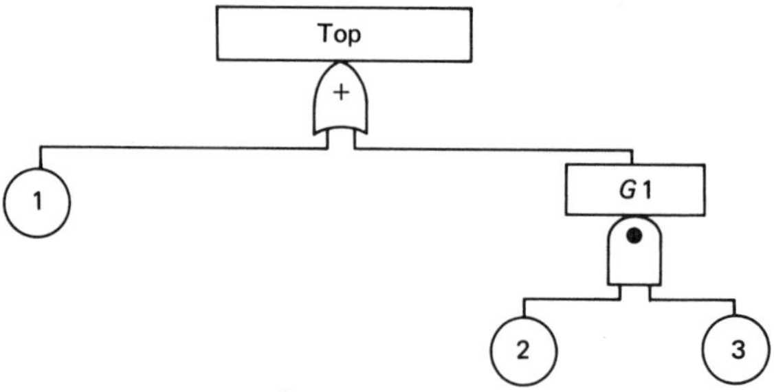

Example 4

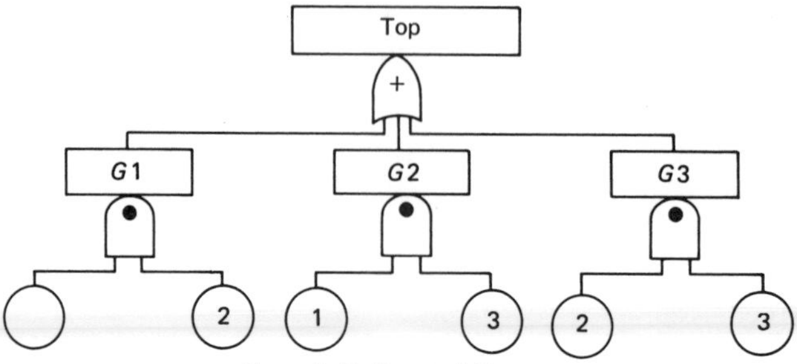

Example 5: 2-out-of-3-system

Example: 1-out-of-2 Twice System. Figure A.1.3 symbolizes a system whose function is to shut down a nuclear power plant in the event of a low coolant pressure. The 2-out-of-2 coincidence unit produces a trip signal provided that the OR unit in both the upper and lower branches simultaneously produces an output signal. Such logic is called *1-out-of-2 twice*. Units c_1 through c_4 are pressure switches. The ith switch will produce an output signal (we call this basic event i) if the pressure p_i drops below a prescribed value, $i = 1, \ldots, 4$. An event tree for this system with top event "trip signal produced" is shown in Figure A.1.4.

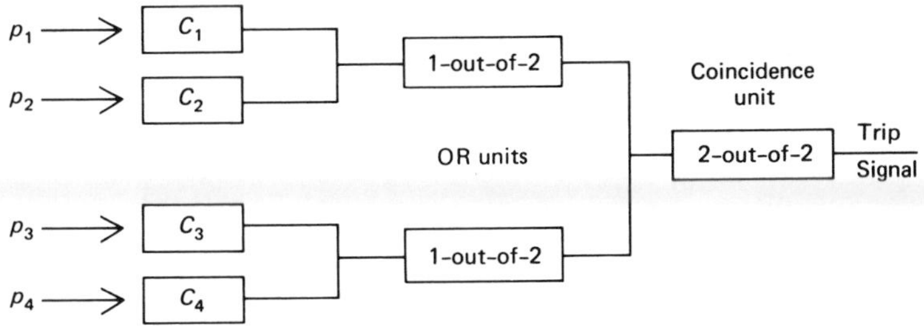

Figure A.1.3.
1-out-of-2 twice system.

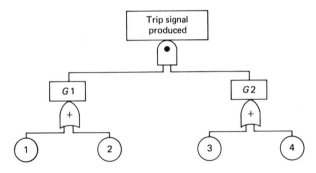

Figure A.1.4.
Event tree for 1-out-of-2 twice system.

Example: Pressure Tank System. Consider the pressure tank engineering diagram in Figure A.1.5. Let the top event (which we wish to prevent) be the rupture of the pressure tank. To start pumping, the switch $S1$ (a push button) is closed and then immediately opened. This allows current to flow in the control branch circuit which activates relay coil $K2$. Relay contacts

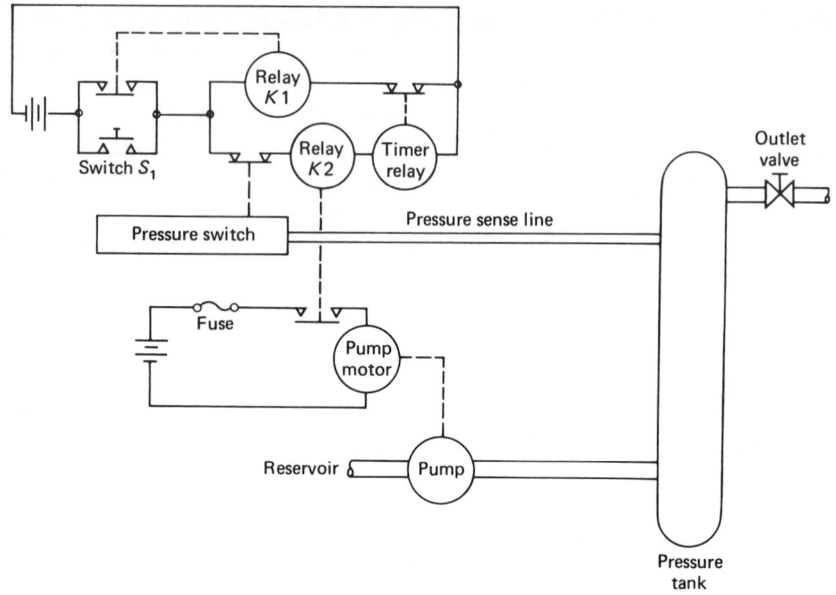

Figure A.1.5.
Pressure tank system.

$K2$ then close and start up the pump motor. After a period of approximately 20 seconds, the pressure switch contacts open (since excess pressure is detected by the pressure switch), deactivating the control circuit which de-energizes the $K2$ coil. The $K2$ contacts then open and shut off the motor. If there is a pressure switch malfunction, then the timer relay contacts open after 60 seconds, de-energizing coil $K2$, and shutting off the pump. The timer resets itself automatically after each cycle.

The event tree drawn in Figure A.1.6 is based on an analysis of the possible failure modes of the system. (When failure rather than success is stressed the event tree is commonly called a *fault tree.*) Circles represent *primary basic events*, while diamonds represent *secondary basic events*. For example, if the $K1$ relay contacts (Figure A.1.5) fail to open under normal operating conditions (that is, within the "design envelope"), this is considered a primary basic event. If the $K1$ relay fails to open because the wrong relay was installed, then this is considered a secondary basic event. A systematic method for drawing fault trees has been developed by David Haasl (1965).

Example: Heart Assist System. An event tree analysis of the electrical shock hazard to a patient using a certain heart assist device was made by Simonaitus, Anderson, and Kaye (1972). Their purpose was to identify and evaluate potential failure modes associated with the device. In order to understand the development of the event tree we first describe the device.

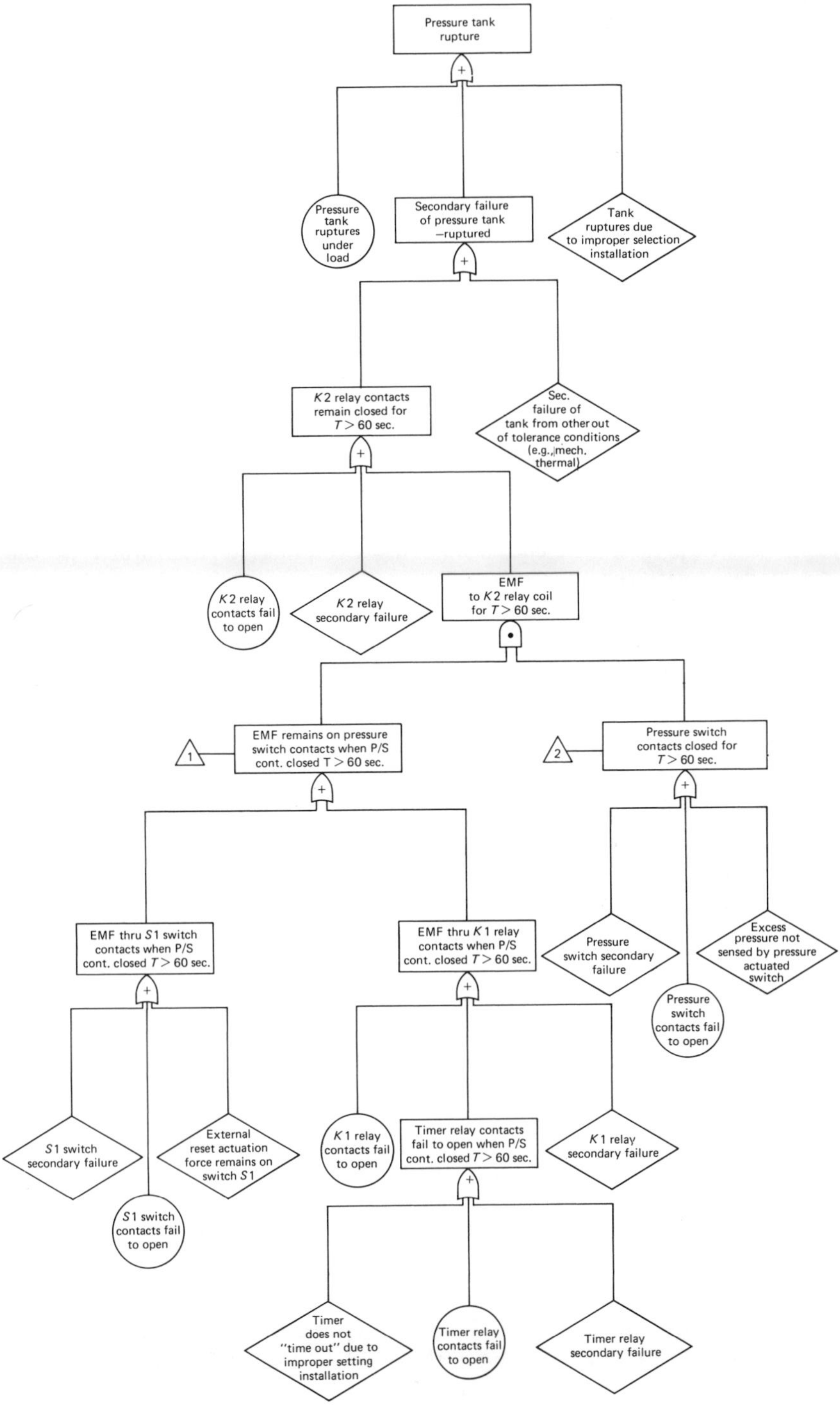

Figure A.1.6.
Pressure tank event tree.

The intra-aortic balloon (IAB) circulatory assist device is intended to provide temporary circulatory assistance following obstruction of blood circulation from the heart. In its application, a balloon-catheter positioned in the thoracic aorta is synchronously inflated and deflated with the action of the heart, resulting in decreased heart work, increased coronary blood flow, and support of the general circulation.

Synchronization of the balloon with the heart is obtained through monitoring of the electrocardiogram (ECG) of the patient by a control console. This console provides all of the control, monitoring display, and alarm functions required for operation of the balloon device. In addition, provision is made for automatic deflation of the balloon in the event of any anomaly in the ECG signal or functioning of the balloon.

A major hazard to the patient is electrical shock, either macroshock or microshock. Macroshock can cause heart failure (ventricular fibrillation) by currents entering the body through connections at the skin. Microshock can cause heart failure by currents having a conductive path directly to the vicinity of the heart.

Other hazards to the patient are balloon inflation during the contraction period of heart function and balloon overpressurization. These hazards are indicated in the following event tree, but are not developed. The connecting symbol

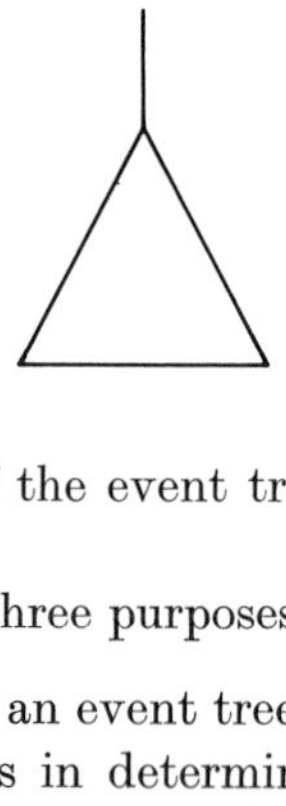

is used to indicate a section of the event tree that is continued in greater detail elsewhere.

Generally, event trees serve three purposes:

(a) In safety analysis, an event tree (or fault tree, as it is called in safety theory) aids in determining the possible causes of an accident. When properly used, the event tree often leads to discovery of failure combinations which otherwise might not have been recognized as causes of the event being analyzed.

(b) The event tree serves as a display of results. If the system design is not adequate, the event tree can be used to show what the weak points are and how they lead to undesirable events. If the design is adequate, the event tree can be used to show that all conceivable causes have been considered.

(c) The event tree provides a convenient and efficient format helpful in the computation of the probability of system success (or failure).

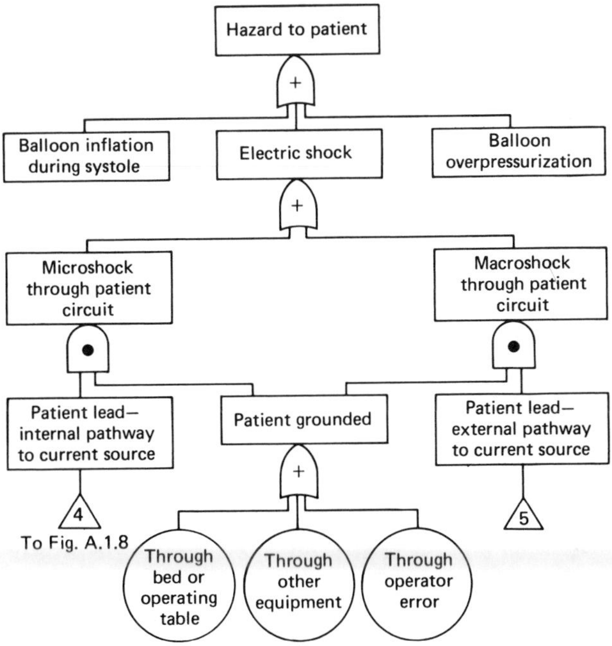

Figure A.1.7.
Intra-aortic balloon device console event tree.

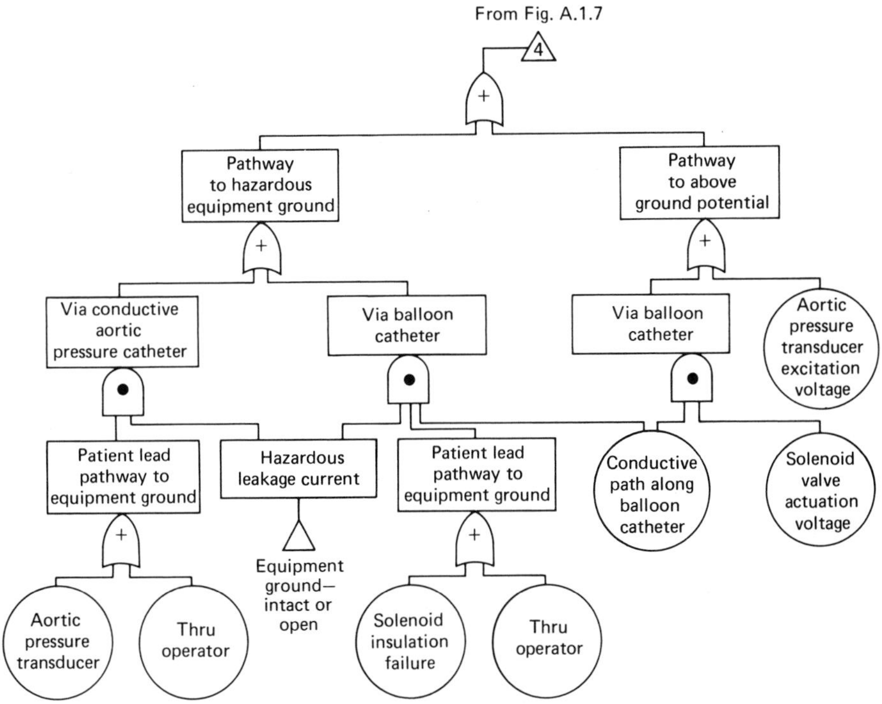

Figure A.1.8.
Microshock event tree.

A.2. MIN CUT SET ALGORITHM FOR EVENT TREES

From an engineering design point of view it is useful to have a listing of the event tree min cut sets and/or min path sets. A *cut set* for an event tree is a set of basic events whose occurrence causes the top event to occur. A cut set is minimal if it cannot be reduced and still insure the occurrence of the top event. If the top event is system failure and basic events correspond to component failures, then this definition of cut set agrees with previous definitions of cut sets for coherent structures.

The simplest and clearest way to explain the min cut set algorithm is to illustrate its operation in an example. Figure A.2.1 is a relabeling of the basic events and gates in the pressure tank event tree described in Figure A.1.6. AND and OR gates are labeled $G1$ through $G8$. The algorithm begins with the gate immediately below the top event, which we label $G0$. If $G0$ is an OR gate, each input is used as an entry in separate rows of a list matrix. If $G0$ is an AND gate, each input is used as an entry in the first row of a list matrix. Since in Figure A.2.1 the gate immediately below the top event is an OR gate, we begin the construction of our list matrix by listing inputs 1, $G1$, and 2 in separate rows as follows:

$$\begin{matrix} 1 \\ G1 \\ 2 \end{matrix}$$

Since any one of these input events can cause the top event to occur, each will be a member of a separate cut set.

The idea of the algorithm is to replace each gate by its input gates and basic events until a list matrix is constructed, all of whose entries are basic events. The rows will then correspond to cut sets.

Since $G1$ is an OR gate, we again replace $G1$ by its input events in separate rows as follows:

$$\begin{matrix} 1 \\ G2 \\ 3 \\ 2. \end{matrix}$$

Since $G2$ is also an OR gate, we replace $G2$ by its input events as follows:

$$\begin{matrix} 1 \\ 4 \\ 5 \\ G3 \\ 3 \\ 2. \end{matrix}$$

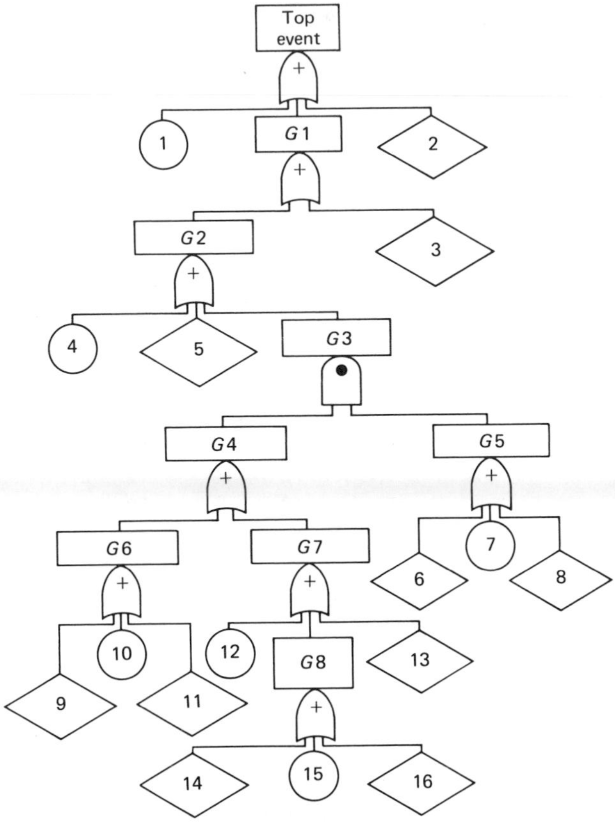

Figure A.2.1.
Pressure tank event tree.

Since $G3$ is an AND gate, we replace the *row* containing $G3$ by its inputs as follows:

1
4
5
$G4$, $G5$
3
2.

Since all inputs to an AND gate must occur to cause the corresponding intermediate event above the AND gate, we see that an AND gate increases the length of its row. An OR gate, on the other hand, increases the number of rows in our list matrix.

Replacing $G4$ by its inputs, we have

$$\begin{array}{c} 1 \\ 4 \\ 5 \\ G6, G5 \\ G7, G5 \\ 3 \\ 2. \end{array}$$

Continuing in this fashion we eventually obtain a list matrix with 29 rows. These are (in a different order),

1	7, 9	8, 9
2	7, 10	8, 10
3	7, 11	8, 11
4	7, 12	8, 12
5	7, 13	8, 13
6, 9	7, 14	8, 14
6, 10	7, 15	8, 15
6, 11	7, 16	8, 16
6, 12		
6, 13		
6, 14		
6, 15		
6, 16		

In the pressure tank event tree (Figure A.2.1), basic events are *not* repeated. For this reason all of our cut sets are minimal cut sets; that is, no one cut set is contained in any other cut set. More generally, with replication of basic events in the event tree, we will *not* obtain only min cut sets by this algorithm. Therefore it will be necessary, in general, to search the list, eliminating cut sets which contain other cut sets. The resulting list will then contain all min cut sets for the event tree.

A.3. DUAL EVENT TREES

In any analysis of an event tree (for example, coherent structure or fault tree) the *dual event tree* should be considered. One reason for this is that the min path sets for an event tree are the min cut sets for the dual event tree. As we saw in Chapter 2, Section 3, we need both min path sets and min cut sets to calculate upper and lower bounds on the probability of system success (or the probability of the top event in the case of event trees).

To draw the dual event tree, replace OR gates by AND gates and AND gates by OR gates in the original event tree. Events are also replaced by their corresponding dual. If the top event is "pressure tank rupture" as in Figure A.1.6, the dual event is "no pressure tank rupture." More generally, dual basic events correspond to the nonoccurrence of the original basic events. The dual event tree for the pressure tank example is drawn in Figure A.3.1. It is easy to verify that there are only two min cut sets for the dual tree. They are

$$\{1', 2', 3', 4', 5', 6', 7', 8'\},$$
$$\{1', 2', 3', 4', 5', 9', 10', 11', 12', 13', 14', 15', 16'\}.$$

Primes indicate that the events considered are dual events.

A *path set* is a set of basic events whose nonoccurrence *insures* the nonoccurrence of the top event. A path set is minimal if it cannot be further reduced and still remain a path set. To find min path sets for an event tree, draw the dual event tree and use the min cut algorithm to find the minimal cuts for the dual event tree. The above min cut sets for the dual event tree in Figure A.3.1 are the *min path sets* for the original event tree of Figure A.2.1.

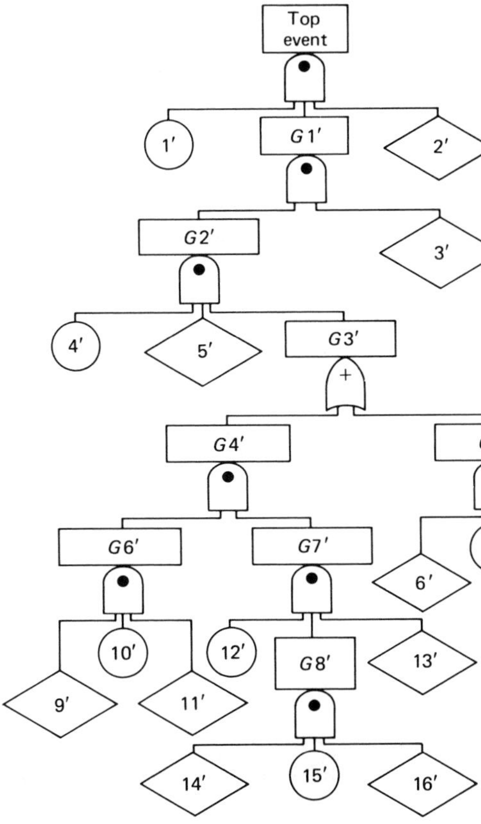

Figure A.3.1.
Pressure tank dual event tree.

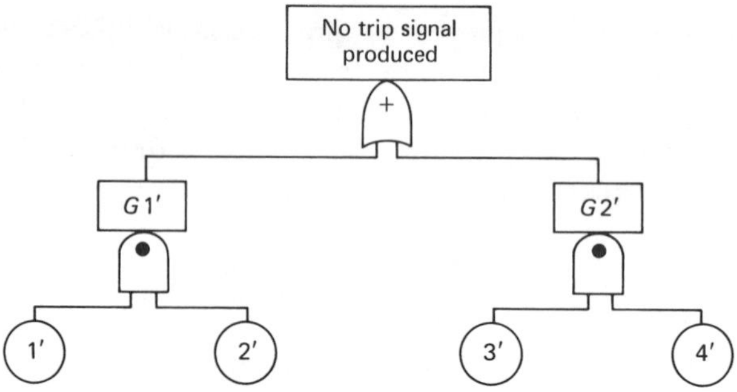

Figure A.3.2.
Dual of 1-out-of-2 twice system event tree.

If all basic events in any path set do not occur, the top event in Figure A.2.1 does *not* occur, that is, the pressure tank does not rupture. If the top event is system failure and basic events correspond to component failures, then this definition of path set corresponds to previous definitions of path set for coherent structures.

Example: The 1-out-of-2 Twice System. The 1-out-of-2 twice system event tree is presented in Figure A.1.4. The min cut sets are {1, 3}, {1, 4}, {2, 3}, and {2, 4}. If the coolant pressure is *not* low, the occurrence of any one of the four min cut set events would produce a *spurious* alarm.

The dual of the event tree is presented in Figure A.3.2. This event tree has two min cuts {1′, 2′} and {3′, 4′}. These are min paths for the original event tree, that is, the nonoccurrence of both basic events 1 and 2 (or 3 and 4) will result in *no* trip signal produced. There are thus only two min path sets in the original event tree which could cause the failure of a trip signal when low coolant pressure is actually present.

From this analysis (which neglects event probabilities), we see that the system has been designed to insure valid alarms when low coolant pressure is present. However, it would appear prone to the production of false alarms since there are four min cut sets, any one of whose occurrence could cause a false alarm. A 2-out-of-3 system, for example, would be less prone to false alarms.

A.4. BOOLEAN REPRESENTATION OF EVENT TREES

Reliability theory, and its foundation in terms of coherent structures, is success-event-oriented. In event tree theory the system event constituting the top event may be system success or system failure or any other system event which can be analyzed. Much of the preceding eight chapters can be easily utilized if we use the following Boolean representation for event trees.

Let $y_i = 1$ if basic event i occurs and $y_i = 0$ otherwise. Let $\mathbf{y} = (y_1, y_2, \ldots, y_n)$ be the vector of basic event outcomes. Define

$$\psi(\mathbf{y}) = \begin{cases} 1 & \text{if the top event occurs for outcome vector } \mathbf{y}, \\ 0 & \text{otherwise.} \end{cases} \tag{A4.1}$$

Assume all basic events are relevant to the event tree; that is, each basic event appears in the union of min cut sets. It is easily verified that ψ is mathematically equivalent to a coherent structure function (Definition 2.1 of Chapter 1). If basic events correspond to component failures and the top event corresponds to system failure, then y_i corresponds to the indicator $1 - x_i$, where $x_i = 1$ means component i is functioning, while $\psi(\mathbf{y})$ corresponds to $1 - \phi(\mathbf{x})$, where ϕ is the coherent structure function. The methods of Section 1 of Chapter 2 can be used to compute the probability of the top event. The min cut algorithm of A.2 together with the inclusion-exclusion principle of Section 1, Chapter 2, is perhaps most efficient for large complex systems. *Reliability importance*, defined in Section 1, Chapter 2, is more appropriately called *probability importance* for event trees. The distribution of time to occurrence of the top event can be calculated using the notation of Chapter 4. The results of Chapters 5, 6, and 7 can be easily translated in terms of the event tree model.

Example: Pressure Tank System. Let $\mathbf{Y} = (Y_1, Y_2, \ldots, Y_{16})$ be the random vector for basic event outcomes in the pressure tank event tree in Figure A.2.1. Let $\psi(\mathbf{Y}) = 1$ if the top event occurs for outcome vector $\mathbf{Y}$; that is, the pressure tank ruptures, and $\psi(\mathbf{Y}) = 0$ otherwise. Then using the min path sets:

$$P_1 = \{1', 2', 3', 4', 5', 6', 7', 8'\}$$

and

$$P_2 = \{1', 2', 3', 4', 5', 9', 10', 11', 12', 13', 14', 15', 16'\}$$

and the min path representation for ψ, we see that

$$\psi(\mathbf{Y}) = \Big(\coprod_{1 \le i \le 8} Y_i\Big)\Big(\coprod_{i \ne 6,7,8} Y_i\Big)$$

$$= \left[1 - \prod_{i=1}^{8}(1 - Y_i)\right]\left[1 - \prod_{i \ne 6,7,8}(1 - Y_i)\right], \tag{A4.2}$$

and $P[\text{top event}] = E\psi(\mathbf{Y})$. Note that (A4.2) differs from the min path representation (3.2) of Chapter 1, Section 3. This is because the Y_i are indicators for failure events, while in (3.2) the X_i are indicators for success events. Also $\psi(\mathbf{Y})$ is the indicator for system failure.

Let $EY_i = q_i$ and assume basic events are *statistically independent.*

Assume that basic event 1 (that is, the pressure tank itself fails) occurs on the average once in 10^8 loading cycles or, in other words, $q_1 = 10^{-8}$. Assume basic event i ($i \neq 1$) occurs on the average once in 10^5 loading cycles or, in other words, $q_i = 10^{-5}$ for $i \neq 1$. Then

$$\begin{aligned} E\psi(\mathbf{Y}) &= 1 - \prod_{i=1}^{8}(1 - q_i) - \prod_{i \neq 6,7,8}(1 - q_i) + \prod_{i=1}^{16}(1 - q_i) \\ &= 1 - (1 - 10^{-8})(1 - 10^{-5})^7 - (1 - 10^{-8})(1 - 10^{-5})^{12} \\ &\quad + (1 - 10^{-8})(1 - 10^{-5})^{15}. \end{aligned}$$

Hence

$$E\psi(\mathbf{Y}) \sim 4 \times 10^{-5}.$$

If it is unreasonable to assume basic events are statistically independent, but it is reasonable to assume they are *associated* (Chapter 2, Section 2), then we can use the min-max bounds (suitably interpreted) of Section 3, Chapter 2 [see (3.12)] to obtain

$$\begin{aligned} E\psi(\mathbf{Y}) &\leq \min_{1 \leq r \leq 2} \coprod_{i \in P_r} q_i \\ &= \min\{1 - (1 - 10^{-8})(1 - 10^{-5})^7, 1 - (1 - 10^{-8})(1 - 10^{-5})^{12}\} \\ &\sim 7 \times 10^{-5}. \end{aligned}$$

On the other hand,

$$E\psi(\mathbf{Y}) \geq \max_{1 \leq s \leq 29} \prod_{i \in K_s} q_i = \max\{10^{-8}, 10^{-5}, 10^{-10}\} = 10^{-5}.$$

The min cut sets, K_s ($1 \leq s \leq 29$) were obtained in Section A.2. Under association, then

$$10^{-5} \leq E\psi(\mathbf{Y}) \leq 7 \times 10^{-5}.$$

Let the random number of cycles to occurrence of basic event i be T_i. Assume $ET_1 = 10^8$ and $ET_i = 10^5$ for $i \neq 1$. Let T be the random time of occurrence of the top event. Since

$$\begin{aligned} T &= \min_{1 \leq s \leq 29} \max_{i \in K_s} T_i \\ &= \max_{1 \leq r \leq 2} \min_{i \in P_r} T_i, \end{aligned}$$

we have

$$\max_{1 \leq r \leq 2} E[\min_{i \in P_r} T_i] \leq ET \leq \min_{1 \leq s \leq 29} E[\max_{i \in K_s} T_i]$$

Assume random variables $T_1, T_2, \ldots, T_{16}$ are associated, with IFRA marginal distributions (see Chapter 4). Then

$$E[\min_{i \in P_r} T_i] \geq [\sum_{i \in P_r} [ET_i]^{-1}]^{-1},$$

so that

$$\begin{aligned} ET &\geq \max\{[10^{-8} + 7 \times 10^{-5}]^{-1}, [10^{-8} + 12 \times 10^{-5}]^{-1}\} \\ &= [10^{-8} + 7 \times 10^{-5}]^{-1} \sim 7^{-1} \times 10^5 \text{ cycles}. \end{aligned}$$

A.5. STRUCTURAL ORDERING OF BASIC EVENTS

In Chapter 1, Section 3, we presented a measure of the structural importance of components in a coherent system. A similar measure of importance of basic events may be defined for event trees. This measure is of course independent of basic event probabilities. We discuss a measure of probability importance of basic events in the following section.

The cut set C_i is a *critical cut set for basic event i* if it is a cut set containing i such that each of its min cut sets contains i. Let $n(i)$ be the number of *critical cut sets for i*. Then we define the *event tree importance of basic event i* by

$$I(i) = 2^{-(n-1)}n(i), \tag{A5.1}$$

where n denotes the number of basic events in the event tree.

To compute $n(i)$, assume the Y_i's are statistically independent, $EY_i = E(1 - Y_i) = \frac{1}{2}$ for $i = 1, 2, \ldots, n$, and use the formula

$$n(i) = 2^{n-1}E[\psi(1_i, \mathbf{Y}) - \psi(0_i, \mathbf{Y})]$$

[see (3.9) in Chapter 1].

Example: The Pressure Tank. We use (A.2) to compute

$$E[\psi(1_j, \mathbf{Y}) \mid EY_i = \tfrac{1}{2}, i = 1, 2, \ldots, n]$$

and

$$E[\psi(0_j, \mathbf{Y}) \mid EY_i = \tfrac{1}{2}, i = 1, 2, \ldots, n].$$

For basic event 1,

$$n(1) = 2^{15}\left\{1 - E\left[1 - \prod_{\substack{i=1\\ i\neq 1}}^{8}(1 - Y_i)\right]\left[1 - \prod_{i\neq 1,6,7,8}(1 - Y_i)\right]\right\}$$

$$= 2^{15}[(\tfrac{1}{2})^7 + (\tfrac{1}{2})^{12} - (\tfrac{1}{2})^{15}] = 263.$$

It is not hard to see that

$$n(1) = n(2) = n(3) = n(4) = n(5) = 263.$$

For basic event 6,

$$n(6) = 2^{15}\left\{E\left[1 - \prod_{i\neq 6,7,8}(1 - Y_i)\right] - E\left[1 - \prod_{\substack{i=1\\ i\neq 6}}^{8}(1 - Y_i)\right]\left[1 - \prod_{i\neq 6,7,8}(1 - Y_i)\right]\right\}$$

$$= 2^{15}[(\tfrac{1}{2})^7 - (\tfrac{1}{2})^{15}] = 255.$$

Also

$$n(6) = n(7) = n(8) = 255.$$

For basic event 9,

$$n(9) = 2^{15}\left\{E\left[1 - \prod_{i=1}^{8}(1 - Y_i)\right] - E\left[1 - \prod_{i=1}^{8}(1 - Y_i)\right]\left[1 - \prod_{i\neq 6,7,8,9}(1 - Y_i)\right]\right\}$$

$$= 2^{15}\{(\tfrac{1}{2})^{12} - (\tfrac{1}{2})^{15}\} = 7.$$

It is not hard to see that

$$n(9) = n(10) = n(11) = n(12) = n(13) = n(14) = n(15) = n(16) = 7.$$

The importance ordering of events is therefore

$$1 \sim 2 \sim 3 \sim 4 \sim 5 > 6 \sim 7 \sim 8 > 9 \sim 10 \sim 11 \sim 12 \sim 13 \sim 14 \sim 15 \sim 16,$$

where "$1 \sim 2$" means 1 and 2 are equally important in the event tree, and "$5 > 6$" means 5 is more important than 6 in the event tree. Table A.1 provides a key to the original example of Figure A.1.5. For example, we see that the pressure tank itself and the $K2$ relay are structurally most important. The pressure switch is next most important, while the timer, the $K1$ relay, and the $S1$ switch are the least important structurally.

Table A.1. Key to Pressure Tank Example

Basic Event i	*Number of Critical Cuts,* $n(i)$, *Containing Basic Event* i	*Description of Basic Events*
1	263	Pressure tank failure
2	263	Secondary failure of pressure tank due to improper selection
3	263	Secondary failure of pressure tank due to out-of-tolerance conditions
4	263	$K2$ relay contacts fail to open
5	263	$K2$ relay secondary failure
6	255	Pressure switch secondary failure
7	255	Pressure switch contacts fail to open
8	255	Excess pressure not sensed by pressure actuated switch
9	7	$S1$ switch secondary failure
10	7	$S1$ switch contacts fail to open
11	7	External reset actuation force remains on switch $S1$
12	7	$K1$ relay contacts fail to open
13	7	$K1$ relay secondary failure
14	7	Timer does not "time off" due to improper setting
15	7	Timer relay contacts fail to open
16	7	Timer relay secondary failure

A.6. PROBABILITY ORDERING OF BASIC EVENTS

In analogy to the definition of reliability importance defined in Chapter 2, Section 1, we define the *probability importance* of basic event j to be

$$I(j) = \sum_{\mathbf{y}} [\psi(1_j, \mathbf{y}) - \psi(0_j, \mathbf{y})] P[\mathbf{Y} = \mathbf{y}]$$

$$= E[\psi(1_j, \mathbf{Y}) - \psi(0_j, \mathbf{Y})].$$

When $P[\mathbf{Y} = \mathbf{y}] = 2^{-n}$ for all $\mathbf{y}$ vectors, then of course

$$I(j) = 2^{-n} \sum_{\mathbf{y}} [\psi(1_j, \mathbf{y}) - \psi(0_j, \mathbf{y})]$$

$$= 2^{-n+1} n(j).$$

Example: Pressure Tank. For the pressure tank example, the outcome of the tank itself failing is the least likely event. For this event, which is labeled 1,

$$I(1) = E\psi(1_1, \mathbf{Y}) - E\psi(0_1, \mathbf{Y})$$

$$= 1 - E\left[1 - \prod_{i=2}^{8} (1 - Y_i)\right]\left[1 - \prod_{i \neq 1,6,7,8} (1 - Y_i)\right]$$

$$\sim 1 - 4 \times 10^{-5}.$$

For the event labeled 2 we have

$$I(2) = 1 - E\left[1 - \prod_{\substack{i \neq 2 \\ i=1}}^{8} (1 - Y_i)\right]\left[1 - \prod_{i \neq 6,7,8,2} (1 - Y_i)\right]$$

$$\sim 1 - 3 \times 10^{-5}.$$

Also it is easy to see that $I(2) = I(3) = I(4) = I(5)$. For the event labeled 6 we have

$$I(6) = 1 - E\left[\prod_{i \neq 6,7,8} (1 - Y_i)\right]$$

$$- E\left[1 - \prod_{\substack{i=1 \\ i \neq 6}}^{8} (1 - Y_i)\right]\left[1 - \prod_{i \neq 6,7,8} (1 - Y_i)\right]$$

$$\sim 9 \times 10^{-5}.$$

Also $I(6) = I(7) = I(8)$.

For the event labeled 9 we have

$$I(9) = E\left[1 - \prod_{i=1}^{8}(1 - Y_i)\right] - E\left[1 - \prod_{i=1}^{8}(1 - Y_i)\right]\left[1 - \prod_{i\neq 6,7,8,9}(1 - Y_i)\right] \sim 1 - 14 \times 10^{-5}.$$

Hence we have

$$I(2) = I(3) = I(4) = I(5) > I(1) > I(9) = I(10) = I(11) = I(12) = I(13) = I(14) = I(15) = I(16) > I(6) = I(7) = I(8).$$

It follows that the $K2$ relay is one of the most important components on the basis of probability ordering, while the pressure tank itself is the second most important component. The pressure switch is the least important component according to this ordering.

A.7. NOTES AND REFERENCES

Section A.1. Fault tree is the name commonly applied to event trees which are failure-oriented. We have used the more general term event tree to show that coherent structures can also be represented in this way. The technique of drawing fault trees was developed by safety engineers at Bell Telephone Laboratories and the Boeing Company. The pressure tank example is due to David Haasl. Howard Lambert (1973) introduced us to this interesting engineering approach. See Fussell (1973a) and (1973b) for fault tree construction methodology.

Section A.2. The min cut algorithm for event trees is due to Fussell and Vesely (1972).

Section A.5. The definition of structural importance is due to Birnbaum (1969). An alternative definition is discussed in Barlow and Proschan (1974).

Section A.6. The definition of probability importance is due to Birnbaum (1969).

REFERENCES

ARNOLD, B. C. 1967. A note on multivariate distributions with specified marginals. *J. Amer. Stat. Assoc.* 62: 1460–1461.

BARLOW, R. E. 1962. Repairman Problems. Chapter 2 in *Studies in Applied Probability and Management Science*, ed. Arrow, Karlin, and Scarf. Stanford, Calif.: Stanford University Press.

BARLOW, R. E. 1965. Bounds on integrals with applications to reliability problems. *Ann. Math. Statist.* 36: 565–574.

BARLOW, R. E., D. J. BARTHOLOMEW, J. M. BREMNER, and H. D. BRUNK. 1972. *Statistical Inference Under Order Restrictions.* New York: John Wiley and Sons.

BARLOW, R. E., and A. W. MARSHALL. 1964. Bounds for distributions with monotone hazard rate, I and II. *Ann. Math. Statist.* 35: 1234–1274.

BARLOW, R. E., and A. W. MARSHALL. 1965. Tables of bounds for distributions with monotone hazard rate. *J. Amer. Statist. Assoc.* 60: 872–890.

BARLOW, R. E., and A. W. MARSHALL. 1967. Bounds on interval probabilities for restricted families of distributions. *Proceedings of the Fifth Berkeley Symposium on Mathematical Statistics and Probability.* III: 229–257.

BARLOW, R. E., A. W. MARSHALL, and F. PROSCHAN. 1963. Properties of probability distributions with monotone hazard rate. *Ann. Math. Statist.* 34: 375–389.

BARLOW, R. E., and F. PROSCHAN. 1964. Comparison of replacement policies, and renewal theory implications. *Ann. Math. Statist.* 35: 577–589.

BARLOW, R. E., and F. PROSCHAN. 1965. *Mathematical Theory of Reliability.* New York: John Wiley and Sons.

BARLOW, R. E., and F. PROSCHAN. 1966a. Inequalities for linear combinations of order statistics from restricted families. *Ann. Math. Statist.* 37: 1574–1592.

Barlow, R. E., and F. Proschan. 1966b. Tolerance and confidence limits for classes of distributions based on failure rate. *Ann. Math. Statist.* 37: 1593–1601.

Barlow, R. E., and F. Proschan. 1973. Availability theory for multicomponent systems. *Multivariate Analysis—III*, ed. P. R. Krishnaiah, pp. 319–335. New York: Academic Press.

Barlow, R. E., and F. Proschan. 1974. Importance of system components and fault tree events. Operations Research Center Report 74–3. University of California, Berkeley. To appear in *Stoch. Proc. Applic.*

Bazovsky, I. 1961. *Reliability Theory and Practice.* Engelwood Cliffs, N.J.: Prentice-Hall.

Bazovsky, I., N. R. MacFarlane, and R. Wunderman. 1962. Study of maintenance cost optimization and reliability of shipboard machinery. United Control Report under ONR Contract Nonr 37400. Seattle, Washington.

Beckenbach, E. F., and R. Bellman. 1961. *Inequalities.* Berlin: Springer-Verlag.

Bessler, S. A., and A. F. Veinott, Jr. 1966. Optimal policy for a dynamic multi-echelon inventory model. *Naval Research Logistics Quarterly.* 13: 355–389.

Billingsley, P. 1968. *Convergence of Probability Measures.* New York: John Wiley and Sons.

Birkhoff, G., and S. MacLane. 1953. *A Survey of Modern Algebra.* Rev. ed. New York: The Macmillan Co.

Birnbaum, Z. W. 1969. On the importance of different components in a multi-component system. *Multivariate Analysis—II*, ed. P. R. Krishnaiah, pp. 581–592. New York: Academic Press.

Birnbaum, Z. W., and J. D. Esary. 1965. Modules of coherent binary systems. *SIAM Journal on Applied Math.* 13: 444–462.

Birnbaum, Z. W., J. D. Esary, and A. W. Marshall. 1966. Stochastic characterization of wearout for components and systems. *Ann. Math. Statist.* 37: 816–825.

Birnbaum, Z. W., J. D. Esary, and S. C. Saunders. 1961. Multicomponent systems and structures, and their reliability. *Technometrics* 3: 55–77.

Bodin, L. D. 1970. Approximations to system reliability using a modular decomposition. *Technometrics* 12: 335–344.

Bruckner, A. M., and E. Ostrow. 1962. Some function classes related to the class of convex functions. *Pacific J. Math.* 12: 1203–1215.

Bryson, M. C., and M. M. Siddiqui. 1969. Some criteria for aging. *J. Amer. Statist. Assoc.* 64: 1472–1483.

Bühlmann, H. 1970. *Mathematical Methods in Risk Theory.* New York: Springer-Verlag.

Chernoff, H., and H. Teicher. 1965. Limit distributions of the minimax of independent identically distributed random variables. *Trans. Amer. Math. Soc.* 116: 474–491.

Chibisov, D. M. 1964. On limit distributions for order statistics. *Theory of Probability and Applications* 9: 142–148.

Cox, D. R. 1962. *Renewal Theory.* New York: John Wiley and Sons.

Cox, D. R., and P. A. W. Lewis. 1966. *The Statistical Analysis of Series of Events.* New York: John Wiley and Sons.

Cox, D. R., and W. L. Smith. 1961. *Queues.* London: Methuen.

Cramér, H. 1951. *Mathematical Methods of Statistics.* Princeton, N.J.: Princeton University Press.

David, H. A. 1970. *Order Statistics.* New York: John Wiley and Sons.

Davis, D. J. 1952. An analysis of some failure data. *J. Amer. Statist. Assoc.* 47: 113–150.

De Haan, L. 1970. *On Regular Variation and Its Application to the Weak Convergence of Sample Extremes.* Mathematical Center, Amsterdam.

Doob, J. L. 1953. *Stochastic Processes.* New York: John Wiley and Sons.

Downton, F. 1970. Bivariate exponential distributions in reliability theory. *J. Royal Statist. Soc.* (B) 32: 408–417.

Drenick, R. F. 1960. The failure law of complex equipment. *J. Soc. Indust. Appl. Math.* 8: 680–690.

Dwass, M., and H. Teicher. 1957. On infinitely divisible random vectors. *Ann. Math. Statist.* 28: 461–470.

Epstein, B., and M. Sobel. 1953. Life testing. *J. Amer. Statist. Assoc.* 48: 486–502.

Epstein, B., and M. Sobel. 1954. Some theorems relevant to life testing from an exponential distribution. *Ann. Math. Statist.* 25: 373–381.

Esary, J. D., and Z. W. Birnbaum. 1965. Modules of coherent binary systems. *J. Soc. Indust. Appl. Math.* 13: 444–462.

Esary, J. D., and A. W. Marshall. 1970. Coherent life functions. *SIAM. J. Appl. Math.* 18: 810–814.

Esary, J. D., A. W. Marshall, and F. Proschan. 1970. Some reliability applications of the hazard transform. *SIAM J. Appl. Math.* 18: 849–860.

Esary, J. D., A. W. Marshall, and F. Proschan. 1971. Determining an approximate constant failure rate for a system whose components have constant failure rates. *Operations Research and Reliability*, ed. D. Grouchko, pp. 195–212. London: Gordon and Breach.

Esary, J. D., A. W. Marshall, and F. Proschan. 1973. Shock models and wear processes. *Ann. Probability.* 1: 627–649.

Esary, J. D., and F. Proschan. 1962. The reliability of coherent systems. *Redundancy Techniques for Computing Systems*, ed. R. H. Wilcox and C. W. Mann, pp. 47–61. Washington, D.C.: Spartan Books.

Esary, J. D., and F. Proschan. 1963a. Coherent structures of nonidentical components. *Technometrics* 5: 191–209.

Esary, J. D., and F. Proschan. 1963b. Relationship between system failure rate and component failure rate. *Technometrics* 5: 183–189.

Esary, J. D., and F. Proschan. 1968. Generating associated random variables. Boeing Scientific Research Laboratories Document D1-82-0696. Seattle, Washington.

Esary, J. D., and F. Proschan. 1970. A reliability bound for systems of maintained, interdependent components. *J. Amer. Statist. Assoc.* 65: 329–338.

Esary, J. D., and F. Proschan. 1972. Relationships among some notions of bivariate dependence. *Ann. Math. Statist.* 43: 651–655.

ESARY, J. D., and F. PROSCHAN. In process. Conservative availability prediction.

ESARY, J. D., F. PROSCHAN, and D. W. WALKUP. 1967. Association of random variables, with applications. *Ann. Math. Statist.* 38: 1466–1474.

FELLER, W. 1966. *An Introduction to Probability Theory and Its Applications*, Vol. II. New York: John Wiley and Sons.

FELLER, W. 1968. *An Introduction to Probability Theory and Its Applications*, Vol. I. 3rd ed. New York: John Wiley and Sons.

FISHER, R. A., and L. H. C. TIPPETT. 1928. Limiting forms of the frequency distribution of the largest or smallest member of a sample. *Proceedings Cambridge Philos. Soc. V.* XXIV: 180–190.

FLEHINGER, B. J. 1962. A general model for the reliability analysis of systems under various preventive maintenance policies. *Ann. Math. Statist.* 33: 137–156.

FRÉCHET, M. 1951. Sur les tableaux de corrélation dont les marges sont données. *Ann. Univ. Lyon Sect. A*, series 3, vol. 14, pp. 53–77.

FRÉCHET, M. 1927. Sur la loi de probabilité de l'écart maximum. *Ann. Soc. Polon. Math. Cracovie* 6: 93.

FUSSELL, J. B., 1973a. A formal methodology for fault tree construction. *Nuclear Science and Engineering.* 52: 421–432.

FUSSELL, J. B. 1973b. Fault tree analysis—concepts and techniques. *Proceedings NATO Advanced Study Institute on Generic Techniques of System Reliability Assessment.* Liverpool, England.

FUSSELL, J. B., and W. E. VESELY. 1972. A new methodology for obtaining cut sets for fault trees. *American Nuclear Society Transactions*, vol. 15, no. 1, pp. 262–263 (June).

GENERAL ELECTRIC CO. 1962. *Tables of the Individual and Cumulative Terms of the Poisson Distribution.* Princeton, N.J.: D. Van Nostrand.

GNEDENKO, B. V. 1943. Sur la distribution limite du terme maximum d'une série aléatoire. *Ann. of Math.* 44: 423–453.

GNEDENKO, B. V., YU. K. BELYAYEV, and A. D. SOLOVYEV. 1969. *Mathematical Methods of Reliability Theory.* New York: Academic Press.

GNEDENKO, B. V., and A. N. KOLMOGOROV. 1954. *Limit Distributions for Sums of Independent Random Variables* (translated from the Russian). Reading, Mass.: Addison-Wesley.

GRIGELIONIS, B. I. 1964. Limit theorems for sums of renewal processes. *Cybernetics in the Service of Communism, Vol. 2, Reliability Theory and Queueing Theory*, ed. A. I. Berg, N. G. Bruevich, and B. V. Gnedenko, pp. 246–266. Moscow-Leningrad: "Energy" Publishing House. In Russian.

GUMBEL, E. J. 1958. *Statistics of Extremes.* New York: Columbia University Press.

GUMBEL, E. J. 1960. A bivariate extension of the exponential distribution. *J. Amer. Statist. Assoc.* 56: 971–977.

GURLAND, J. 1955. Distribution of the maxima of the arithmetic mean of correlated random variables. *Ann. Math. Statist.* 26: 294–300.

HAASL, D. 1965. Advanced concepts in fault tree analysis. *System Safety Symposium*, sponsored by the University of Washington and The Boeing Company. Seattle, Washington.

HAIGHT, F. A. 1967. *Handbook of the Poisson Distribution.* New York: John Wiley and Sons.

Handbook Reliability Engineering. 1968. NAVAIR 00-65-502, NAVORD OD-41146. Naval Air Systems Command.

HARDY, G. H., J. E. LITTLEWOOD, and G. PÓLYA. 1952. *Inequalities.* London: Cambridge University Press.

HARRIS, R. 1970. A multivariate definition for increasing hazard rate distributions. *Ann. Math. Statist.* 41: 713–717.

HARRIS, R. 1970. An application of extreme value theory to reliability theory. *Ann. Math. Statist.* 41: 1456–1465.

HEWITT, E., and K. STROMBERG. 1965. *Real and Abstract Analysis.* New York: Springer-Verlag.

HOLLANDER, M., and F. PROSCHAN. 1972. Testing whether new is better than used. *Ann. Math. Statist.* 43: 1136–1146.

JORDAN, C. 1950. *Calculus of Finite Differences.* New York: Chelsea Publishing Co.

JUNCOSA, M. L. 1949. On the distribution of the minimum in a sequence of mutually independent random variables. *Duke Math. J.* 16: 609–618.

KAMINS, M. 1962. Rules for planned replacement of aircraft and missile parts. RAND Memo, RM-2810-PR (abridged).

KAO, J. H. K. 1958. Computer methods for estimating Weibull parameters in reliability studies. *IRE Transactions on Reliability and Quality Control,* PGRQC 13: 15–22.

KARAMATA, J. 1930. Sur un mode de croissance régulière des functions. *Mathematica* (Clug) 4: 38–53.

KARLIN, S. 1964. Total positivity, absorption probabilities and applications. *Trans. Amer. Math. Soc.* III: 33–107.

KARLIN, S. 1968. *Total Positivity,* Vol. I. Stanford, Calif.: Stanford University Press.

KARLIN, S., and A. NOVIKOFF. 1963. Generalized convex inequalities. *Pacific J. Math.* 13: 1251–1279.

KARLIN, S., and F. PROSCHAN. 1960. Pólya type distributions of convolutions. *Ann. Math. Statist.* 31: 721–736.

KARLIN, S., F. PROSCHAN, and R. E. BARLOW. 1961. Moment inequalities of Pólya frequency functions. *Pacific J. Math.* 11: 1023–1033.

KETTELLE, J. D., JR. 1962. Least-cost allocation of reliability investment. *Operations Research* 10: 249–265.

KHINCHIN, A. IA. 1956a. Streams of random events without aftereffect. *Theory of Probability and Its Applications* (SIAM translation of Russian journal) 1: 1–15.

KHINCHIN, A. IA. 1956b. On Poisson streams of random events. *Theory of Probability and Its Applications* (SIAM translation of Russian journal) 1: 248–255.

LAMBERT, H. E. 1973. System safety analysis and fault tree analysis. Lawrence Livermore Laboratory Technical Report. Livermore, California.

LARRIBEAU, JR. 1969. The automatic calculation of the reliability of coherent structures. ORC 69-4. Operations Research Center. University of California, Berkeley.

LARSEN, R. I. 1969. A new mathematical model of air pollutant concentration averaging time and frequency. *J. of the Air Pollution Control Association* 19: 24–30.

LEADBETTER, M. R., and W. A. SMITH. 1963. On the renewal function for the Weibull distribution. *Technometrics* 5: 393–396.

LEHMANN, E. L. 1966. Some concepts of dependence. *Ann. Math. Statist.* 37: 1137–1153.

LIEBLEIN, J., and M. ZELEN. 1956. Statistical investigation of the fatigue life of deep-groove ball bearings. *J. Res., Nat. Bureau Stand.* 57: 273–316.

LOÈVE, M. 1956. Ranking limit problem. *Third Berkeley Symposium on Mathematical Statistics and Probability*, pp. 177–194. Berkeley, Calif.: University of California Press.

LOMNICKI, Z. A. 1966. A note on the Weibull renewal process. *Biometrika* 53: 375–381.

MARCUS, M., and M. PINSKY. 1969. On the domain of attraction of $e^{-e^{-x}}$. *J. of Math. Analysis and Applications* 28: 440–449.

MARSHALL, A. W., and I. OLKIN. 1967a. A multivariate exponential distribution. *J. Amer. Statist. Assoc.* 62: 30–44.

MARSHALL, A. W., and I. OLKIN. 1967b. A generalized bivariate exponential distribution. *J. Appl. Prob.* 4: 291–302.

MARSHALL, A. W., I. OLKIN, and F. PROSCHAN. 1967. Monotonicity of ratios of means and other applications of majorization. *Inequalities*, ed. O. Shisha, pp. 177–190. New York: Academic Press.

MARSHALL, A. W., and F. PROSCHAN. 1970. Mean life of series and parallel systems. *J. Appl. Prob.* 7: 165–174.

MARSHALL, A. W., and F. PROSCHAN. 1972. Classes of distributions applicable in replacement, with renewal theory implications. *Proceedings of the 6th Berkeley Symposium on Mathematical Statistics and Probability*, vol. I, ed. L. LeCam, J. Neyman, and E. L. Scott, pp. 395–415. Berkeley, Calif.: University of California Press.

MEJZLER, D. 1965. On a certain class of limit distributions and their domain of attraction. *Trans. Amer. Math. Soc.* 117: 205–236.

MOLINA, E. C. 1942. *Poisson's Exponential Binomial Limit.* Princeton, N.J.: D. Van Nostrand Co.

MOOD, A. M. 1950. *Introduction to the Theory of Statistics.* New York: McGraw-Hill.

MOORE, E. F., and C. E. SHANNON. 1956. Reliable circuits using less reliable relays. *J. of the Franklin Institute*, vol. 262, part I, pp. 191–208, and vol. 262, part II, pp. 281–297.

MUNROE, M. E. 1953. *Introduction to Measure and Integration.* Cambridge, Mass.: Addison-Wesley.

OSOSKOV, G. A. 1956. A limit theorem for flows of similar events. *Theory of Prob. and Its Appl.* 1: 248–255.

PALM, C. 1943. Intensitatsschwankungen im fernsprechverkehr. *Ericsson Technics* 44: 3–189.

PARZEN, E. 1962. *Stochastic Processes.* San Francisco: Holden-Day, Inc.

PEARSON, E. S., and H. O. HARTLEY. 1958. *Biometrika Tables for Statisticians*, Vol. I. 2d ed. London: Cambridge University Press.

RÉNYI, A. 1953. On the theory of order statistics. *Acta Math. Acad. Science Hungar.* 4: 191–206.

ROSS, S. M. 1970. *Applied Probability Models with Optimization Applications.* San Francisco: Holden-Day, Inc.

ROSS, S. M. 1972. *Introduction to Probability Models.* New York: Academic Press.

SARKAR, T. K. 1969. Some lower bounds of reliability. Technical Report No. 124. Departments of Operations Research and of Statistics, Stanford University. Stanford, California.

SAUNDERS, S. C. 1957. Reliability, failure rate and life exceedence expectancies when the underlying distribution is Gamma. Boeing Mathematical Note No. 169. Seattle, Washington.

SEAL, H. 1969. *Stochastic Theory of a Risk Business.* New York: John Wiley and Sons.

SETHURAMAN, J. 1965. On a characterization of the three limiting types of the extreme. *Sankhyā A*, pp. 357–364.

SIMONAITUS, D. F., R. T. ANDERSON, and M. P. KAYE. 1972. Reliability evaluation of a heart assist system. *Proceedings of the 1972 Annual Reliability and Maintainability Symposium.* San Francisco.

SMIRNOV, N. V. 1952. Limit distributions for the terms of a variational series. *Trans. Amer. Math. Soc. Series*, vol. 1, no. 67, pp. 1–64.

SMITH, W. L. 1958. Renewal theory and its ramifications. *J. Roy. Statist. Soc.*, series B, vol. 20, pp. 243–302.

SMITH, W. L., and M. R. LEADBETTER. 1963. On the renewal function for the Weibull distribution. *Technometrics* 5: 393–396.

SOLOVYEV, A. D. 1964. On standby redundancy without renewal. *Cybernetics in the Service of Communism, Vol. 2, Reliability Theory and Queueing Theory*, ed. A. I. Berg, N. G. Bruevich, and B. V. Gnedenko, pp. 83–121. Moscow.

SOLOVYEV, A. D., and I. A. USHAKOV. 1967. Some estimates for systems of "aging" components. *Automatika i Vycislitelnaja Tehnika* 6: 38–44. In Russian.

STRAUB, E. 1970. Application of reliability theory to insurance. ASTIN Colloquium. Randers, Denmark.

SUKHATME, P. V. 1936. On the analysis of k samples from exponential population with especial reference to the problem of random intervals. *Statist. Res. Mem.* 1: 94–112.

SUKHATME, P. V. 1937. Tests of significance for samples of the χ^2-population with two degrees of freedom. *Ann. Eugen. Lond.* 8: 52–56.

TEICHER, H. 1954. On the multivariate Poisson distribution. *Skand. Aktuarietidskr.* 37: 1–9.

THOMPSON, W. A., and E. C. BRINDLEY, JR. 1972. Dependence and aging aspects of multivariate survival. *J. Amer. Stat. Assoc.* 67: 822–830.

VAN ZWET, W. R. 1964. *Convex Transformations of Random Variables.* Amsterdam: Mathematisch Centrum.

VEINOTT, A. F. 1965. Optimal policy in a dynamic, single product, nonstationary inventory model with several demand classes. *Operations Research* 13: 761–778.

Von Mises, R. 1964. La distribution de la plus grande de *n* valeurs. *Selected Papers of Richard von Mises, II*. Providence, R.I.: American Mathematical Society.

Watson, G. S. 1954. Extreme values in samples from m-dependent stationary stochastic processes. *Ann. Math. Statist.* 25: 798–800.

Weibull, W. 1939. A statistical theory of the strength of materials. *Ing. Vetenskaps Akad. Handl.*, no. 151, pp. 1–45.

INDEX